The Infra-red Spectra of Complex Molecules

Infra-red Spectra of Complex Molecules, Volume 2, has the title: *Advances in Infra-red Group Frequencies.*

The Infra-red Spectra of Complex Molecules

L. J. BELLAMY

C.B.E., B.Sc., Ph.D.

CHAPMAN AND HALL

LONDON

First published 1954
by Methuen and Co. Ltd.,
Second edition, 1958
Reprinted six times
Third edition, 1975
published by Chapman and Hall Ltd.
11 New Fetter Lane, London EC4P 4EE

Typeset by Santype (Coldtype Division)
Salisbury, Wiltshire

Printed in Great Britain by
Lowe & Brydone Ltd.
Thetford, Norfolk

© 1975, L. J. Bellamy

ISBN 0 412 13850 6

Distributed in the U.S.A.
by Halsted Press, a Division
of John Wiley and Sons, Inc., New York

Library of Congress Catalog Card Number 75-2435

Preface to the Third Edition

The full revision of this text has presented a number of problems. The basic data have changed little since the second edition, although they have been much extended in depth and in detail. To some extent this has helped to shorten the present text as many controversial issues which needed to be presented at some length have now been resolved so that only the final conclusions need to be given. However, it remains the case that very few new group frequencies have emerged over the past fifteen years and the emphasis of group-frequency studies has tended to shift away from the identification of specific groups towards their other possible uses in the solution of structural and chemical problems. The study of the interplay of mechanical and electronic effects in determining the directions and extents of group-frequency shifts which result from changes in the substituents has told us much about chemical mechanisms themselves, and has opened up new possibilities for the use of group frequencies in areas such as the study of rotational isomerism, the measurement of bond angles, and the use of frequency relationships to derive other chemical and physical properties.

For this reason, when the revision of this book was last considered in 1967, I decided that as the basic facts had changed but little, I would do better to write instead a new and complementary text which would concentrate on providing as far as was possible, an explanation of the known facts about the behaviour of group frequencies in chemical and physical terms, using for this purpose all those new references which would otherwise have been used in a revision of this text. This book *Advances in Infrared Group*

Frequencies, I regarded as essentially Volume 2 in that it sought to complete and to round off the first. One book sought to set out the known facts, and the other to set out why these facts were as they were, and to provide the basic understanding which is essential to the proper use of group frequencies. However, the time has now come for a revision of the present book, and it is intended that the two will now be issued as originally intended, as Volumes 1 and 2. I have not therefore repeated all the references of 'Volume 2' in this text, but I have duplicated a few of the most important and those giving extensive bibliographies so that this book can if required stand on its own. For the same reasons I have eliminated the final chapter on the origins of group-frequency shifts as this is the subject matter of the second volume.

A second problem of this revision is the vast literature which has grown up in the twenty years since this book was first written. At that time it was possible, at least to aspire, to include all the relevant references. To try to do so now would require a bibliography as large as the book itself. Moreover a number of specialist texts have appeared in which the whole book is devoted to the material which has to be presented here in one chapter, and these already offer the complete documentation that a specialist may need. Thus, two reviews by Katritzky and his coworkers deal exclusively with the group frequencies of heterocyclics and between them cite over 1500 references. Varsanyi has provided a comparable review of the aromatics, Adams of coordination compounds and Bentley *et al.* of the far infrared. There is also a wide coverage on inorganics by Nakamoto and by Lawson, whilst the supplementary Raman group frequency data has recently been very excellently covered by Dollish, Fateley and Bentley. I have therefore confined myself to those references which are most relevant and contented myself with reference to these more detailed sources where appropriate.

The spectra which replace those of earlier editions were prepared by Dr A. R. Tatchell of Thames Polytechnic, and I am most grateful to him for allowing me to make use of them. They were originally selected for use in a teaching course on group frequencies, which we gave together, and were therefore chosen for their ability to illustrate to beginners in this field many of the more important group frequencies.

Contents

Contents ix

Part Two. Vibrations involving mainly C—O and O—H linkages

Correlation Charts

1. Hydrogen stretching and triple-bond vibrations. $3750-2000$ cm^{-1}
2. Double-bond vibrations etc. $2000-1500$ cm^{-1}
3. Carbonyl vibrations. $1900-1500$ cm^{-1}
4. Single-bond vibrations etc. (I) $1500-650$ cm^{-1}
5. Single-bond vibrations etc. (II) $1500-650$ cm^{-1}

Infra-red Spectra

1. *n*-Hexane. 3500–1300 cm^{-1}. Thin film
2. 2-Methylpentane. 3500–1300 cm^{-1}. Thin film
3. Dec-1-ene. 3500–1300 cm^{-1}. Thin film
4. *trans*-Stilbene. 1300–400 cm^{-1}. KBr disc
5. Styrene. 3500–1300 cm^{-1}. Thin film
6. Styrene 1300–400 cm^{-1}. Thin film
7. Phenyl Acetylene. 3500–1300 cm^{-1}. Thin film
8. Phenyl Acetylene. 1300–400 cm^{-1}. Thin film
9. Aromatic substitution patterns. 2000–1600 cm^{-1}
10. *para*-Cresol. 3500–1300 cm^{-1}. I.M. CCl$_4$ solution
11. *tert*-Butyl methyl ketone. 4000–650 cm^{-1}. Thin film
12. *n*-Heptaldehyde. 4000–650 cm^{-1}. Thin film
13. Di-*iso*propyl ether. 4000–650 cm^{-1}. Thin film
14. Acetic Anhydride. 4000–650 cm^{-1}. Thin film
15. Propionic acid. 4000–650 cm^{-1}. Thin film
16. Propionic acid. 3500–2000 cm^{-1}. Solution 0.005M CCl$_4$
17. Methyl salicylate. 4000–650 cm^{-1}. Thin film
18. *n*-Butylamine. 4000–650 cm^{-1}. Thin film
19. Benzamide. 3500–1300 cm^{-1}. KBr disc
20. Methionine. 4000–650 cm^{-1}. KBr disc
21. Benzonitrile. 3500–1300 cm^{-1}. Thin film
22. Benzonitrile. 1300–400 cm^{-1}. Thin film
23. *N*-Methyl acetamide. 3500–650 cm^{-1}. Thin film
24. Methyl acrylate. 4000–650 cm^{-1}. Thin film
25. Benzoyl chloride. 4000–650 cm^{-1}. Thin film
26. Triphenyl phosphate. 3500–1300 cm^{-1}. Melt
27. Triphenyl phosphate. 1300–400 cm^{-1}. Melt
28. Di-*iso*propyl sulphone. 4000–650 cm^{-1}. Melt
29. Nitrobenzene. 4000–650 cm^{-1}. Thin film
30. Dimethyl sulphoxide. 4000–650 cm^{-1}. Thin film
31. Polymeric silicone. 4000–650 cm^{-1}. Thin film
32. Calcium sulphate Dihydrate. 4000–650 cm^{-1}. KBr disc

1

Introduction

Infra-red spectroscopy is widely employed in commerce and in organic chemistry for the recognition and the quantitative analysis of structural units in unknown compounds. Almost all the spectrometers employed in industry are used for this purpose, although, in the hands of theoretical physicists, the spectra can also be used to obtain fundamental data on the mechanics of simple molecules. The latter studies are extremely useful to the analyst, in that the particular motions associated with the various characteristic frequencies are determined, so that it is possible to assess to some extent the likelihood of frequency shifts occurring with changes in the local environment of the group. Nevertheless, these studies are necessarily restricted to simple molecules in which the frequencies associated with various structural units are often out of line with those found for more complex materials.

In the interpretation of spectra, therefore, the analyst must rely upon the empirical data which have been accumulated relating infra-red absorption bands with structural units. This involves a very complete knowledge of all the widely scattered work on this subject, and this is not easy to obtain. A number of correlation charts summarising this information have been published from time to time and, properly used, these are of great value. However, as their authors point out, the correlations are of unequal value and, in some cases, are based on the study of limited groups of compounds, so that their incautious use can lead to wholly misleading results. Furthermore, if the frequency ranges given for a specific grouping, such as the carbonyl vibration, were to be extended to cover all the known cases in which interference or interaction effects occur, the

1

final range would be too wide to be of any value; whereas, in fact, the particular point at which a carbonyl absorption appears within this range will often throw a good deal of light on the nature of the adjoining structure.

The present work is, therefore, an attempt to present a critical review of the data on which infra-red spectral correlations are based, indicating the classes of compounds which have been studied in each case and the known factors which can influence the frequencies or intensities of the characteristic bands. In doing so, especial attention has been paid to publications dealing with the groups of compounds containing common structural units and although a considerable number of fundamental studies on single molecules have been covered, no attempt has been made to provide a complete bibliography of these, as the absorption of structural units of very simple compounds are often not typical of those found in larger molecules. For ease of reference the chemical names given in the original publications have been retained throughout, although this has involved occasional departures from the accepted British nomenclature.

Many of the correlations discussed, particularly those in the high-frequency region, are capable of giving structural information of great value and precision, and the position and intensities of the absorption bands can be used to confirm the presence of a particular group and to obtain information as to its environment. Others, particularly in the region of skeletal vibrations, are subject to considerable frequency alterations with structural change. These can only be employed with caution and cannot safely be applied to structures widely different from those on which the correlations are based. Nevertheless, even these can be of value in indicating possible structures, whilst the absence of any bands in the appropriate region is usually a good indication of the absence of the particular grouping from the molecule. In any case, in work of this type, the spectroscopist is expected to indicate not only the presence of groupings of which he can be reasonably certain, but also to discuss the various structures which may be present but which cannot be identified with certainty. It is in assisting him to determine the proper weight which can be given to any particular identification that it is hoped this book will prove to be of value.

The subject-matter of this book has been strictly confined to the empirical interpretation of infra-red spectra. No attempt has been made to cover the many related aspects of practical spectroscopy, such as sample preparation, cell construction, quantitative analysis, instrumentation, etc. These topics have all be adequately discussed in a number of existing texts.

In any book of this kind which deals with a mass of isolated empirical observations about a variety of different structural units, a purely arbitrary arrangement of subject-matter must be employed, and I have followed the general lines of the well-known correlation charts which have appeared in the literature from time to time [1–8]. The various types of linkage are roughly classified into four main classes: –

I. Carbon–carbon and carbon–hydrogen links. II. Carbon–oxygen and oxygen–hydrogen links. III. Carbon–nitrogen and nitrogen–hydrogen links. IV. Linkages involving other elements, or which are related to inorganic structures. This arrangement has not however been followed too rigidly. The discussion on amides, for example, cannot be divided between parts II and III; and it has been included at the head of part III, where it follows directly upon the discussion of carbonyl frequencies in part II. Similarly, the chapter on nitro-compounds and related structures heads part IV, where it follows the discussion of carbon–nitrogen linkages.

At the beginning of each chapter a brief outline is given of the correlations to be discussed, together with a table giving the various frequency ranges, whilst at the end of this chapter a series of charts is given in which the correlations are summarised in the usual line drawing form. The first is intended to give the reader an outline of the particular correlations which exist before discussing any one of them in detail, whilst the second is designed to enable him to see readily any other structural units which might be expected to absorb in any specific region and which might therefore interfere. It cannot be too strongly emphasised that the indiscriminate use of either of these summaries for correlation work without reference to the detailed work on which they are based can only lead to error. An endeavour has been made to make the tables and discussion as complete as possible, but they are intended only for use in conjunction with the more detailed accounts of their origin and reliability. In both the charts and the tables a rough indication of the absorption intensity is given by the symbols (s) – strong; (m) – medium; (w) – weak and (v) – variable.

A small group of spectra, and partial spectra, have been included at the end of each of the four parts. These have been chosen to illustrate as far as possible some of the correlations which have been discussed, and the individual assignments for the various structural units involved have been indicated on them. As before, these include some doubtful correlations inserted for completeness, and the identifications are therefore not certain in all cases. These illustrations may also be of value to workers newly entering this field, for practice purposes.

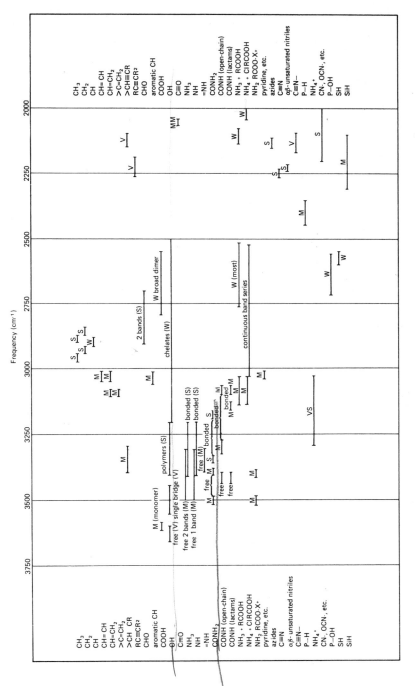

Correlation Chart No. 1. Hydrogen stretching and triple-bond vibrations. 3750–2000 cm^{-1}

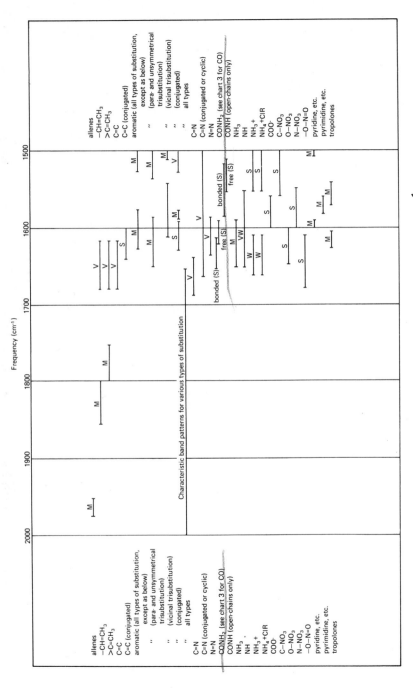

Correlation Chart No. 2. Double-bond vibrations etc. 2000—1500 cm^{-1}

6

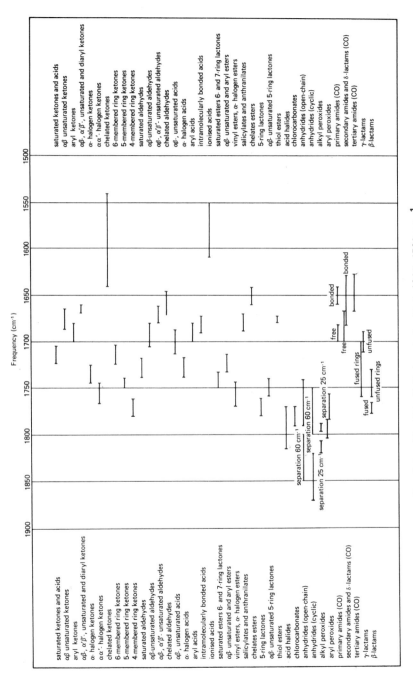

Correlation Chart No. 3. Carbonyl vibrations. 1900–1500 cm^{-1}

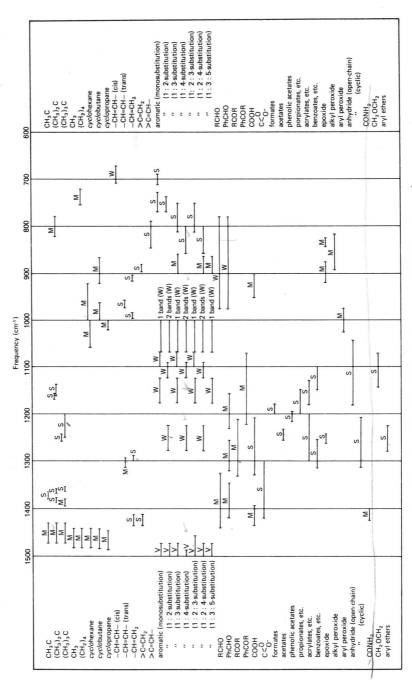

Correlation Chart No. 4. Single-bond vibrations etc. (I) 1500—650 cm⁻¹

8

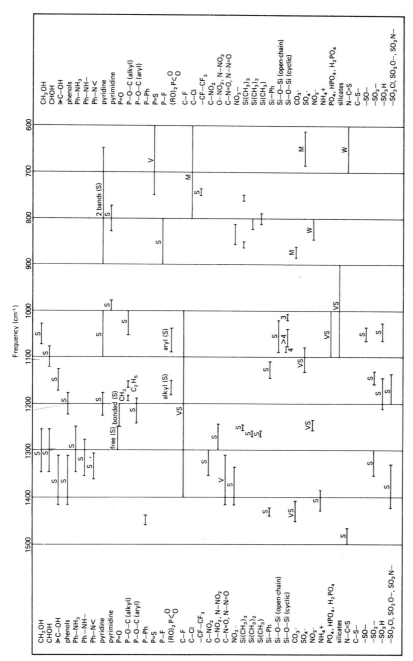

Frequency (cm⁻¹)

Correlation Chart No. 5. Single-bond vibrations etc. (II) 1500—650 cm⁻¹

9

Part One | Vibrations of C—C and C—H Linkages

2

Alkanes

2.1. Introduction and table

The study of the characteristic absorption bands arising from various types of hydrocarbon structures has produced a number of reliable correlations between the band positions and the types of grouping present. Of these, the correlations involving C–H stretching and bending modes are the most specific, but correlations involving skeletal C–C frequencies are also available for certain types of branched-chain structures. The C–H stretching mode absorptions are especially useful in differentiating saturated and unsaturated materials and, when combined with a study of the deformation modes, it is also possible to obtain an approximate indication of the proportions of CH_3, CH_2 and CH groups present in hydrocarbons.

The symmetrical CH_3 deformation is also valuable in diagnostic work in the recognition of both the $C–CH_3$ link and branched-chain structures, whilst assignments of the latter can be checked by reference to skeletal frequencies.

Correlations for the $–(CH_2)_4–$ group and for saturated cyclic hydrocarbons have also been considered in this section.

The correlations discussed are listed in Table 2.

2.2. C–H Stretching vibrations

2.2(a). Position. In the region of 3000 cm^{-1} infra-red absorption bands arise from the C–H stretching modes of organic materials. This was recognised as early as 1905 by Coblentz [1], whilst Bonino [2] in 1929 pointed out that the characteristic CH bands of aromatic and aliphatic compounds in this region occurred at different recognisable

Table 2

CH *Stretching Frequencies*

CH_3	2962 and 2872 cm^{-1} ± 10 cm^{-1} (s)
CH_2	2926 and 2853 cm^{-1} ± 10 cm^{-1} (s)
CH (tertiary)	2890 ± 10 cm^{-1} (w)

CH *Deformation Frequencies*

$C-CH_3$	1450 ± 20 cm^{-1} (m) (asymmetrical)
$-CH_2-$	1465 ± 20 cm^{-1} (m)
$C-CH_3$	1380−1370 cm^{-1} (s) (symmetrical)
$-C-(CH_3)_2$	1385−1380 (s) 1370−1365 cm^{-1} (s) (approx. equal intensity)
$-C-(CH_3)_3$	1395−1385 (m) 1365 (s) cm^{-1}
$-CH-$	Near 1340 cm^{-1} (w)

Skeletal Vibrations

$(CH_3)_3-C-R$	1250 ± 5 cm^{-1} (s) 1250−1200 cm^{-1} (s) near 415 cm^{-1}
$(CH_3)_2-C$	1170 ± 5 cm^{-1} (s) 1170−1140 cm^{-1} (s) near 800 cm^{-1}
$-(CH_2)_4-$	750−720 cm^{-1} (m)
$-(CH_2)_4-O-$ (or other functional group)	742−734 (s) (not applicable to hydrocarbons)

Ring Systems

*Cyclo*propane	1020−1000 cm^{-1} (m)
*Cyclo*butane	920−866 cm^{-1} or 1000−960 cm^{-1}
*Cyclo*pentane	Nil } (m)
*Cyclo*hexane	1005−925 cm^{-1} and 1055−1000 cm^{-1}

positions. The study of the nature of these bands and the correlation of their position with various types of structure is due to Fox and Martin [3−6]. In a series of papers between 1937 and 1940 they were able to show that, in hydrocarbons, the position of the absorption peaks from CH stretching modes is determined almost entirely by the nature of the linkage involved and is virtually independent of the rest of the structure. Although this work was done many years ago it was obviously done with great care as the frequency ranges have remained virtually unaltered despite much later more detailed work at higher resolution. Snyder *et al.* [19] for example give values in the same range. Thus, in saturated materials, a close group of absorption peaks arises from CH_3, CH_2 and CH groups; in aromatics and unsaturated compounds there is a rise to higher frequencies which becomes even more marked in the alkynes. Makic *et al.* [7] have correlated the positions of the CH_2 stretching frequencies with the degree of s character of the carbon orbital, and give the expressions $v_{as} = 16.5 \times \%s + 2542$ and $v_s = 20.3 \times \%s + 2360$. They also report a good correlation with spin-spin coupling constants in the NMR. A similar relation with J values and also with the dissociation energies and with the bond lengths has been reported by McKeen *et al.* [115]. Using the relatively low dispersion of a sodium chloride prism it is not possible to do more than recognise the main classes, but with higher resolution the dispersion is sufficient to enable the differentiation of the various types of CH groups within the main classes, so that CH_3, CH_2 and CH groups can be recognised individually. The absorption peaks arising from unsaturated and aromatic vibrations are considered in Chapters 3 and 5, and this discussion is limited to the bands arising from CH stretching modes in compounds in which the carbon atom concerned is saturated.

CH_3 *group.* Fox and Martin [6] examined a large number of hydrocarbons containing methyl groups, and found that in all cases two strong bands occurred at $2962\ cm^{-1}$ and $2872\ cm^{-1}$, corresponding to asymmetrical and symmetrical stretching modes. In all the compounds examined, the variation from these values was not more than $+10\ cm^{-1}$. In addition, two weaker intermediate frequencies were noted in some cases. Snyder and Schachtschneider [19] have shown that the higher frequency peak is split into a doublet with a separation of about $10\ cm^{-1}$ from the near degenerate v_{as} modes of the two symmetry classes. The two main frequencies were not altered in position within the range given by the presence of a double bond adjacent to the methyl group, but in these cases the lower CH_3 frequency ($2872\ cm^{-1}$) was split into two by a resonance effect. The mean value of this pair, however, remained within a few wave numbers of $2872\ cm^{-1}$.

CH_2 *group.* This group gives rise to two characteristic bands at 2926 and $2853\ cm^{-1}$ corresponding to the in-phase and out-of-phase vibrations of the hydrogen atoms [4, 6]. As before, the mean values are remarkably constant in position, being within $\pm 10\ cm^{-1}$ over a range of hydrocarbon materials including not only aliphatic materials but also the CH_2 vibrations of compounds such as benzyl alcohol and diphenylmethane.

In the case of compounds containing the group $CH_2 - C = C$, a resonance splitting effect, similar to that with CH_3 groups, is observed. In this case, also, the mean frequency is close to the normal value.

CH *group*. The absorption intensity of $\geqslant$CH groups is extremely low in comparison with those of CH_2 and CH_3 groups, but a weak band, which Fox and Martin [6] believe to be characteristic, can be traced at 2890 cm^{-1} in such compounds as triphenylmethane. The assignment is, however, of very limited value. In certain cases when the group is directly linked to oxygen the intensity is considerably enhanced, but frequency shifts due to the presence of the oxygen atom complicate the picture, and Pozefsky and Coggeshall [50] were unable to identify any specific absorption from this cause in oxygenated materials.

All the above correlations have been fully substantiated by later workers [8, 9], and no cases are known of true saturated hydrocarbons in which there is any significant departure from them. However, variations in the force field immediately around the C—H group will affect the stretching frequency, and it is because of this that the stretching frequency of =C—H is different from that of

—C—H. Similarly, substituents such as halogens which are directly joined to the C—H group, or the inclusion of CH_2 groups in strained rings, will affect the frequencies of the stretching modes and the correlations will not then apply.

2.2(b). Factors influencing the characteristic C—H stretching frequency. As has been indicated above, the correlations derived for C—H stretching frequencies are directly applicable only to unstrained hydrocarbons in which the groups concerned are linked to other carbon atoms. With polar substituents the frequencies move outside the quoted ranges. With $CH_3 X$ compounds for example a strongly electron attracting substituent as X generally raises both frequencies, so that higher values are given by the methyl halides and by compounds such as acetonitrile. However other factors are also involved as is shown by the fact that the frequencies of methyl chloride are higher than those of methyl fluoride. The general upward trend of the frequencies is broadly related to the Taft σ values of the substituents, and this is also reflected in the intensities of the ν_s bands which fall progressively as the frequency rises [81]. In acetonitrile for example the intensity of the symmetric CH stretch is only one fourth of the value found in normal alkanes.

Exceptions to the general rule that polar groups raise both frequencies are found with oxygen, nitrogen and sulphur atoms when these carry an undelocalised lone pair of electrons. Interactions then occur between the lone pair and that CH bond which lies in a *trans* position to it. The result is that the CH bonds at the same carbon atom are no longer equivalent [81, 82, 113, 115–118]. In dimethyl ether for example deuteration studies have shown that the individual CH frequencies can differ by as much as $100 \ cm^{-1}$. There is therefore a reduction in the coupling between the two stretching modes, but the overall result is to throw the two bands further apart. The antisymmetric mode appears in the normal range, but the symmetric band is at very significantly lower frequencies [115]. The same effect is observed in methyl ethyl ether where it influences both the CH_3 and CH_2 bands.

These low values of ν_s provide valuable group frequencies for the identification of OCH_3 and NCH_3 groups. They have been known for some time but have been generally thought to arise from a Fermi resonance interaction with a CH_3 deformation mode [119]. However, the deuteration studies have shown that although Fermi resonance can occur, the impact is relatively very small compared with the effects of the lone pair interaction. The correlations must be applied with caution, bearing in mind that the effect does not occur when the lone pair is delocalised or donated into a vacant orbital as in coordination compounds. Even the involvement of the lone pair in a strong hydrogen bond can be sufficient to cause a frequency shift of these bands. The frequency ranges are as follows:

CH_3O	$2820–2810 \ cm^{-1}$
$(CH_3)_2N$	$2825–2810$ and $2775–2765 \ cm^{-1}$
CH_3N	$2805–2780 \ cm^{-1}$
Aryl N CH_3	$2830–2810 \ cm^{-1}$

Bohlmann [82] has shown that a similar effect occurs with CH_2 groups attached to nitrogen, and studies on the deuterated ethers have shown that this is also true for attachment to oxygen. Apart from this, the presence of oxygen or nitrogen atoms further along the chain has only a minimal effect upon the frequencies [50], although intensity effects are relayed through many carbon atoms. The special case of a methyl group attached to a carbonyl link is interesting from this point of view. Partial deuteration studies show that even these have small differences between the individual CH bonds, although the mechanism of interaction must clearly be different from that of OCH_3 systems. These effects are insufficient to cause any very significant frequency shifts but they are very strongly reflected in the intensities. The CH stretching bands are reduced in intensity by a factor of four [125] whilst the symmetrical deformation vibration increases in intensity by a factor of thirteen. This provides a useful guide to the identification of $CH_3 CO$ groups in simple compounds such as acetic anhydride. The intensities of the stretching and bending bands of CH_2 groups adjacent to a carbonyl link are similarly affected.

Nolin and Jones [79], studied all the various partly deuterated diethyl ketones. $C_2 H_5 COC_2 H_5$ has bands at 2977 cm^{-1}, 2936 cm^{-1}, 2883 cm^{-1} and 2902 cm^{-1}. The first three of these are absent from the spectrum of $CD_3 CH_2 COCH_2 CD_3$, and must therefore be associated with the CH_3 group. The appearance of only one CH_2 stretching frequency in these compounds is probably due to the considerable intensity changes which occur when this group is adjacent to a carbonyl structure, and a second, very weak band is believed to occur at 2955 cm^{-1}.

Another factor which can occasionally cause a shift in CH stretching frequencies is ring strain. The C–H stretching frequencies of unstrained rings are normal, and Plyler and Acquista [41] have shown that in nineteen *cyclo*pentane materials there is no very great shift of the CH_2 vibrations away from the normal value. With smaller rings, however, the effect of the ring size upon the bond lengths becomes apparent, and there are some frequency shifts. Roberts and Chambers [51] have shown that the frequencies of the CH_2 bands increase and the intensities decrease progressively as the ring size becomes smaller. They conclude, therefore, that the electrical character of the C–H bonds is a relatively continuous function of ring size. In this work they examined the CH_2 vibrations of three-, four-, five- and six-membered rings substituted with a single chlorine or bromine atom, and in each series they found that the values for six-membered rings were normal and for five-membered rings very nearly so; with further decrease in ring size, however, the frequency rose considerably, so that the CH_2 vibrations of bromo-*cyclo*propane were at 3077 cm^{-1} and 2985 cm^{-1} in contrast to the values of 2924 and 2841 cm^{-1} in bromo-*cyclo*hexane. In *cyclo*butane [80] the increase is still insufficient to raise the C–H stretching frequency above 3000 cm^{-1}, but *cyclo*propane derivatives can be readily identified [62, 73] if unsaturation is absent. The methylene stretching frequency also remains high in steroids and triterpenes containing bridged structures, including a *cyclo*propane ring, and absorption in the 3060–3040-cm^{-1} region has been used for their identification [74]. Epoxy structures can be similarly identified [114], provided they are not fully alkylated, and these absorb in the range 3056–2990 cm^{-1}. Force constant calculations show that these rises in the CH stretching frequencies are associated with real increases in the strengths of the CH bonds and are not a result of changes in coupling due to angle effects [124].

The high CH stretching frequencies of three-membered rings of this type reflects their well-known 'unsaturated character', and for the same reason they are sometimes difficult to differentiate from aromatic or olefinic absorptions. Absorption above 3000 cm^{-1} can also arise in special cases in four-membered rings which contain polar groups able to influence the hybridisation of the CH bonds. β-butyrolactone [121] is such a case with CH bands at 3014 and 2933 cm^{-1}. Other examples are 3-methylene oxirane [122] and cyclobutanone [123].

2.2(c). Factors influencing the intensity. Rose [13] in 1938, in work on the overtones of the CH fundamentals, noted a considerable degree of constancy in the intensity of absorption of various molecular units, and this has been confirmed by Fox and Martin [6] working on the fundamentals themselves. Using the integrated absorption area as a measure of intensity they find that, in long-chain paraffins, the intensities of the CH_2 and CH_3 bands are directly related to the proportions of these groups present and the intensities of the CH_2 bands change by a steady increment for each unit increase in chain length. With branched chains the agreement was less good but, if account is taken of all the CH_3 bands in assessing the integrated absorption areas, it is reasonably satisfactory. With 2 : 3-dimethylbutane, 2 : 2 : 3-trimethylbutane and 2 : 3 : 4-trimethylbutane the ratios of CH_2 groups are found as 4.42 : 5 : 5.21, as against 4 : 5 : 5.

When a double bond is adjacent to the CH_3 group, the method fails if measurements are confined to the normal frequencies due to the band-splitting effect, but again, an integrated absorption area value based on all the methyl groups is reasonably good. For 1-butylene, *cis-* and *trans-*2-butylene and trimethylethylene they obtained values of 1 : 1.76 : 1.89 : 3.50, as compared with 1 : 2 : 2 : 3 for the ratio of the numbers of methyl groups in the molecules.

A similar degree of constancy was found for $=CH$ and $=CH_2$

group absorption intensities, and these authors quote the following relative intensity values based on the integrated absorption areas summed for all the bands associated with the group:

$$\diagdown\!\!\!/\text{CH } 410; \quad =CH_2 1800; \quad -\diagup\!\!\!\!\!\diagdown CH 120; \quad \diagdown\!\!\!/\text{CH}_2 2400; \quad \diagdown\!\!\!/\text{C}-CH_3 2230;$$

$$\diagdown\!\!\!/\text{CH}-CH_3 3800.$$

The reduction in intensity of CH_2 and CH_3 groups adjacent to double bonds is apparent from these values, as is the low relative intensity of the $-CH$ group absorption.

A somewhat similar study has been made by Francis [9], although he has used the integrated intensities of the 2900 cm^{-1}, 1460 cm^{-1} and 1375 cm^{-1} bands to obtain values for the unit structural group intensities of CH_3, $-CH_2-$ and $-CH-$ groups. He was able to show that the calculated absorption intensities of twelve aliphatic hydrocarbons at each of these frequencies generally stay close to the observed values and that it was only with very complex branched-chain materials that there was any significant departure from the additive values obtained by treating the hydrocarbon as a mixture of non-interacting CH_3, $-CH_2-$ and $-CH-$ units. In a subsequent paper [72] he proposes the use of integrated intensities between 790 and 685 cm^{-1} for the determination of paraffinic methylene groups. cycloPentyl and cyclohexyl residues can be estimated separately using bands at 2957 cm^{-1} and 2926 cm^{-1} after these have been corrected for the interfering methyl and methylene absorptions.

Quantitative work in the 3000 cm^{-1} region therefore affords a method for the estimation of the relative numbers of CH_3, CH_2 and CH groups in hydrocarbons. Hastings et al. [61] have used a somewhat similar method, including also measurements in the 800–700 cm^{-1} region, for the determination of methyl and methylene groups in paraffin and methylene groups in substituted naphthalenes in complex mixtures.

It has already been noted, however, that Pozefsky and Coggeshall [50] find that extinction coefficients of the CH_3 and CH_2 groups in the 2890–2850 cm^{-1} region are not the same in oxygen-containing materials, so that these findings cannot be safely applied to non-hydrocarbons without supplementary basic work. This has been confirmed by Francis [71], who finds a marked fall in the integrated absorption intensities of methyl or methylene groups adjacent to an oxygen atom or to a carbonyl group. Mirone and Fabbri [83] have also noted that the overall intensity of the CH_2 band in secondary alcohols and tertiary alcohols is lower than in the corresponding primary compounds.

Following the work of Rose [13], similar estimations are possible in the overtone region; Hibbard and Cleaves [14] have used the second overtone region for this purpose, and Tuot and Barchewitz [15] have noted that the intensities of CH_3 and CH_2 bands in this region are functions of the number of groups present and have employed them for analysis. Gauthier [52] has employed a similar technique in the investigation of the proportion of branched chains in liquid hydrocarbons, working in the $8000-5000$ cm^{-1} region. Evans, Hibbard and Powell [53] have also investigated this region extensively and have shown that it can be applied to relatively large molecules of molecular weight up to 500. They estimate that they are able to determine the various groups to within 0.40 group or less in paraffins, 1.2 group or less in naphthenes and 0.9 group or less in aromatics. Valuable reviews of characteristic frequencies and analyses in the overtone region has been given by Kaye [84], Whetsel [127], Tosi and Pinto [126] and others [85].

This region has many advantages for this type of work, especially as normal optics can be employed and detectors of very high sensitivity are available.

2.3. CH_3-C and CH_2 deformation vibrations

2.3(a). Position. It was recognised quite early in the study of infra-red spectra that CH_2 and CH_3 groupings give rise to absorption near 1460 cm^{-1} due to hydrogen bending vibrations [16, 17, 18]. It is now accepted that CH_2 deformations give rise to absorption very close to 1465 cm^{-1} and asymmetrical CH_3 deformations to absorption at 1450 cm^{-1}. Variations of more than $10-20$ cm^{-1} from these values are extremely rare except in cases in which the presence of strongly electronegative atoms causes a change in the charge distribution on the carbon atom.

The data on which these assignments are based are primarily derived from the spectra of the relatively large number of simple hydrocarbons which have now been subjected to mathematical analysis by various workers. In the case of the CH_3 group, two frequencies arise from the bending of the hydrogen atoms about the carbon atom, the asymmetrical mode giving rise to the 1460 cm^{-1} absorption, whilst the symmetrical mode is responsible for the band near 1375 cm^{-1}. The corresponding $-\overset{|}{\underset{|}{C}}-H$ bending band is less well defined. This is due to its inherently low intensity in comparison with the bands from CH_3 and CH_2 groups and also to the fact that less compounds in which this group occurs have been fully studied. It appears as a weak band close to 1340 cm^{-1} in hydrocarbons [10, 63, 69], but is generally not sufficiently intense to be of much value in structural analysis although, in certain cases in which the attachment is to oxygen or nitrogen atoms, a considerable increase in intensity occurs.

The bands in the 1460 cm^{-1} region arising from CH_2 and CH_3 groups lie too close together to enable the two groupings to be readily differentiated, especially in view of the small shifts which may be induced by changes of state, but the overall intensity in the 1460 cm^{-1} region is directly proportional to the number of such groupings present and, as mentioned, has been employed by Francis [9] in conjunction with other bands for the calculation of the relative proportions of CH_3, CH_2 and CH groupings present in a molecule.

The symmetrical deformation mode of the hydrogen atoms of a methyl group results in an absorption band in the range 1385–1370 cm^{-1} which is extremely stable in position provided that the methyl group is attached to another carbon atom. This is a most valuable correlation, especially as splitting of the band, in certain circumstances, allows the identification of branched-chain systems.

The intensity of this band is considerably greater for each unit of structure than that for the corresponding methylene and asymmetrical methyl deformation and is directly proportional to the number of units present. Even when splitting of the band occurs, as in branched chains, Francis [9] has shown that the integrated absorption in the 1380 cm^{-1} region is still approximately related to the number of CH_3 groups and may be used in conjunction with the data from the other regions for the determination of the proportion of methyl groups present. This absorption has been similarly used by Freeman [65] in the determination of the degree of branching of fatty acids.

Luther and Czerwany [86] have carried out an extensive study of the integrated absorption areas of CH absorption bands in normal paraffins up to C_{32} and in many branched-chain materials. They quote values for the integrated absorption coefficients of CH_2 deformation and wagging modes, and for CH_3 asymmetric and symmetric deformations. As more data become available on the variations in such intensities with structural alterations, studies of this type will be a very valuable aid in correlation work.

2.3(b). Factors influencing CH_2 and CH_3 bending frequencies and intensities.

As with the CH stretching frequencies, changes in the state of aggregation can lead to shifts of up to $10 \ cm^{-1}$. Richards and Thompson [21], for example, found shifts of this order on examination of a number of hydrocarbons in the liquid and solid states. However, these shifts were accompanied by other changes, and they noted that bands obtained from the solids were sharper and that they doubled in many cases. This effect has been confirmed by Sutherland [12], who finds the $1460 \ cm^{-1}$ absorptions to be double below the transition point in solid paraffins, but single above this and in the liquid state. This arises from the interaction between the CH_2 deformation frequencies in neighbouring hydrocarbon chains. Sheppard and Simpson [64] quote similar data for many other hydrocarbons.

Changes in these CH_2 deformation frequencies are also induced by the introduction of electro-negative groups in the immediate vicinity. Thus, in the case of unsaturated materials, the CH_2 deformation frequency of a vinyl type $=CH_2$ group is shifted to $1420-1410 \ cm^{-1}$, and this fact is of value in the identification of this type of double bond (Chapter 3). In allene [22], the frequency falls further to $1389 \ cm^{-1}$. On the other hand the influence of unsaturated groups is mainly limited to the carbon atoms forming the bond, and adjacent CH_2 groups are relatively unaffected. However, when the unsaturated link occurs in a cyclic system a small shift to a lower frequency of the adjacent methylene vibration sometimes occurs. Thus in many sterols the CH_2 deformation frequency of the group $-C=C-CH_2-$ occurs [66] at $1438 \ cm^{-1}$. Changes which occur due to the substitution of halogens are again limited to the CH bending vibrations at the particular carbon atom to which they are attached. In methylene fluoride [23] the frequency is raised to $1508 \ cm^{-1}$, whilst it is $1429 \ cm^{-1}$ in the chloride [24]; similarly the asymmetrical deformation mode of methyl fluoride absorbs at $1471 \ cm^{-1}$ and the iodide [25] at $1441 \ cm^{-1}$. A typical illustration of the extent of such shifts is given by Brown and Sheppard [26] in discussing the spectra of some brominated hydrocarbons; 1 : 2-dibromoethane absorbed at $1435 \ cm^{-1}$, whereas n-butyl bromide and n-propyl bromide absorbed at 1435 cm^{-1} and also at 1460 cm^{-1}; 1 : 4-dibromo-butane gave two CH_2 deformation frequencies at 1433 and 1440 cm^{-1}. Shifts also occur when the grouping is attached to other atoms such as nitrogen or sulphur, but the extent of these is not well defined and depends to some extent on the nature of the other attachments to these atoms.

With carbonyl substitution, the intensity of the CH_2 and CH_3 deformations of adjacent methylene groups is considerably increased and there is a shift towards lower frequency similar to that of vinyl groupings. Acetone for example absorbs at 1431 cm^{-1} and acetic acid at 1418 cm^{-1}. Francis [71] has assigned the deformation frequency of a methylene group adjacent to a carbonyl at 1410 cm^{-1} and this has been confirmed by other workers for a variety of compounds. Thus in acids containing the group $-CH_2COOH$ Hadži and Sheppard [20] have observed a strong band at this point which is absent from other acids and which is quite distinct from the 1420 cm^{-1} band originating from the carboxyl group itself (Chapter 10), whilst studies of the ketones $CH_3 CH_2 CO$ $CH_2 CH_3$, $CH_3 CD_2 CO CD_2 CH_3$ and $CD_3 CH_2 CO CH_2 CD_3$ have indicated very clearly [67, 79] that the CH_2 deformation occurs at 1415 cm^{-1} and the CH_3 symmetrical deformation at 1380 cm^{-1}. A very large number of steroids have also been examined from this point of view, and it has been found that in these, active methylene groups adjacent to carbonyls in the 3- or 17-position or in the side-chain absorb in the range $1418-1408 \text{ cm}^{-1}$. However, a carbonyl group at the 4-, 6-, 7-, 11- or 12-position appears to have somewhat less influence on adjacent CH_2 groups which absorb at 1434 cm^{-1}, close to the frequencies of cyclic methylene groups activated by a double bond [66, 67, 87]. These findings have been confirmed by deuteration studies.

In steroid acetates the influence of the CO grouping upon the CH_3 symmetrical mode is small, and although splitting into two bands at 1375 cm^{-1} and 1365 cm^{-1} has been found in 3- and 17-acetates, this is probably due to steric effects similar to those responsible for the splitting of the C—O stretching absorption (Chapter 11) rather than to any influence of the carbonyl group. When the methyl group is directly attached to the CO group, however, its influence is, of course, more pronounced [66], and the symmetrical CH_3 defor-mation of sterols with a $COCH_3$ group in the side-chain has been assigned at 1356 cm^{-1}.

Strained ring systems can also lead to small alterations in the charge distribution at methylene groups, with a corresponding alteration in the characteristic frequency. The CH_2 frequencies for ethylene oxide, imide and sulphide are 1500, 1475 and 1471 cm^{-1} respectively [27]. It will be seen that the effect is not large, and with larger rings with less strain, the CH_2 vibrations tend towards the normal values. In six-membered steroid rings this frequency falls between 1456 and 1448 cm^{-1}, which is very slightly lower than the value for methylene groups in the side-chain (1468—1466 cm^{-1}), so that in favourable cases it is possible to differentiate between the two types [66, 67]. When *cyclo*propane rings are included within larger systems, such as triterpenes, the CH deformation frequencies are not sufficiently distinctive to enable them to be identified directly [74]. However, the location and type of such rings can be found by a study of the changes in CH_3 and CH_2 deformation frequencies which take place when the rings are opened with hydrogen chloride and deuterium chloride [88].

The shifts in the position of the symmetrical CH_3 absorption band due to changes of state have been investigated by Richards and Thompson [21], who found that they were relatively small and rarely more than 10 cm^{-1}. Changes in the type of substitution on the carbon atom to which the methyl group is attached also have very little influence upon the band position except in cases where a second methyl group is attached. This is illustrated by the following values assigned to the symmetrical CH_3 mode in molecules subjected to mathematical analysis [29]: acetaldehyde 1370 cm^{-1}, methyl *iso*cyanate 1377 cm^{-1}, methylacetylene 1379 cm^{-1}, acetic acid 1381 cm^{-1} and methyl cyanide 1396 cm^{-1}. In cases in which a second methyl group is attached to the same carbon atom, resonance splitting occurs and two separate bands appear [17, 54, 64]. The relative intensities of these bands are usually a direct indication of the extent of branching of the chain. The simplest case, ethane [30], shows this effect and absorbs at 1374 and 1379 cm^{-1}. With the *iso*propyl grouping $(CH_3)_2 CH-$ or $(CH_3)_2 -CH-$, the bands split further apart and appear at about an equal distance on each side of the mean value at 1385—1350 cm^{-1}, with approximately equal intensities. This effect was first noted by Thompson and Torkington [17], who found doubling of the CH_3 vibration in a number of 2 : 2-dimethyl-substituted hydrocarbons.

The splitting appears to depend to a very considerable degree upon the bond angle between the methyl groups. In systems in which the methyl groups are attached to an unsaturated carbon atom and are therefore at an angle of 120° to each other the symmetric deformation band does not split. In *iso*propyl groups with a normal tetrahedral angle the splitting is considerable and well defined, and the separation of the two bands increases progressively as the bond angle is narrowed further. With the tertiary butyl grouping $(CH_3)_3-C-$ the splitting appears to be very slightly further apart, with bands at 1395 cm^{-1} and 1365 cm^{-1}, whilst the intensity of the second band is about twice that of the first. The positions and relative intensities of bands in this region therefore afford a very valuable indication of the presence or absence of various types of branched chains in a molecule. These correlations have been very fully discussed by Sheppard and Simpson [64], and by McMurry and Thornton [60]. A typical illustration of their use is the work of Sobotka and Stynler [31] in the identification of normal, *iso*- and *neo*-palmitic and stearic acids by studies in this region. It must be remembered, however, that the intensity pattern may be influenced by the presence of other methyl groups elsewhere in the molecule, which will contribute their own specific absorptions in this region. The characteristic skeletal absorptions of various types of branched chain should, therefore, also be sought in the identification of *iso*propyl and *tert.*-butyl groups. One other instance in which structural features influence the CH_3 symmetrical deformation frequency has been found in the steroid series [66]. A considerable number of sterols containing two re-entrant methyl groups situated between rings have been shown to absorb at 1384 cm^{-1} and 1378 cm^{-1}, whereas thirteen other steroids in which only one such methyl group is present show only one of these bands. However, these frequencies may be influenced to some extent by stereo-chemical factors. Characteristic frequencies for methyl and methylene groups in various environments in steroids have been summarised by Jones and Herling [111].

The deformation frequency of the single CH bond of the *iso*propyl group is too weak to be of value for identification in the infra-red. However it gives rise to a band in the 1350—1330 cm^{-1} region in the Raman spectrum which is readily identified.

2.4. Deformation vibrations of methyl groups attached to elements other than carbon

The position of the CH_3 symmetrical deformation frequency when the methyl group is attached to an element other than carbon is determined largely by the electro-negativity of the element, and by its position in the periodic table [89, 90]. The magnitude of the resulting shifts can be very considerable. The methyl halides, fluoride, chloride, bromide and iodide absorb at 1475, 1355, 1305 and 1255 cm^{-1} respectively [29]. Examples of other frequencies with different elements are given by methanol (1456 cm^{-1}) and dimethyl ether (1466 cm^{-1}) for the CH_3-O group, dimethyl sulphide (1323 cm^{-1}) for the CH_3-S group, and methylamine (1418 cm^{-1}) for the CH_3-N group. Similarly, silicones with the CH_3-Si group absorb constantly at 1250 cm^{-1} and phosphorus compounds with the CH_3-P link may absorb near 1280 cm^{-1}. When the methyl group is attached to an aromatic ring it shows a very specific rocking frequency [91] near 1042 cm^{-1}. This correlation must be used with causion, however, as *meta*-substituted compounds have a ring frequency near this point, and some interactions can occur in *ortho*-substituted materials.

Bellamy and Williams [89] have pointed out that in any CH_3X compound the individual CH frequencies are all directly related to one another and to the HX-stretching frequency.

2.5. Skeletal vibrations

2.5(a). *iso*Propyl and *tert.*-butyl structures. Skeletal vibrations are usually very sensitive to changes in the immediate environment of the vibrating group and hence considerable frequency shifts can be caused. However, these considerations do not apply to branched units such as the structures

$$\begin{array}{ccc} CH_3 & CH_3 & CH_3 \\ \diagdown & \diagup & \diagdown \\ & C & \quad \text{and} \quad \\ \diagup & \diagdown & \diagup \\ CH_3 & C & CH_3 \end{array} \quad CH-C,$$

each of which might be expected to exhibit characteristic frequencies of their own. Simpson and Sutherland [32, 33] have been able to calculate the expected frequencies ν_2 and ν_3 of these groupings on the basis of a relatively simple force field. For 2 : 2-dimethyl-substituted paraffins containing the group $(CH_3)_3CR$, they were able to show that no great variation in either of these frequencies is to be expected from alteration in the mass of the R group from methyl (15) up to heptyl (99). Throughout this series the ν_2 frequency would be expected to decrease slowly by only 37 cm^{-1} and the ν_3 frequency by only 97 cm^{-1}.

Experimentally they have found that in paraffins up to and

including the nonanes, strong absorption bands occur near 1250 cm^{-1} and 1200 cm^{-1} whenever the 2 : 2-dimethyl group is present. The former is almost constant in position, is well seen also in the Raman spectrum, and does not vary by more than a few wave-numbers throughout the series, whilst the second band coincides with the first in *neo*pentane, and then falls steadily throughout the series to 1200 cm^{-1}. The greater consistency in the position of the 1250 cm^{-1} band is to be expected as this arises from the ν_2 vibration, which is a perpendicular mode of the $(CH_3)_3C$ group and so will be much less affected than the ν_3 parallel vibration when the rest of the hydrocarbon chain varies. The C–C skeletal modes can sometimes be differentiated from the various CH_2 wagging modes by a comparison of the spectra of hydrocarbons with those of corresponding compounds with polar substituents. The CH_2 modes often become intensified in the second cases, and in this way further support has been obtained for the skeletal assignments quoted above [92].

Similar calculations by the same authors for the *iso*propyl type skeletons indicate that the ν_2 and ν_3 frequencies should occur near 1170 and 1145 cm^{-1} and experimental evidence again confirms this. In the series 2-methylpropane to 2-methylnonane, they find one frequency to be constant in the range $1170-1167 \text{ cm}^{-1}$, whilst the second falls steadily from 1170 cm^{-1} in the first to 1142 cm^{-1} in the last. The structure $(CH_3)_2C{<}$ with no free hydrogen atom on the central carbon gives only a single band in this region [54] at 1195 cm^{-1}. A second band near 1210 cm^{-1} has been observed in some cases, but it is more variable in position [64].

These correlations have also been fully substantiated by the observations of Rasmussen [28]. Whilst differing in the interpretation of these bands, he has given empirical data obtained on a large number of branched-chain paraffins which confirm that bands at the positions indicated can be associated with these different types of branched-chain structure. These correlations appear to hold, also, for many non-hydrocarbons provided no oxygen atoms are attached

near the chain branching. The structure

$$\begin{array}{c} C \qquad C \\ \diagdown\;\diagup \\ \diagup\!\!\!\!C\!\!\!\!\diagdown \\ C \qquad O \end{array}$$

absorbs near

900 cm^{-1} (Chapter 7).

In addition to these correlations, Simpson and Sutherland have pointed out that a third skeletal vibration (ν_4) can also be predicted for each of the branched structures discussed. For *tert.*-butyl compounds this occurs at 415 cm^{-1} and for *iso*propyl compounds near 850 cm^{-1}. These are, however, less valuable than the others. Very little information is available as regards the first of these, whilst the second is variable within the range 840–793 cm^{-1} in the series of 2-methyl compounds quoted.

A considerable number of correlations have also been worked out for a series of highly specialised hydrocarbon structures. These have been developed by Sheppard and Simpson [64] and by McMurry and Thornton [19, 60] and they include correlations for CH_2 wagging, rocking and twisting modes, and for C–C stretching vibrations as well as for various patterns of methyl-group substitution. As they are not generally applicable they have not been included here and the original publications must be consulted for them. In many cases they include details of the number of compounds on which any particular correlation is based, together with some indication of the intrinsic intensity of the bands in question.

2.5(b). Paraffinic chains. As has been mentioned above, correlations for open-chain skeletal vibrations are frequently unreliable and cover a wide range of frequencies. Correlations involving single bond stretchings of ethyl and *n*-propyl groups have been developed by Kohlrausch [34] and by Sheppard and Simpson [64], and are included in Colthup's frequency-correlation chart [10], and in the chart issued by the Spectroscopic Group of the Institute of Petroleum [55]. However, the bands are frequently weak and cover such a wide frequency range as to be of little practical value for work on complex molecules. There is, however, one correlation arising from an open-chain vibration which has proved to be extremely valuable in practice. This is that concerned with a long chain of the type $-(CH_2)_n-$ ($n = 4$ or more) which gives rise to a strong absorption band near 720 cm^{-1}. This band arises from a rocking mode of the (CH_2) groups and is not strictly a skeletal mode, but it can conveniently be considered here, as it arises from a structural group rather than from a single unit.

Tuot and Lecomte [35] in 1943 gave data on a series of thirty alcohols in which they noted the appearance of a band near 720 cm^{-1} when at least four adjacent straight-chain methylene groups were present, and Thompson and Torkington [17] assigned the doublet at 721 and 732 cm^{-1} in polythene to this cause. The vibration has been shown to be due to a rocking mode of the CH_2

groups by Sheppard and Sutherland [36] working on fully deuterated paraffins and by Sutherland and Vallence Jones [37], who
used polarised radiation to show that the vibrations were at right
angles to the hydrocarbon chains. The former workers suggested the
range $760-720$ cm^{-1} for this vibration in hydrocarbons.

It is clear, therefore, that a number of directly linked $-CH_2-$
groups will give rise to absorption in this region, but it is not to be
expected that any very clear-cut line of demarcation will be possible
as regards the number of CH_2 groups necessary for its production.
Generally the number is given as four, and no hydrocarbon
containing four or more CH_2 groups, directly linked, has been
reported in which this band is not present in the spectra. On the other
hand it appears in some, but not all, compounds in which only three
(CH_2) groups are linked, and there is a suggestion of a gradual shift
towards higher frequency as the number of groups is
reduced [38, 60]. The correlation can, therefore, be used positively
for recognising the absence of the $-(CH_2)_4-$ group but must be
interpreted with caution when the band is present.

Recently Wiberley and Bassett [39] have suggested a more precise
correlation for the n-butyl group itself. In a series of eighteen n-butyl
derivatives in which the butyl group is attached to oxygen, nitrogen
or some other functional group, this frequency was found to fall
always within the range $742-734$ cm^{-1}. They would, therefore,
restrict this correlation to compounds of this type and they point
out that with this restricted frequency range it is not applicable to
hydrocarbons. In this connection they quote Smith as having pointed
out to them that of seven hundred A.P.I. hydrocarbon spectra, two
hundred and twenty absorb in the range $750-720$ cm^{-1}; of these,
thirty-four lie in the $742-734$ cm^{-1} range, of which only six
contain butyl groups, whilst seventeen contain propyl groups.
Similarly six compounds containing the n-butyl group absorp
between 728 and 730 cm^{-1}, whilst others absorb above 742 cm^{-1}.
This analysis is confirmed by McMurry and Thornton [60] and
emphasises the need for caution in the use of this band for diagnostic
purposes. A useful discussion of this band has also been given by
Bomstein [93].

One factor, which is of value in identifying a band in the
$750-720$ cm^{-1} region as being due to this type of methylene-group
vibration, is its behaviour when the material is examined in different
states. Like the CH_2 deformation frequency, this band is always
double below the transition point and in the solid state, and single in

the liquid state and in solution [12, 68], and in this property will enable this band to be recognised among others in the same spectral region which arise from other causes and do not show this behaviour. This phenomenon has been extensively studied by many workers [77, 99—100] using such techniques as polarised radiation, temperature dependence, etc. It is now certain that the doubling of this band in the crystalline material arises from interactions between neighbouring molecules in the crystalline phase. Studies of the relative intensities of these bands therefore afford a convenient method for the determination of the crystalline/amorphous ratio. It may also be possible to obtain some tentative confirmation of the presence of long-chain methylene groups by the examination of the 1300—1200 cm^{-1} region of the spectrum. Brown, Sheppard and Simpson [40] have found weak bands near 1300 cm^{-1} and near 1240 cm^{-1} in a number of long-chain paraffins which they attribute to CH_2 wagging vibrations. This region is complicated by the presence of C–C stretching modes coupled with CH_2 wagging frequencies, but in the compounds examined the 1300 cm^{-1} band appeared in all the even-numbered carbon atom paraffins from four to fourteen carbon toms, and the 1240 cm^{-1} vibration in all from six carbon atoms upwards. A number of workers have studied this region in detail [19, 75, 100—104, 112], and the various frequencies can be disentangled and assigned in many simple hydrocarbons. The number of bands increases progressively with the chain length, and is reduced by the introduction of substituents which lowers the number of adjacent methylene groups. However, these absorptions are weak in normal hydrocarbons, and it is only in compounds with polar end groups that they become sufficiently intense to be identified without difficulty. There are, however, possibilities at least for the use of this region in the determination of chain lengths, and some success in this direction has been achieved in the fatty acid series (Chapter 10).

2.5(c). Cyclic systems. The saturated cyclic compounds form another group from which characteristic frequencies arising from ring deformations might be expected, and a number of papers have been published in this field [41, 51, 54]. These will be considered under individual ring systems, but it should be appreciated that, for the larger systems at least, the correlations quoted are not of much practical value. The most characteristic of the vibrations of cyclic systems is the ring-breathing frequency, which is unfortunately inactive in the infra-red but is readily identified in Raman spectra [105].

Cyclo*propane*. Bartleson, Burk and Lankelma [42] in 1946 noted that *cyclo*propane derivatives appeared to absorb near 1026 cm^{-1} and 866 cm^{-1}. This observation was followed up by Derfer, Pickett and Boord [43], who examined fourteen different *cyclo*propane derivatives. They found that a strong band at 866 cm^{-1} was not consistently present, but that in all cases there was a band at 1020–1000 cm^{-1}. The only exception to this was *cyclo*propane itself, which absorbs just outside this region. In three hundred other hydrocarbon spectra this band was not present more often than it would be expected to occur by chance. They, therefore, regarded this band as being typical of the *cyclo*propane structure and suggested that it arises from a ring deformation mode. The band was present in the spectra of the compounds of Bartleson *et al.*, and Marrison [44] has since pointed out that it is also present in the spectrum of vinyl *cyclo*propane [45]. Slabey [58] has confirmed this and found the band also in di*cyclo*propyl. In later papers [59, 73] he reports results of studies on large numbers of *cyclo*-propane derivatives. Thus thirty-four compounds with this structure absorbed in the range 1048–1017 cm^{-1}, of which twenty-nine absorbed within 5 cm^{-1} of 1021 cm^{-1}. Wiberley and Bunce [62] also confirm this frequency in nine *cyclo*propane compounds, and Bridson-Jones *et al.* [69, 70] have observed it in seven other cases. The occurrence of this band within a very narrow frequency range is not confined to simple structures and it appears, for example, in the spectrum of *i*-cholestane [46] in which the *cyclo*propane ring is part of a bridged-ring system of a six-membered ring, and also in carene [47] and car.3.-ene, in which one side of the *cyclo*propane ring is supplied by a side of another six-membered ring. The occurrence of a similar band in the spectrum of artenol, a triterpenoid ketone, has been used by Barton [46] as evidence for the presence of a *cyclo*propane ring system. However, as Cole has pointed out [74], it is necessary to be very cautious in the use of this correlation in oxygenated compounds. Whilst he observes a band near 1010 cm^{-1} in a number of triterpenes containing *cyclo*propane rings, he does not consider it sufficiently characteristic without reference also to the CH stretching region. This correlation is also supported by Sheppard's work on naphthene hydrocarbons [54]. *Cyclo*propane rings can be readily recognised in the Raman spectra as they give rise to a band in the range 1220–1200 cm^{-1} in mono and in 1:2-dialkyl derivatives. This band moves to 1320 cm^{-1} in 1:1-disubstituted *cyclo*propanes.

Cyclo*butane*. The position of *cyclo*butane correlations is less clear cut than is the case with *cyclo*propane partly because the characteristic bands are considerably less intense. Wilson [48] assigned bands at 920 and 903 cm^{-1} in *cyclo*butane to two CH$_2$ rocking frequencies, and Derfer *et al.* [43] followed this by the observation that seven substituted *cyclo*butanes all absorbed in the narrow range 920–910 cm^{-1}. They pointed out, however, that if the assignment to CH$_2$ modes was correct, this band would not be expected to appear in fully substituted materials. Reid and Sack [49] later examined a series of eight substituted *cyclo*butanes in which all the carbon atoms were substituted. Their samples were examined as solids and all absorbed between 888 and 868 cm^{-1} and none between 920 and 910 cm^{-1}. Marrison [44] has pointed out that neither methylene nor octafluoro *cyclo*butane absorbs in the 920–910 cm^{-1} region but that all the *cyclo*butanes in the A.P.I. series absorb between 1000 and 960 cm^{-1}. The value of this latter correlation is partly discounted by the additional observation that many *cyclo*pentane derivatives also absorb in this region. Sheppard [54] supports the 910 cm^{-1} correlation but does not state the compounds studied. A band near this position occurs in the spectra of a number of *cyclo*butane derivatives described by Roberts and Chambers [51, 56] and by Roberts and Simmons [57]. *Cyclo*butane compounds show a strong and very characteristic C–C stretching band in their Raman spectra, near 933 cm^{-1}.

Cyclo*pentane*. As stated, Marrison [44] has noted that the spectra of *cyclo*pentane derivatives contain a band whose average position is 977 cm^{-1}. The variation from compound to compound is, however, too wide to afford this observation any significance. Sheppard [54] regards bands near 930 and 890 cm^{-1} as being characteristic and the latter is a particularly useful diagnostic band in the Raman spectrum.

Cyclo*hexane*. Marrison [44] has noted that the published spectra of all *cyclo*hexane derivatives (with two exceptions) show bands in the ranges 1005–952 cm^{-1} and 1055–1000 cm^{-1}. This relates to fifty published spectra, and the two exceptions (*cyclo*hexane and β-gammexane) each show one of these bands. In addition, he quotes a further nine materials he has examined which absorb in these regions. These tentative correlations do not, however, apply to *cyclo*hexanone derivatives in which the introduction of the carbonyl group would be expected to alter any characteristic ring deformation frequencies. Sheppard [54] quotes values near 1260 cm^{-1} and 890 cm^{-1} for *cyclo*hexanes but again gives no data.

2.5(d). Low-frequency absorptions. Hydrocarbons have a number of weak absorption bands due to skeletal vibrations in the region below 650 cm^{-1}. These are usually extremely weak, and can be studied only in relatively thick films. Plyler [106] has given data on a number of such materials over the range $650{-}400 \text{ cm}^{-1}$, and Borello and Mussa [107] have suggested there is a characteristic methyl-group frequency in this region which can be used in the estimation of CH_3 groups in polyethylene.

Studies at lower frequencies have been made by Gates [108], and more recently Donneaud [109, 110] has found some interesting differences between straight- and branched-chain compounds in this region and discussed their possible analytical applications.

2.6. Bibliography

1. Coblentz, *Investigations of Infra-red Spectra* (Carnegie Institute Washington, Part 1, 1905).
2. Bonino, *Trans. Faraday Soc.*, 1929, **25**, 876.
3. Fox and Martin, *Proc. Roy. Soc.*, 1937, **A162**, 419.
4. Fox and Martin, *ibid.*, 1938, **A167**, 257.
5. Fox and Martin, *J. Chem. Soc.*, 1939, 318.
6. Fox and Martin, *Proc. Roy. Soc.*, 1940, **A175**, 208, **234**.
7. Makic, Meic and Randic, *J. Mol. Spectroscopy*, 1972, **I2**, 482.
8. Saier and Coggeshall, *Analyt. Chem.*, 1948, **20**, 812.
9. Francis, *J. Chem. Phys.*, 1950, **18**, 861.
10. Colthup, *J. Opt. Soc. Amer.*, 1950, **40**, 397.
11. Jones, McKay and Sinclair, *J. Amer. Chem. Soc.*, 1952, **74**, 2575.
12. Sutherland, *Discuss. Faraday Soc.*, 1950, **9**, 274.
13. Rose, *J. Res. Nat. Bur. Stand.*, 1938, **20**, 129.
14. Hibbard and Cleaves, *Analyt. Chem.*, 1949, **21**, 486.
15. Tuot and Barchewitz, *Bull. Soc. Chim.*, 1950, 851.
16. Barnes, Gore, Liddel and Van Zandt Williams, *Infra-red Spectroscopy* (Reinhold, 1944).
17. Thompson and Torkington, *Proc. Roy. Soc.*, 1945, **A184**, 3.
18. Thompson and Torkington, *Trans. Faraday Soc.*, 1945, **41**, 246.
19. Snyder and Schachtschneider, *Spectrochim. Acta*, 1963, **19**, 117.
20. Hadži and Sheppard, *Proc. Roy. Soc.*, 1953, **A216**, 247.
21. Richards and Thompson, *Proc. Roy. Soc.*, 1948, **A195**, 1.
22. Herzberg, *Infra-red and Raman Spectra of Polyatomic Molecules* (Van Nostrand, 1945), p. 339.
23. Pitzer, *J. Chem. Phys.*, 1944, **12**, 310.
24. Herzberg, *op. cit.*, p. 317.
25. Herzberg, *op. cit.*, p. 314.
26. Brown and Sheppard, *Discuss. Faraday Soc.*, 1950, **9**, 144.
27. Thompson and Cave, *Trans. Faraday Soc.*, 1951, **47**, 946, 951.
28. Rasmussen, *J. Chem. Phys.*, 1948, **16**, 712.

29. Cf. Randall, Fowler, Fuson and Dangl, *Infra-red Determination of Organic Structures* (Van Nostrand, 1949).
30. Herzberg, *op. cit.*, p. 344.
31. Sobotka and Stynler, *J. Amer. Chem. Soc.*, 1950, **72**, 5139.
32. Sutherland and Simpson, *J. Chem. Phys.*, 1947, **15**, 153.
33. Simpson and Sutherland, *Proc. Roy. Soc.*, 1949, **A199**, 169.
34. Kohlrausch, *Ramanspektren* (Akad. Verlags. Becker-Erler, Leipzig, 1943).
35. Tuot and Lecomte, *C.R. Acad. Sci. Paris*, 1943, **216**, 339.
36. Sheppard and Sutherland, *Nature*, 1947, **159**, 739.
37. Sutherland and Vallence Jones, *Nature*, 1947, **160**, 567.
38. Fellgett, Harris, Simpson, Sutherland, Thompson, Whiffen and Willis, *Inst. Petroleum Reports*, XI, 1946.
39. Wiberley and Bassett, *Analyt. Chem.*, 1950, **22**, 841.
40. Brown, Sheppard and Simpson, *Discuss. Faraday Soc.*, 1950, **9**, 261.
41. Plyler and Acquista, *J. Res. Nat. Bur. Stand.*, 1949, **43**, 37.
42. Bartleson, Burk and Lankelma, *J. Amer. Chem. Soc.*, 1946, **68**, 2513.
43. Derfer, Pickett and Boord, *J. Amer. Chem. Soc.*, 1949, **71**, 2482.
44. Marrison, *J. Chem. Soc.*, 1951, 1614.
45. Van Volkenburgh, Greenlee, Derfer and Boord, *J. Amer. Chem. Soc.*, 1949, **71**, 3595.
46. Barton, *J. Chem. Soc.*, 1951, 1444.
47. Pliva and Herout, *Coll. Trav. Chim. Tchecosl*, 1950, **15**, 160.
48. Wilson, *J. Chem. Phys.*, 1943, **11**, 369.
49. Reid and Sack, *J. Amer. Chem. Soc.*, 1951, **73**, 1985.
50. Pozefsky and Coggeshall, *Analyt. Chem.*, 1951, **23**, 1611.
51. Roberts and Chambers, *J. Amer. Chem. Soc.*, 1951, **73**, 5030.
52. Gauthier, *C. R. Acad. Sci. Paris*, 1950, **231**, 837.
53. Evans, Hibbard and Powell, *Analyt. Chem.*, 1951, **23**, 1604.
54. Sheppard, *J. Inst. Pet.*, 1951, **37**, 95.
55. "The Spectroscopic Panel of the Hydrocarbon Research Group," *J. Inst. Pet.*, 1951, **37**, 109.
56. Roberts and Chambers, *J. Amer. Chem. Soc.*, 1951, **73**, 5034.
57. Roberts and Simmons, *J. Amer. Chem. Soc.*, 1951, **73**, 5487.
58. Slabey, *J. Amer. Chem. Soc.*, 1952, **74**, 4930.
59. Slabey, *ibid.*, p. 4928.
60. McMurry and Thornton, *Analyt. Chem.*, 1952, **24**, 318.
61. Hastings, Watson, Williams and Anderson, *Analyt Chem.*, 1952, **24**, 611.
62. Wiberley and Bunce, *Analyt. Chem.*, 1952, **24**, 623.
63. Pitzer and Kilpatrick, *Chem. Reviews*, 1946, **39**, 435.
64. Sheppard and Simpson, *Quarterly Reviews*, 1953, **7**, 19.
65. Freeman, *J. Amer. Chem. Soc.*, 1953, **75**, 1859.
66. Jones and Cole, *J. Amer. Chem. Soc.*, 1952, **74**, 5648.
67. Jones, Cole and Nolin, *ibid.*, p. 5662.
68. Robert and Favre, *C.R. Acad. Sci. (Paris)*, 1952, **234**, 2270.
69. Bridson-Jones, Buckley, Cross and Driver, *J. Chem. Soc.*, 1951, 2999.
70. Bridson-Jones and Buckley, *ibid.*, p. 3009.
71. Francis, *J. Chem. Phys.*, 1951, **19**, 942.
72. Francis, *Analyt. Chem.*, 1953, **25**, 1466.
73. Slabey, *J. Amer. Chem. Soc.*, 1954, **76**, 3604.
74. Cole, *J. Chem. Soc.*, 1954, 3807, 3810.
75. Tschamler, *J. Chem. Phys.*, 1954, **22**, 1845.
76. Keller and Sandeman, *J. Polymer. Sci.*, 1955, **15**, 133.
77. Tobin and Carrano, *J. Polymer. Sci.*, 1957, **24**, 93.
78. Stein and Sutherland, *ibid.*, 1953, **12**, 370.

79. Nolin and Jones, *J. Amer. Chem. Soc.,* 1953, **75**, 5626.
80. Rathgens, Freeman, Gwinn and Pitzer, *ibid.,* 1953, **75**, 5634.
81. Higuchi, Kuno, Tanaka and Kamason, *Spectrochim. Acta,* 1972, **28A**, 1335.
82. Bohlmann, *Chem. Ber,* 1958, **91**, 2157.
83. Mirone and Fabbri, *Gazz. chim.,* 1954, **84**, 187.
84. Kaye, *Spectrochim. Acta,* 1954, **6**, 257.
85. Lauer and Rosenbaum, *App. Spectroscopy,* 1952, **6**, 29.
86. Luther and Czerwany, *Z. Phys. Chem.,* 1956, **6**, 286.
87. Jones, Nolin and Roberts, *J. Amer. Chem. Soc.,* 1955, **77**, 6331.
88. Barton, Page and Warnhoff, *J. Chem. Soc.,* 1954, 2715.
89. Bellamy and Williams, *ibid.,* 1956, 2753.
90. Sheppard, *Trans. Faraday Soc.,* 1955, **51**, 1465.
91. Randle and Whiffen, *J. Chem. Soc.,* 1955, 3497.
92. Sheppard and Simpson, *J. Chem. Phys.,* 1955, **23**, 582.
93. Bomstein, *Analyt. Chem.,* 1953, **25**, 512.
94. Keller and Sandeman, *J. Polymer. Sci.,* 1954, **13**, 511.
95. Stein and Sutherland, *J. Chem. Phys.,* 1954, **22**, 1993.
96. Krimm, *ibid.,* 1954, **22**, 567.
97. *Idem, Phys. Rev.,* 1954, **94**, 1426.
98. Novak, *Zhur. Tekh Fiz.,* 1954, **24**, 18.
99. Nikitin and Pokrovskiy, *Doklady, Akad Nauk, S.S.S.R.,* 1954, **95**, 109.
100. Krimm, Liang and Sutherland, *J. Chem. Phys.,* 1956, **25**, 549.
101. Primas and Günthard, *Helv. Chim. Acta,* 1953, **36**, 1791; 1956, **39**, 1182.
102. Brown, Sheppard and Simpson, *Phil. Trans.,* 1954, **A247**, 35.
103. *Idem, J. Phys. Radium,* 1954, **15**, 593.
104. Mizushima, *The Structure of Molecules* (Academic Press, 1954), pp. 98 *et seq.*
105. Bateuv, *Izvest Akad. Nauk. S.S.S.R. Otdel Khim Nauk,* 1947, **1**, 3.
106. Plyler, *J. Opt. Soc. Amer.,* 1947, **37**, 746.
107. Borello and Mussa, *J. Polymer Sci.,* 1954, **13**, 402.
108. Gates, *J. Chem. Phys.,* 1949, **17**, 393.
109. Donneaud, *Compt. Rend. Acad. Sci. Paris,* 1954, **239**, 1480.
110. *Idem, Rev. Inst. Franc. Petrole,* 1955, **10**, 1525.
111. Jones and Herling, *J. Org. Chem.,* 1954, **19**, 1252.
112. Brini, *Bull. Soc. Chim. Fr.,* 1955, 996.
113. Henbest, Meakins, Nicholls and Wagland, *J. Chem. Soc.,* 1957, 1462.
114. Henbest, Meakins, Nicholls and Taylor, *ibid.,* 1459.
115. McKean, *Chem. Comm.,* 1971, 1373.
116. McKeen, Duncan and Batt, *Spectrochim. Acta,* 1973, **29A**, 1037.
117. Kreuger and Jan, *Can. J. Chem.,* 1970, **48**, 3329, 3236.
118. Kreuger, Jan and Wieser, *J. Mol. Structure,* 1970, **15**, 575.
119. Bellamy, *Advances in Infra-red Group Frequencies* (Methuen, London, 1968).
120. Gates and Steel, *J. Mol. Structure,* 1968, **1**, 349.
121. Durig and Morrissey, *J. Mol. Structure,* 1968, **2**, 377.
122. Durig and Morrissey, *J. Chem. Phys.,* 1966, **45**, 1269.
123. Danti, Lafferty and Lord, *J. Chem. Phys.,* 1960, **33**, 294.
124. Galakov and Simov, *J. Mol. Structure,* 1972, **11**, 341.
125. Vakhleuva, Finkel, Sverdlov and Zaitseva, *Optics and Spectroscopy,* 1968, **25**, 161.
126. Tosi and Pinto, *Spectrochim. Acta,* 1872, **28A**, 585.
127. Whetsel, *Applied Spectroscopy Reviews,* 1968, **2**, 1.

3

Alkenes

3.1. Introduction and table

Infra-red absorption bands associated with the various types of double bond have been extensively studied, so that recognition of the presence and type of unsaturation in an unknown compound is usually possible with reasonable certainty, although this depends to some extent on the amount of substitution at the double bond. This is due to the fact that the most characteristic frequencies available for identification arise from CH stretching and deformation vibrations, so that in materials of the type $R_1 R_2 C = CHR_3$ the number of characteristic frequencies is reduced. Unsaturation in compounds of the type $R_1 R_2 C = CR_3 R_4$ is extremely difficult to detect other than in the Raman spectra, as the only remaining characteristic frequency is the $C = C$ stretching mode, the intensity of which is considerably reduced by the symmetry around the bond. A very full review of fundamental work on simple molecules in this field has been given by Sheppard and Simpson [47]. Other useful reviews are those of Lecomte and Naves [80], Gruzdev [81], O'Connor [82], and Dollish *et al.* [122].

Many of the characteristic frequencies of double bonds occur in the region of C—O and C—C stretching bands, so that the occurrence of an absorption band at a specific frequency can not alone be taken as evidence for the presence of a certain type of double bond, although from the absence of a band we may conclude the absence of the group with fair confidence.

On the other hand, examination of all the regions in which specific absorption frequencies occur usually allows a fairly certain identification of the double bond and its type, at least for the less heavily substituted materials. The relatively constant intensity of the specific bands arising from CH stretching and CH out-of-plane deformations in relation to the molecular weight has also resulted in the development of numerous quantitative methods for the analysis of complex hydrocarbons and polymers. It has been suggested by some workers that a statistical relationship connects the C=C and CH frequencies of olefines [83]. This may well hold good for any limited series of related hydrocarbons, but is unlikely to apply very widely, as the C=C stretching frequencies are more sensitive to changes of mass than are the deformation modes.

The correlations relating bond structure and frequencies are listed in Table 3.

Table 3

C=C *Stretching Vibrations*

Non-conjugated	$1680-1620$ cm^{-1} (intensity very variable)
Phenyl conjugated	Near 1625 cm^{-1} (intensity enhanced)
CO or C=C conjugated	Near 1600 cm^{-1} (intensity enhanced)

C–H *Stretching and Deformation Vibrations*

—CH=CH— (*trans*)	$3040-3010$ cm^{-1} (m)	CH stretching
	$970-960$ cm^{-1} (s)	CH out-of-plane deformation
	$1310-1295$ cm^{-1} (s-w)	CH in-plane deformation
—CH=CH— (*cis*)	$3040-3010$ cm^{-1} (m)	CH stretching
	Near 690 cm^{-1}	CH out-of-plane deformation (correlation uncertain)
CH=CH$_2$ (Vinyl)	$3040-3010$ cm^{-1} (m)	CH stretching
	$3095-3075$ cm^{-1} (m)	CH stretching
	$995-985$ cm^{-1} (s)	CH out-of-plane deformation
	$915-590$ cm^{-1} (s)	CH$_2$ out-of-plane deformation
	$1856-1800$ cm^{-1} (m)	Possible overtone of the above
	$1420-1410$ cm^{-1} (s)	CH$_2$ in-plane deformation
	$1300-1290$ cm^{-1} (s-w)	CH in-plane deformation
CH$_1$R$_2$=CH$_2$	$3095-3075$ cm^{-1} (m)	CH stretching
	$895-885$ cm^{-1} (s)	Out-of-plane deformation
	$1800-1750$ cm^{-1} (m)	Possible overtone of the above
	$1420-1410$ cm^{-1} (s)	CH$_2$ in-plane deformation
CR$_1$R$_2$=CHR$_3$	$3040-3010$ cm^{-1} (m)	CH stretching
	$840-790$ cm^{-1} (s)	CH out-of-plane deformation

3.2. C=C Stretching vibrations

3.2(a). Position and intensity. In contrast to the strong C=O absorption the C=C stretching vibration gives rise only to weak bands in the infra-red in non-conjugated compounds. The position of the absorption peak is modified to some extent by the nature of the substituents on the two carbon atoms, but the main factors influencing both the frequency and the intensity are symmetry considerations, conjugation and fluorine substitution. The influence of each of these factors is discussed below.

In compounds which contain only an isolated double bond the absorption peak normally occurs within the range 1680–1620 cm^{-1}, with the majority absorbing in the 1660–1640 cm^{-1} region. Barnes *et al.* [1] list about twenty compounds in which the absorption appears in the 1660–1640 cm^{-1} range, whilst numerous fundamental studies on hydrocarbons and small molecules have also resulted in the assignment of the C=C stretching vibration to this region. Thus, Sheppard [2] finds a band at 1645 cm^{-1} in butene-1, which he assigns to this mode, and Thompson and Torkington [3] have reported a band near 1650 cm^{-1} in a series of allyl halides and allyl alcohol. Randell *et al.* [4] also include a number of examples of structural assignments of the C=C in this region, such as *iso*butylene at 1637 cm^{-1} and propylene and butene-2 at 1647 cm^{-1}.

This correlation has been further studied by Sheppard and Sutherland [5], and, as a result of studies on hydrocarbons, they have suggested that it is possible to differentiate between the double-bond types $RR_1C=CH_2$ and $RR_1C=CHR_2$ by means of the position of the absorption peak in this region. They quote ranges of 1655–1645 cm^{-1} and 1680–1670 cm^{-1}, respectively, for these two types of linkage. These two linkages can also be differentiated by the intensity in this region, as the terminal double bond gives rise to a considerably more intense band than that from a double bond included in the chain. This differentiation has been supported by Thompson and Whiffen [6], who point out that it is also in line with the results of Raman spectra of similar compounds [7], and it has also been applied by Thompson and Torkington [8] in assigning the 1645 and 1690 cm^{-1} bands of polymerised 2 : 3-dimethylbutadiene to the two types of double bond. No characteristic band has been assigned to the $R_1R_2C=CR_3R_4$ linkage, but from Raman data it is probable that this will also absorb in the 1690–1680 cm^{-1} region [47, 58]. The intensity of this band in the infra-red will, however, always be low, and often it will not be observed at all. The classified studies of the A.P.I. spectra by McMurry and Thornton [58] support this assignment, as do the summaries of Sheppard [55] and of Sheppard and Simpson [47], of the results of studies of simple olefinic compounds. The latter workers quote the following average values for the C=C vibration in simple olefines of the type studied:

$$CHR=CH_2, 1643 \text{ cm}^{-1}; CHR^1=CHR^2 \text{ (trans)}, 1673 \text{ cm}^{-1};$$
$$CHR^1=CHR^2 \text{ (cis)}, 1657 \text{ cm}^{-1}; CR^1R^2=CH_2, 1653 \text{ cm}^{-1};$$
$$CR^1R^2=CHR^3, 1670 \text{ cm}^{-1}; CR^1R^2=CR^3R^4, 1670 \text{ cm}^{-1}$$

This division of the correlation into groups has been valuable in hydrocarbon studies and in work with relatively simple molecules. However, it is not intended for use with oxygenated or other olefines with polar groupings, as minor disturbances of the C=C modes then occur. Vinyl compounds [74] with an oxygen or a carbonyl group at the double bond absorb between 1652 and 1611 cm^{-1}, with normal vinyl esters [102] near 1640 cm^{-1} and acrylates and crotonates in the range [62] 1638–1633 cm^{-1}. The corresponding vinylidene derivatives absorb between 1670 and 1632 cm^{-1}. In general, with the exception of fluorine which is discussed below, the attachment of a polar atom lowers the C=C stretching frequency, so that vinyl chloride absorbs at 1610 cm^{-1} and vinyl bromide at 1593 cm^{-1}. This is probably also related to the strength of the C–X bond as the attachment of elements such as silicon or sulphur [111] gives low frequencies also. Similarly, small differences in C=C frequencies occur even with the minor differences in strain which occur in the various steroid rings. Jones et al. [9] have studied a large number of such materials and have found that the position of the C=C absorption band is closely related to the relative position of the unsaturated unit in the ring system. Thus ten different Δ^5-unsaturated steroids absorbed in the range 1672–1664 cm^{-1}, whereas a number of Δ^{11}-steroids absorbed between 1628 cm^{-1} and 1624 cm^{-1}. Similar correlations were found for other positions of the double bond, and they concluded that the peak frequency was dependent on the double-bond position and that it was relatively unaffected by other structural features. In all these steroids the absorption was uniformly weak. In none of the unconjugated materials (twenty-nine in all), however, did the absorption fall outside the overall range 1680–1620 cm^{-1}. Henbest et al. [77, 108] and Cole and Thornton [109] have obtained similar data for cis double bonds in steroids and triterpenes.

Studies of the influence of greater degrees of ring strain have been made by Lord and Walker [75]. As the size of the ring diminishes and strain is increased, the C–H stretching frequency rises and the –C=C– frequency falls. The C=C frequency of *cyclo*heptene is 1651 cm^{-1} and *cyclo*butene 1566 cm^{-1}. The fusion of an additional ring increases the degree of strain, so that the five-membered rings of *biscyclo*(2-2-1)-hept-2-ene absorb at 1568 cm^{-1} close to *cyclo*butene. The comparable bridged system with a six-membered ring likewise absorbs close to *cyclo*pentene. Supporting evidence that fusion of a second ring does in fact increase the ring strain is given by reactivity studies [106], which show a similar effect. However, with the last member of the series the effects of angle change are reversed, and *cyclo*propene (1647 cm^{-1}) absorbs at a higher frequency than *cyclo*butene (1566 cm^{-1}). The reasons for this are well understood; coupling with the adjacent C–C single bonds is minimal in *cyclo*butene with 90° angles. In *cyclo*propene, the C=C stretching must be accompanied by a corresponding extension of the single bonds and the frequency is therefore raised.

The substitution of methyl groups on the double bond reduces the effects on the C=C frequency of changes in ring size from six- to four-membered rings, and all of these absorb at much the same frequency (1685 cm^{-1}). In these cases the reductions in coupling which result from changes to smaller ring angles are balanced by corresponding increases in coupling with the C–CH$_3$ bonds which are also changing in angle in the opposite direction. However, in the substituted *cyclo*propenes these two effects cease to cancel and now reinforce each other. The result is a dramatic jump in the double bond frequency. Methyl*cyclo*propene absorbs at 1784 cm^{-1} and 1 : 2-dichlorpropene [123] at over 1800 cm^{-1}. The origins and significance of these effects have been discussed by Lord and Miller [84].

With exocyclic double bonds the effect of strain is in the opposite direction, leading to rises in the C=C frequencies as the ring strain is increased. Methylene groups attached to three-, four-, five- and six-membered ring systems absorb respectively [84] at 1736 cm^{-1} [124], 1678 cm^{-1}, 1657 cm^{-1} and 1651 cm^{-1}. Studies on related molecules include, 1 : 2-dimethylene*cyclo*pentane [110], trimethylene*cyclo*propane [126] and tetramethylene*cyclo*butane [125]. The effects of ring strain on C=C intensities have not been studied in detail, but it is noteworthy that some bridged-ring *cyclo*heptene compounds [108] studied by Henbest *et* *al.*, in which the high C—H frequencies indicate considerable strain, did not show any detectable C=C absorptions.

In general, therefore, the C=C absorption will be found as a weak band in the range 1680—1620 cm^{-1}. In simple hydrocarbons and in steroids [85, 108] further correlations are available within this range which will give valuable additional information. Reference to the other regions of the spectrum in which CH vibrations occur and a consideration of the relative intensities of all the bands connected with double-bond structure are essential in the recognition of the type of linkage involved, especially in symmetrical non-conjugated C=C compounds in which the C=C frequency may not appear. In cases of doubt the changes following hydrogenation, bromination or other chemical attack upon the double bond can be studied. One such application by Leonard and Gash [78] is valuable in the identification of αβ-unsaturated tertiary amines. These compounds epimerise to the partial structure $\begin{array}{c}\\[-0.5em]\overline{C}H-C=\overset{+}{N}\\\end{array}$ on treatment with perchloric acid, with a consequent shift of the double-bond absorption of 20—50 cm^{-1} towards higher frequencies. βγ-Unsaturated amines, on the other hand, are not affected and continue to show absorption in the original 1665—1640-cm^{-1} range.

As indicated below, C=C absorptions vary considerably in intensity, depending upon the symmetry of the bond and on factors such as conjugation, etc. No detailed studies of absolute intensities have been made, but Jones and Sandorfy [86] have compiled a useful summary of intensity data in the A.P.I. series, and Davison and Bates [74] report extinction coefficients for a variety of oxygenated olefines.

3.2(b). The influence of symmetry. It can be predicted that no C=C stretching vibration will appear in the infra-red from compounds with a *trans* double bond at a centre of symmetry. This arises from the fact that infra-red absorption takes place only when there is a change of dipole moment, and there is no appreciable change involved in the C=C stretching vibration of a *trans* symmetrical molecule. On the other hand, some change will occur in the dipole moment in the case of compounds with *cis* double bonds, whilst both types will be active in Raman spectra [47, 55]. The intensity of the C=C double bond in the infra-red would therefore be expected to diminish when it is moved from the end of a chain towards the centre and the molecule becomes more symmetrical. This affords a ready explanation of the enhanced intensity of this group absorption in terminal $R_1R_2C=CH_2$ structures as against compounds of the type $R_1R_2C=CHR_3$.

Numerous examples of this effect are known. Thus, ethylene and *trans*-dichloroethylene are both inactive in the infra-red in this region, but emit strongly at 1623 cm^{-1} and 1577 cm^{-1}, respectively, in their Raman spectra [10]. *cis*-Dichloroethylene [11], on the other hand, shows a strong infra-red absorption at 1590 cm^{-1}. Similarly, Flett [12] has found no C=C absorption in fumaric acid, and this is, again, associated with the presence of a centre of symmetry.

Many cases have been reported in which the intensity of the C=C band falls as the tendency to symmetry increases. Kletz and Sumner [13] drew attention to the fall in intensity of this band in a series of trimethylpentenes in which the symmetry was gradually increased, and they have suggested that observations on the relative intensity may give an indication of the type of linkage involved. Thus $2:4:4$-trimethylpent-1-ene with a terminal $=CH_2$ link shows the strongest absorption of this series. Assigning this an arbitrary value of 1, the value for the optical density of the corresponding $2:4:4$-tri-methylpent-2-ene is only 0.35. The same value is given by $3:4:4$-trimethylpent-2-ene, which also has the $R_1R_2C=CHR_3$ structure, and there is a further decrease in intensity with increase of symmetry in $2:3:4$-trimethylpentene to 0.14 when the double bond is of the type $R_1R_2C=CR_3R_4$. Shreve *et al.* [14] have been unable to detect the C=C stretching vibration in *trans* long-chain unsaturated fatty acids and alcohols in which the double bond is situated towards the centre of the chain. On the other hand, weak indications of this link could be found with comparable *cis*-compounds. In contrast, Δ^{10}-undecenoic acid and its methyl ester both show relatively strong C=C absorption due to the terminal vinyl grouping.

Intensity studies on the C=C linkage will therefore yield some indication of the nature of the double bond involved, and in cases where this is known they can be used for the study of *cis—trans* isomerism. A typical example of the latter use is that of Bernstein and Powling [15] in studies on *cis*- and *trans*-1 : 2-dichloro-1-propene and their deutero-derivatives. After separation of the two isomers they could be readily identified by the relatively high intensity of the C=C absorption at 1606 cm^{-1} (D) and 1614 cm^{-1} (H) in the *cis*-compound, against the low intensity of the bands at

1605 cm^{-1} (D) and 1615 cm^{-1} (H) in the *trans*-compound. Additional evidence in these cases can also be obtained from the overall number of bands in the spectrum. These are considerably fewer in the more symmetrical isomer, and in the case quoted only about half the number of absorption bands appear in the *trans*- as compared with the *cis*-compound. It is also a general — but obviously not invariable — rule that the C=C absorption of *cis*-isomers is a little lower than that of the corresponding *trans*-compounds [47, 112–114]. The small but significance shifts which occur in this way are well illustrated in the various isomers of the compound $CH_3 CH=CH-CH=CH-COOCH_3$ which have been studied by Allan, Meakins and Whiting [87]. The *trans–trans*-compound absorbs at 1642 and 1614 cm^{-1}, and the *cis–cis*- at 1623 and 1587 cm^{-1}, whilst the mixed isomers show intermediate values. This effect has been used, for example, in the differentiation of *cis*- and *trans*-propenylbenzenes [73]. In all cases the C=C frequency is of course very intense in the Raman spectrum, and this is the most positive method of identification. It is particularly useful also in cases where it is difficult to differentiate between a C=C and a C=O band in the infra-red as the latter is always relatively weak in the Raman spectrum.

3.2(c). The influence of conjugation. *Aliphatic Conjugation.* When aliphatic conjugation of C=C bonds occurs, splitting of the two double-bond absorption bands usually results [16, 17]. This arises partly by mechanical interaction and partly because the nature of the normal modes has been altered, and, for two identical bonds, the modes of the in-phase and opposite-phase simultaneous vibration of the two bonds will come into play. Nevertheless, in simple conjugated systems one band is usually stronger than the other, and this usually occurs at about 30 cm^{-1} lower frequency than the corresponding non-conjugated material. Barnes *et al.* [1] quote a number of examples of absorption near 1600 cm^{-1} arising from conjugated C=C systems, and 1 : 3-butadiene (1597 cm^{-1}) [18] and vinyl acetylene [2] (1600 cm^{-1}) are other instances. In all cases a very considerable enhancement of the intensity is observed in the infra-red.

The fullest investigation of the effects of a number of conjugated double links upon each other is due to Blout, Fields and Karplus [19]. They examined a series of long-chain conjugated materials, including polyenes containing two, three, four and five conjugated C=C linkages. As they expected, the absorption pattern was complex in the 1650–1600 cm^{-1} region, but they noted that the diene had two, the triene three, and the tetraene four double bands in this region. The strongest band in all cases remained near 1650 cm^{-1} throughout the series. They also examined polyene azides and polyene aldehydes containing two, four, six, eight and ten conjugated double linkages. In these cases, in which conjugation with a different type of unsaturated link is also involved, the assignment becomes even more complicated, but the C=C vibrations appeared to shift progressively towards lower frequencies as the number of bonds in conjugation was increased. They point out that their results cannot be properly interpreted without mathematical analysis, but their findings confirm the general statement that conjugation results in a shift of the main C=C absorption band towards lower frequencies. The extent and occurrence of this shift will, however, be conditioned by many other factors which are not yet fully understood. Most conjugated dienes take a *trans* configuration about the C=C–C=C link and no bands due to a *cis* form are seen. However, in highly substituted dienes there is the possibility of the occurrence of a *cis/trans* form which will give rise to additional bands, and in rare cases this occurs with simple dienes. 1 : 3-Penta-diene is an example of this, giving 4 C=C bands rather than 2 due to the contribution of the *cis/trans* form. The differences between the C=C frequencies of *s cis* and *s trans* ketones are not sufficient in themselves to enable the two to be differentiated, but this can be done on the basis of the separation from the C=O band. In *s trans* compounds the separation of these bands is less than 60 cm^{-1} but it is more than this with the *s cis* series. There are also characteristic intensity differences [127].

Conjugation with acetylenic links has been studied by Allan *et al.* [87] and many others [128]. The frequency effects are similar, except that doubling of the –C=C– band does not, of course, occur. The intensity of the absorption is also increased, but the effect is smaller than that arising from conjugation with a carbonyl group. Thus the two compounds

$$CH_3-C\equiv C-CH=CH-COOCH_3 \text{ and } CH_3-CH=CH-C\equiv C-COOCH_3$$

can be readily differentiated by the reversal of the relative intensities of the C=C and C≡C absorptions, depending upon which is adjacent to the carbonyl group. Conjugation effects in olefines have also been discussed by Scrocco and Salvetti [92], and other examples are available in papers by Sörensen and Sörensen [115] and by Bohlmann and Mannhardt [116].

In the sterol series Jones *et al.* [9] have examined a limited number of conjugated dienes, all of which give two maxima in the $C=C$ stretching region. A $\Delta^{3:5}$-sterol, for example, absorbs at 1618 and 1578 cm^{-1}. However, marked interaction effects appear to arise when ester groups are also present in the molecule, and two $\Delta^{3:5}$ esters absorb at 1670 and 1639 cm^{-1}, although the ester group is not close to the double bonds. In this connection it should be noted that doubling of the $C=C$ absorption is a common feature both of vinyl ethers and of acrylates [74]. This is probably associated with the appearance of an overtone frequency or results from rotational isomerism, but the presence of more than one $C=C$ absorption does not necessarily imply that more than one olefinic group is present. Conjugation effects in steroids and triterpenes have also been studied by Henbest *et al.* [108] and by Cole and Thornton [109].

Aromatic conjugation. When a double bond is conjugated with an aromatic ring the $C=C$ frequency shift is generally less than that occurring in full aliphatic conjugation. This is to be expected from the weaker nature of the double bond of the aromatic system. Absorption therefore appears in the neighbourhood of 1625 cm^{-1}. Barnes *et al.* [1] quote several examples of this shift occurring with absorption close to 1625 cm^{-1}. Cinnamic acid, as a typical example, has its main $C=C$ absorption near 1626 cm^{-1}. Bearing in mind, however, the overall frequency range of 1680—1620 cm^{-1} of the unconjugated $C=C$ link, it will be seen that absorption near 1625 cm^{-1} does not necessarily indicate conjugation with an aryl group. However, consideration of the relative intensity of this band and of those of the aromatic ring, which also become enhanced on conjugation, will frequently give a useful lead as to whether this type of conjugation is present. There is, also, a third aromatic $C=C$ vibration which appears near 1590 cm^{-1} when the ring is in conjugation, which is also of considerable assistance in interpretation (see Chapter 5).

αβ unsaturated ketones, etc. Conjugation of a double bond with a carbonyl grouping results in a low-frequency shift of the $C=C$ absorption which is similar to, but smaller than, that arising from a normal pair of conjugate double bonds. Thus Rasmussen *et al.* [20] place the $C=C$ absorption of isophorone at 1639 cm^{-1}, and state that other unpublished studies on this type of conjugated ketone show that the $C=C$ band falls in the range 1647—1621 cm^{-1}.

Similarly, they find acetyl acetone acetate absorbs strongly at 1633 cm^{-1}, and Blout *et al.* [19] find the strongest bands in the C=C region for crotonaldehyde and 2 : 4-hexadienal at 1638 cm^{-1} and 1642 cm^{-1} in solution. In the solid state these compounds give values about 10 cm^{-1} higher. Raman data give essentially similar results for such systems [21]. As in the case of conjugated dienes, however, the degree of shift can be considerably influenced by other factors, such as inclusion in ring systems. In unsaturated keto-steroids, for example, the frequency in some cases falls as low as 1588 cm^{-1}, its actual value being determined almost entirely by its position in the system. Jones *et al.* [9] and also Furchgott *et al.* [22] have examined numbers of such compounds and have been able to correlate the observed frequencies with ring position; $\Delta^{16.20}$ ketones, for example, absorb at 1592–1588 cm^{-1}, whilst $\Delta^{4.3}$ ketones absorb at 1617 cm^{-1}. The C=C frequency is only slightly reduced in $\alpha\beta$ unsaturated acids. Freeman [37] quotes 1653–1631 cm^{-1} for this vibration in a series of 2-hexenoic acids. Data on the molecular extinction coefficients of a number of compounds of this type are now available in the literature [74, 87].

In esters the C=C frequency is close to 1640 cm^{-1}. Indeed McManis quotes 51 vinyl esters [102] as absorbing within 2 cm^{-1} of this point except where there is an electronegative atom at the α-carbon when there is a small frequency rise. Acrylates and crotonates absorb [62] between 1638 and 1633 cm^{-1}.

3.2(d). Influence of fluorine substitution. Thompson and Torkington [23] have drawn attention to the remarkable increase in the C=C stretching frequency which occurs when the fluorine atoms are substituted on one of the carbon atoms constituting the double bond. The C=C absorption of ethylene at 1626 cm^{-1} shows only a small frequency shift to 1650–1645 cm^{-1} with a single fluorine substituent, but rises to 1730 cm^{-1} in $CH_2=CF_2$ and $CCl_2=CF_2$. The ethylene frequency quoted is, of course, derived from Raman data, but the remainder show sufficient asymmetry for this vibration to be active in the infra-red. The reasons for this shift are not known, but must indicate a shortening of the double bond under the influence of the fluorine substitution. These findings have been confirmed by Hatcher and Yost [64], and by others [65, 66], and cases of a greater frequency shift on further fluorine substitution are known. Edgell [24] finds the C=C stretching absorption of fully fluorinated propene at 1798 cm^{-1}, and Park, Lycan and

Lacher [25] have published a spectrum of trifluoroethylene in which a strong band is shown near 1780 cm^{-1}. The latter workers do not make any specific assignment for this absorption, but it is significant that the band is absent from the corresponding saturated dichloride $CF_2 Cl{-}CFClH$ and also from the dibromide. Perfluorobutene-1 also absorbs [88] at 1792 cm^{-1}, and it is significant that the frequency falls to 1733 cm^{-1} in perfluorobutene-2. Other comparable cases have been discussed by Brice *et al.* [88] and by Weiblen [118]. Haszeldine [67] has tabulated the $C{=}C$ frequencies for variously substituted ethylenes and has shown that when the fluorine is not directly attached to the double bond, as in the group $-CH{=}CH{-}CF_3$, its influence is virtually negligible.

3.3. $=$ CH Stretching vibrations

One of the most valuable means of detection of double bonds through infra-red spectra is the examination of the region near 3000 cm^{-1} with high-dispersion. The C–H valence stretching absorptions occur in this region, and these are virtually independent of the nature of the associated structure apart from the valency state of the carbon atom. The groupings $=CH_2$ and $=CRH$ therefore exhibit characteristic CH stretching frequencies which can be recognised and are easily distinguished from the $-CH_2$ and $-CH_3$ frequencies occurring at longer wave-lengths. Fox and Martin [26] have shown that the C–H stretching frequencies of a number of hydrocarbons containing the $=CH_2$ group occur in the range $3092-3077 \text{ cm}^{-1}$. All of these, except ethylene, showed a second band in the region $3025-3012 \text{ cm}^{-1}$, which is also shown by five other unsaturated hydrocarbons, and is ascribed to the $=CH-$ vibration. By the use of these bands it is therefore possible to differentiate the structures $-CH{=}CH_2$ (both bands), $-\overset{|}{C}{=}CH_2$ (3079 cm^{-1} only), and $-\overset{|}{C}{=}CH-$ (3019 cm^{-1} only). Sheppard and Simpson [47], for example, record the CH stretching absorptions for a considerable number of substituted ethylenes. The $=CH_2$ absorption is found in the range $3095-3075 \text{ cm}^{-1}$ (thirteen examples) and the $=CH$ absorption in the range $3030-3000 \text{ cm}^{-1}$ (sixteen examples). The normal CH vibrations of saturated structures occur at frequencies below 3000 cm^{-1}, and do not interfere. In the case of some of the higher hydrocarbons, however, the methyl-group absorption band can result in masking of the lower frequency $=CH$ band.

These findings have been confirmed for hydrocarbons by Saunders and Smith [63] and applied by Saier and Coggeshall [27] to the quantitative estimation of complex hydrocarbon mixtures. They have analysed successfully mixtures of the isomers 2 : 4 : 4-trimethyl 1- and 2-pentenes, using the $=CH_2$ band of the former near 3090 cm^{-1} and mixtures of n-octane, 2-octene and 1-octene, using bands at 3067, 3012 and 2857 cm^{-1}. The $=C-H$ stretching vibrations of aromatic rings occur near 3030 cm^{-1}, and these also have been used in quantitative studies such as in the analysis of benzene n-heptane mixtures and in the determination of benzyl alcohol in n-octanol and methyl ethyl ketone. Several other instances are given [27] of this type of analysis, all of which give recovery figures within 1.0 per cent of the theoretical amounts. These absorptions can also be used in kinetic studies, as in the thermal polymerisation of styrene [90].

The identification of double-bond types in sterols and triterpenes has also been carried out by Jones *et al.* [9, 28] and others [108, 109] using the high-frequency region. They found that in saturated steroids the first absorption maximum in this region was at 2970 cm^{-1}, whereas steroids containing an ethylenic double bond show a band in the 3040–3030 cm^{-1} range. In some cases the insolubility of the material made the observation of the absorption peak difficult in solution, but a band in the appropriate region has been found in a considerable number of unsaturated steroids, and appears to show little change of position with ring structure or with ketonic conjugation. Vinyl testosterone also shows a weak band near 3085 cm^{-1}, arising from the terminal vinyl grouping, whilst four ethynyl derivatives absorb at 3310 cm^{-1} due to the $C{\equiv}C-H$ group. The positions of these bands are very little affected by the nature of other substituents, indeed, the $N=CH_2$ group in formaldoxime absorbs at 3098 cm^{-1}. With terminal $=CH_2$ groups the 3085 cm^{-1} band arises from the antisymmetric stretch, and a second absorption can be traced at lower frequencies which is due to the symmetric mode. This falls below 3000 cm^{-1} and is usually obscured by absorptions from alkyl groups, but it can be identified at high resolution. Duncan [129] has demonstrated a very useful relationship between the separation of the antisymmetric and symmetric stretching bands and the bond angle between the two hydrogen atoms and the carbon. This is in accord with theory as at a bond angle of 90° the mechanical coupling will be reduced to zero. Nevertheless it is satisfying to find such excellent agreement with the bond angles of accurately measured molecules. This kind of relationship may well exist in many other related series and the development of parallel correlations would provide a valuable new use for group frequency data.

Although the occurrence of the olefinic CH stretches above 3000 cm^{-1} provides a very valuable differentiation from alkyl CH stretches which fall below, it should be borne in mind that there are a few instances, such as chloroform and *cyclo*propane derivatives, in which high frequencies are given by saturated CH links.

The C–H stretching bands give rise to a first harmonic in the $8000-5000 \text{ cm}^{-1}$ region, and these have been employed in a similar way by Rose [29] and others [130], in the identification of structural units.

3.4. Out-of-plane = CH deformation vibrations

One of the most widely used methods of detecting carbon–carbon double bonds and of differentiating the various types is through a study of the out-of-plane deformation of the attached hydrogen atoms. These gave rise to highly characteristic absorptions in the region $1000-800 \text{ cm}^{-1}$.

The first correlation of these absorptions as arising from unsaturated centres was due to Lambert and Lecomte [30, 31], but more precise rules were later formulated by Thompson and Torkington [8], Gore and Johnson [32], Rasmussen and Brattain [16, 33], and Sheppard [55]. The detailed assignment of the various vibrational modes producing these characteristic bands has been described by Sheppard and Sutherland [34], and by Sheppard and Simpson [47], and Torkington [45] has given a mathematical treatment which enables observed frequency shifts to be correlated with the electronegativity of other substituents.

3.4(a). Di-substituted ethylenes. –CH=CH– (*trans.*). *trans*-Ethylenic double bonds give rise to a medium to strong band at $990-965 \text{ cm}^{-1}$. This has been shown by Kilpatrick and Pitzer [3] to be due to the hydrogen atoms which are out of plane at the double bond. In consequence, alteration in the weight of either of the substituents on the carbon atoms has very little effect on the position of the band, and the intensity varies inversely with the molecular weight. This follows from the fact that the absolute intensity from each double bond is constant, but the proportion of double bonds per unit volume falls as the chain length is increased. Rasmussen and Brattain [33] have shown that this absorption band appears only with *trans* double bonds, and this is valuable in elucidating problems of *cis*- and *trans*-structure.

Typical cases of absorption within the 980–965 cm^{-1} range are listed by Shepparᴅ and Sutherland [34], by Sheppard [55] and by Sheppard and Simpson [47] for a number of olefines and ethylenes, and Hampton [36] lists another series of hydrocarbons with additional data. In almost all these cases the frequency lies close to 965 cm^{-1}. Long-chain *trans*-unsaturated fatty acids and alcohols all absorb [38] at 965 cm^{-1}, but if the double bond is $\alpha\beta$ to the carbonyl group, some small displacement to higher frequencies occurs. Crotonic, cinnamic, *o*-coumaric and acrylic acids [12] absorb in the range 980–974 cm^{-1}, and many other examples are known [70, 71, 87, 93, 112]. This effect is a general one for all types of carbonyl groups, including amides and similar materials. Substitution as the α-carbon atom of CN, OH or OR groups has no effect upon this frequency [76, 87, 93], but a rise to 978 cm^{-1} occurs in some methyl-substituted ethylenes. The presence of an α-chlorine atom in an olefine cannot therefore be detected directly from this frequency, but a new characteristic band of unknown origin is found near 1250 cm^{-1} in such materials [76, 87]. The direct substitution of a halogen on to the double bond has a more pronounced effect. Haszeldine [67] quotes values of 935 cm^{-1} for the CH deformation of $-CH=CHBr$ and $-CH=CHCl$, and Kitson [76] confirms this. *trans*-1 : 2-Dialkylvinyl ethers also absorb about 30 cm^{-1} below the normal level [93]. In general an oxygen or halogen atom directly attached to the double bond reduces the frequency by about 35 cm^{-1}. This effect is additive so that the shift in *trans* dichloroethylene is 70 cm^{-1}.

The influence of conjugation upon this frequency has been extensively studied, particularly in relation to multiple conjugations such as occur in vegetable oils, etc., and the available data have been very well reviewed by O'Connor [82], who has a very full bibliography. The effects are again small, but there is usually a shift towards higher frequencies, particularly in long conjugated chains. Allan *et al.* [87] describe many such compounds in which a steady increase in frequency occurs as the conjugated chain is lengthened, and in which the frequency reaches a limiting value of about 1000 cm^{-1} when the chain is also conjugated with acid or ester groups. Some attempts have been made [82, 94, 107] to subdivide the 1000–965 cm^{-1} range into regions characteristic of various steric patterns, in long-chain fatty acids. Thus, *trans–trans*-compounds absorb near 980 cm^{-1}, *cis–trans*- near 984 cm^{-1}, *trans–trans–trans*- near 995 cm^{-1}, etc. However, whilst this approach may be useful within the fatty acid series, it cannot be indiscriminately used elsewhere. For example, Chapman and Taylor [96] instance a number of cases, such as β-carotene with eleven conjugated double bonds, in which the CH deformation frequency is essentially unaltered at 968 cm^{-1}.

Conjugation with aromatic rings, as in the stilbenes [39, 97], does not influence the position of the 965 cm^{-1} band, but it is displaced to slightly lower frequency when conjugated with acetylenic links [87]. The intensity of this absorption in other substituted stilbenes has been studied by Orr [119], who finds significant alterations in band width where steric effects are likely.

The extinction coefficient of this band has been measured by a number of workers, and despite the difficulty in comparing values obtained on different instruments, it is clear that it has a fairly consistent value. This is also of value in confirming that an absorption band at 965 cm^{-1} is due to the —CH=CH— (*trans*) structure. As the band is situated in a spectral region in which very many other vibrations occur, its absence can be regarded as conclusive evidence for the absence of this type of linkage, but when the band is present, confirmation must be sought from intensity measurements and from other spectral regions, such as the C—H stretching and C=C vibrations.

Anderson and Seyfried [40] have given a value for the extinction coefficient of the *trans*-ethylenic double-bond absorption at 967 cm^{-1} in terms of a functional group absorption coefficient. The value agrees reasonably well with those quoted by Hampton [36] and with those from the A.P.I. series. The latter (on the basis of the whole molecular weight of the compounds) range from 132 to 141, whilst Shreve *et al.* [38] give 140–156 for elaidic acid and related materials. The variation from compound to compound is appreciable, but is small in relation to other materials, such as, for example, methyl oleate and triolein, for which Shreve gives values of 12.2 and 74.4, respectively.

The relative stability of the intensity of this band in relation to the proportion of double bonds has been used by many workers for quantitative studies. In the fatty acid series the intensity of this band appears to remain roughly additive even in conjugated materials [94, 98], but Allan *et al.* [87] note considerable variations when conjugation is with COOR or with acetylenic links. The work of Shreve already mentioned was designed to give a method for the determination of *trans*-octadecenoic acids, esters and alcohols in complex mixtures, and similar methods have been used for the estimation of *trans*-olefines in lubricating oils [41, 68, 95] and in gasoline [42]. An especially valuable application has been in the quantitative and qualitative study of polymerisation reactions. These include applications to low-temperature polymers of butadiene and styrene [36, 43], natural and synthetic rubbers [44, 60, 69] and to the vulcanisation of natural rubber [59]. In the terpene field [46], and in sterols [48, 61] also, studies on the absorption at 965 cm^{-1} coupled with work on the hydrogen deformations of other double-bond types have given valuable structural information. All these applications provide further evidence for the constancy of this frequency in a wide range of different materials.

The evidence that this band is specific for *trans* types of ethylenic links rests, of course, upon its absence in comparable materials of the *cis* series [16]. The corresponding *cis*-compounds are often difficult to obtain completely free from the *trans*-isomer, but the band at 965 cm^{-1} has been effectively eliminated in many cases. In others the intensity at 965 cm^{-1} is extremely low — as in oleic acid, for example — and differentiation between the two isomers has frequently been made on this basis. Typical cases of such a differentiation are those between *cis*- and *trans*-phenylbutadiene [49], polyurethanes [50], long-chain unsaturated enols [51], and the isomers of crotonyl chloride [52]. It is, however, necessary to exercise some caution in dealing with this assignment. For example, the group —CH=C(CH$_3$)— in the *trans*-configuration can also give rise to an absorption in the 1000—980 cm^{-1} region when conjugated with a carbonyl group [80].

3.4(b). Di-substituted ethylenes (*cis*). Owing to the relatively small numbers of *cis*-isomers available in pure form, much less information is available on the out-of-plane hydrogen deformations [34] of *cis*-di-substituted ethylenes. Several spectra of such compounds are available in the American Petroleum Institute series and elsewhere [47, 55], and these suggest that the absorption peak due to this vibration lies near 690 cm^{-1}, although it appears to be much more variable in position than the corresponding band from the *trans* form. Thus, *cis*-2-pentene absorbs at 698 cm^{-1} and 4-methyl-*cis*-2-pentene at 719 cm^{-1}. Sheppard and Simpson [47] quote a range of 728–675 cm^{-1} for a number of simple ethylene derivatives, but even with these some of the assignments are uncertain. Support for this assignment has been forthcoming from many workers [76, 79, 87, 99, 112, 120]. In hydrocarbons the frequency usually falls near 730 cm^{-1}, but it is very much more sensitive to the nature of the surrounding structure than is the corresponding 965 cm^{-1} *trans* band. The substitution of chlorine [76], methyl [76, 79, 99] or an oxygenated group [87] at the α-position results in a marked shift to higher frequencies. 1-Chloro-2-butene (*cis*) absorbs [76] at 769 cm^{-1} and 1 : 4-dichloro-2-butene (*cis*) at 787 cm^{-1}, and the shifts from the other substituents listed are of the same order of magnitude.

Conjugation of the double bond also causes an upward frequency shift, the extent of which depends upon whether or not conjugation is on both sides, and on the number of double bonds involved [79]; and shifts up to 780 cm^{-1} are known [100, 101]. On the other hand, acetylenic conjugation results in only a small frequency shift to slightly lower frequency, and a fairly stable value of 720 cm^{-1} is found [87]. Conjugation with carbonyl groups has a very marked effect, and the group –CH=CH–COOR (*cis*) absorbs near 820 cm^{-1} with sufficient regularity for this to be a useful assignment, especially as this value is not altered significantly by further conjugation. This absorption is usually much weaker in intensity than that from the *trans*-series, but it is usually considerably enhanced on conjugation.

In complex systems, such as the steroids and triterpenes, etc., *cis*-olefines of this type give rise to a multiple series of bands in the 800–650 cm^{-1} region. Useful information can be obtained from the study of these patterns in relation to the substitution arrangements of the rings [77, 108, 109], but it is usually not possible to identify any particular band of the series with the –CH=CH– *cis*-absorption, and considerable disturbances result from the introduction of conjugated substituents. This emphasises the caution which is necessary in the identification of *cis*-olefines in this region, and it should not be forgotten that many other structures, including even some *trans*-olefines, absorb here also. *trans*-Phenylpropenyl carbinol [54] and *trans*-crotonyl chloride [53], for example, absorb in the 690 cm^{-1} region.

3.4(c). Mono-substituted ethylenes. The vinyl type double bond gives rise to two strong bands near 990 cm^{-1} and 910 cm^{-1}. The first of these is connected with the hydrogen deformation mode of the

$$\begin{array}{c}H \\ \diagdown \\ \diagup \end{array} C = C \begin{array}{c} \diagup \\ \diagdown \\ H \end{array}$$
structure [34], and is absent from asymmetrically

di-substituted ethylenes. The second band arises from out-of-plane deformations of the hydrogens of the =CH$_2$ group and occurs in both types of linkage.

As with the $-$CH=CH$-$ link, these frequencies are remarkably constant in position in hydrocarbons. Anderson and Seyfried [40], for example, report twelve vinyl compounds as absorbing at 995 ± 2 cm^{-1} and 910 ± 2 cm^{-1}, and Hampton [36] lists a further nine materials taken from the A.P.I. series, in which the frequency range for the second band is 912–909 cm^{-1}. McMurry and Thornton [58] list fifteen such compounds, whilst Sheppard and Sutherland [34], and Sheppard and Simpson [47] also quote numbers of vinyl compounds absorbing in the frequency ranges 995–985 cm^{-1} and 915–905 cm^{-1}. Small shifts occur in structures RCH=CH$_2$, in which the α carbon atom of the R grouping is fully substituted, when the frequency moves to 1005–995 cm^{-1}.

The presence of polar groups can cause significant shifts away from these ranges and their effects on the two bands are not the same. The 990 cm^{-1} band originates in a CH=CH twisting mode similar to that which is responsible for the 965 cm^{-1} band of *trans* ethylenes, and it responds to substitution effects in exactly the same way. Thus it is little influenced by conjugation, the attachment of carbonyl groups, or by polar atoms at the α carbon. For substituents such as CH$_2$Cl, CH$_2$CN, CH$_2$OR, CH$_2$SR, CH$_2$OH, C=O *etc.* the band remains in the 990–980 cm^{-1} range [63, 74, 76, 93, 103], although it does rise to 1000 cm^{-1} with tertiary carbons at the double bond. However, the direct attachment of oxygen, nitrogen or a halogen to the double bond does have a very marked effect and the frequency falls. Vinyl ethers absorb near 960 cm^{-1} and vinyl esters near 950 cm^{-1} [74, 89, 93].

The 910 cm^{-1} band is more sensitive to changes at the α carbon atom, which lead to upward frequency shifts of about 12 cm^{-1} for polar atoms, with larger shifts with the CH$_2$OCO group (932 cm^{-1}) and the carbonyl group (965–950 cm^{-1}). The direct attachment of polar atoms results in large downwards shifts. The 910 cm^{-1} band moves to 810 cm^{-1} in vinyl ethers [89, 93] and in the vinylidene compounds with two attached oxygens the shift of the corresponding band is doubled (720 cm^{-1}).

Halogen substituents, again, show some interference when substituted on one of the carbon atoms of the double bond, but possibilities are limited in this case to vinyl fluoride, chloride and iodide. These have been examined by Thompson and Torkington [56], who found the expected lowering in frequencies with absorptions at 860, 895 and 902 cm^{-1}, respectively, corresponding to the 910 cm^{-1} band in alkyl compounds.

Both of the bands of this correlation show the same consistency in intensity in relation to molecular weight, which is a feature of out-of-plane hydrogen deformations at a double bond. For quantitative work the 910 cm^{-1} band has generally been employed, as it is the more useful of the two for the analysis of olefine contents of polymers and hydrocarbons. Anderson [40] finds a constant value for the intensity of the 910 cm^{-1} band within about 7 per cent, after allowance for molecular weight, and the A.P.I. series, quoted by Hampton [36], all show extinction coefficients within the range 143−153 on a molecular weight basis. The most detailed study is due to Johansen [121], who showed the integrated area of this band was reasonably constant although the band width varied considerably. Variations in the observed extinction coefficients with various polar substituents are also recorded by Davison [74].

Applications of these frequencies to qualititative and quantitative analysis are, of course, included in the studies referred to under the discussion of the −CH=CH− link, as they were all primarily concerned with the detection and estimation of the various double-bond types in mixtures [36, 38, 40−46]. This correlation, also, is probably applicable within rather wider limits to non-hydrocarbons.

In addition to this pair of characteristic frequencies, vinyl compounds nearly always show a medium intensity band in the range 1850−1800 cm^{-1} which may be an overtone associated in some way with the bands at 910 and 990 cm^{-1}. The suggestion is supported by the fact that with the double-bond types $R_1 R_2 C = CH_2$ in which the out-of-plane hydrogen deformations occur at a lower frequency than 900 cm^{-1}, the possible overtone band is also observed at a lower frequency (near 1800−1750 cm^{-1}). The intensity of these 'overtone' bands is often considerably greater than would be expected, and the frequencies are unusual in falling above the expected value rather than below (negative anharmonicity). The reasons for this are not known, but the bands afford a useful confirmation of the presence of the $=CH_2$ grouping in many cases.

3.4(d). Asymmetric di-substituted ethylenes. $R_1 R_2 C = CH_2$.

The hydrogen out-of-plane deformations in compounds of the type $R_1 R_2 C = CH_2$ occur at 890 cm^{-1}. The consistency of frequency is of the same order as with the double-bond types already described. Sheppard and Simpson [47], for example, quote a range of 892—887 cm^{-1} for seven different compounds with this group. The influences of polar groups are essentially the same as with the corresponding frequency of vinyl compounds. The groups Cl, CN, OR and OH substituted on an α-carbon atom usually give a small shift [74, 76, 93,] to 900 cm^{-1}, but this does not always occur, and 2 : 3-dichloro-1 : 3-butadiene, for example, absorbs in the normal position [105] at 892 cm^{-1}. Carbonyl groups directly on the double bond [74, 76, 93, 104] raise the frequency to 930 − 945 cm^{-1}, and similar values are given with a directly attached nitro group [103]. The direct attachment of oxygen, as before, leads to a major shift [89, 93] to 795 cm^{-1}, with a further fall to 710 cm^{-1} in diethoxy compounds [93]. Conjugation with an olefinic group raises this frequency slightly in some cases [104], but the effect is more marked with acetylenic conjugation (908 cm^{-1}) or with the direct attachment of a nitrile group (934 cm^{-1}). With halogen substitution, the frequency falls as before. Thompson and Torkington [56] quote shifts of 5, 11 and 71 cm^{-1} for this frequency in vinylidene bromide, chloride and fluoride respectively. The intensity of this band is a little more variable, and Anderson [40] quotes deviations from the average of up to 30 per cent. Nevertheless, absorption intensity analyses based on this frequency can be made with reasonable accuracy in many cases. Typical illustrations of the utilisation of absorption in analysis have already been quoted [36, 38, 40—46].

As with the vinyl compounds, a band also occurs in the range 1800—1750 cm^{-1}, which may be due to an overtone of the 890 cm^{-1} frequency and which is often useful in providing confirmation of the presence of the asymmetric ethylene linkage.

3.4(e). Tri-substituted ethylenes. $R_1 R_2 C = CHR_3$.

With tri-substitution at the double bond the specificity of the CH out-of-plane deformation is much reduced, and the frequency shows more variation with the nature of the substituents. These changes are probably connected with the nature of the substituents R_1, R_2 and R_3, which can now cause a greater proportional variation in the force constants than in the mono- and di-substituted ethylenes. The general frequency range for this structure is $840-800$ cm^{-1}, although a few cases have been reported with absorptions as low as 790 cm^{-1}. As before, the bulk of the data on which this assignment is based are to be found in the A.P.I. series of spectra [58], but a number of examples are quoted by Sheppard and Sutherland [34], Thompson and Torkington [3] and others. The variation in frequency of this band renders it of less use for analytical purposes than those from other types of double bond, but it has, nevertheless, been successfully employed for qualitative identifications in hydrocarbons and for quantitative estimations of olefine mixtures. One example of a suitable application is that of Ruzicka *et al.* [57], who have used the presence or absence of a band in the $840-800$ cm^{-1} region to differentiate between sesquiterpenes containing the $R_1 R_2 C{=}CHR_3$ group and the fully substituted types $R_1 R_2 C{=}CR_3 R_4$, in which no band can arise from hydrogen deformations about the double bond. In hydrocarbons in which the presence of a trisubstituted ethylene link is already known it has been suggested that the precise position of the CH deformation can be used to differentiate between configurations in which the hydrogen atom is *trans* to the longest chain, and those where it is not. However the differences are small [54]. 3-Methyl *trans*-2-hexene absorbs at 825 cm^{-1} and 3-methyl *cis*-2-hexene at 810 cm^{-1}. Hirschmann [72] has found this band near 800 cm^{-1} in a large number of unsaturated sterols and has noted a second band near 812 cm^{-1} which appears to occur only in Δ^5-3β-acetoxy compounds. The influence of halogen substitution at the double bond has not been fully studied, but would be expected to cause reduction in the frequency; neoprene, which contains the structure $R_1 ClC{=}CHR_2$, absorbs at 826 cm^{-1}. However, this frequency is not a particularly trustworthy one, and it can be used only with great caution. A band in the $1000-980$ cm^{-1} region has been identified with the group $\underset{CH_3}{\overset{\diagdown}{\diagup}}C{=}CH-$ (*trans*) in certain irones by Lecomte and Naves [80], this may be a CH_3-C stretching frequency, and a somewhat similar correlation has been suggested by Gunzler *et al.* [117] for fully substituted ethylenes carrying a methyl substituent. In these cases a reasonably constant absorption occurs near 1155 cm^{-1}.

3.5. In-plane =CH deformation vibrations

The in-plane deformation frequencies of hydrogen atoms about a double bond have been less used for analytical work. This is principally due to the fact that symmetry considerations either forbid the appearance of the normal vibration in certain cases or result in a weakening of the intensity. Nevertheless, these frequencies have a similar precision to those arising from out-of-plane deformations, and in double-bond types, which give rise to strong bands, the detection of the in-plane frequencies can provide useful confirmation of the presence of the double bond, despite the fact that the bands normally occur in the region associated also with C—C stretching modes and with saturated —CH deformations. These bands appear with medium or strong intensity in the Raman spectra and then afford a reliable method of structural diagnosis [122], particularly in cases such as *cis* disubstituted ethylenes where the infra-red correlations are not reliable.

3.5(a). Di-substituted ethylenes (*trans*). *The* v_1 in-plane hydrogen deformation in symmetrically substituted ethylenes should appear only in the Raman spectrum and is forbidden in the infra-red. Sheppard and Sutherland [34] have shown that, in fact, a strong band appears in the region 1310—1290 cm^{-1} in both the Raman and infra-red spectra of a series of hydrocarbons of this type. They have assigned these to the two in-plane deformations v_1 and v_2 with different symmetry classes. A band in this region can be traced in the infra-red spectrum of all hydrocarbons of this type, but the intensity appears to be somewhat variable and the band is occasionally weak in intensity [34, 47, 55].

3.5(b). Di-substituted ethylenes (*cis*). *cis*-Di-substituted ethylenes show a strong Raman line near 1260 cm^{-1} which is assigned to the v_1 vibration [47, 55, 122]. This band would be permitted by symmetry considerations to appear also in the infra-red spectrum, but no consistent absorption can be found in this region in the infra-red. However, a medium intensity band is found near 1405 cm^{-1} in many cases. This is also attributed to a CH in-plane deformation [47].

3.5(c). Mono-substituted ethylenes. As in the case of the out-of-plane hydrogen deformations, the vinyl-type compounds give rise to two frequencies from the corresponding planar modes. The ν_1 frequency arises from deformations of the hydrogen of the $=CH_2$ structure, whilst ν_2 rises, chiefly, from the $-CH=C-$ structure. Two bands are found in both the infra-red and Raman spectra, in the ranges $1420-1410$ cm^{-1} and $1300-1290$ cm^{-1}. The higher frequency band has been assigned to the $=CH_2$ deformations [34, 47, 55] in analogy with $R_1 R_2 C=CH_2$ structures in which it is also present, and with the corresponding CH_2 deformation near 1450 cm^{-1} in saturated hydrocarbons. The lower-frequency band is in the expected region and is found also in *trans*-symmetrically di-substituted ethylenes.

The high-frequency band near 1415 cm^{-1} is usually strong in the infra-red and is very stable in position, so that it serves as a useful guide in analysis. The substitution of CH_2OR and CH_2OCOR groups has little or no effect upon this frequency, but in the case of acrylates and methacrylates it shows a slight fall [74] to 1405 cm^{-1}. The lower-frequency band is also relatively stable in position, but appears to be more variable in intensity. Both bands appear with medium intensity in the Raman spectra.

3.5(d). Asymmetric di-substituted ethylenes. This type of structure would be expected, from the above, to show strong absorption in the range $1420-1410$ cm^{-1} in both Raman and infra-red spectra, due to the planar deformation of the $=CH_2$ group. This is found experimentally and the frequency shows the expected degree of consistency of position within a few cm^{-1} [34, 47, 55].

3.5(e). Tri-substituted ethylenes. The in-plane deformation of the single hydrogen atom of these structures gives rise to a weak Raman line in the region $1360-1320$ cm^{-1}. Although allowed in the infra-red, the corresponding band is weak in most cases and, as such, is of little value for work on structural analysis.

3.6. Bibliography

1. Barnes, Gore, Liddel and Williams, *Infra-red Spectroscopy* (Reinhold, 1944).
2. Sheppard, *J. Chem. Phys.*, 1949, **17**, 74.
3. Thompson and Torkington, *Trans. Faraday Soc.*, 1946, **42**, 432.

4. Randall, Fowler, Fuson and Dangl, *Infra-red Determination of Organic Structures* (Van Nostrand, 1949).
5. Sheppard and Sutherland, *J. Chem. Soc.*, 1947, 1540.
6. Thompson and Whiffen, *ibid.*, 1948, 1412.
7. Hibben, *The Raman Effect and its Chemical Applications* (Reinhold, 1939), p. 166.
8. Thompson and Torkington, *Trans. Faraday Soc.*, 1945, 41, 246.
9. Jones, Humphries, Packard and Dobriner, *J. Amer. Chem. Soc.*, 1950, 72, 86.
10. Herzberg, *Indra-red and Raman Spectra of Polyatomic Molecules* (Van Nostrand, 1945), p. 325.
11. *Idem, ibid.*, p. 329.
12. Flett, *J. Chem. Soc.*, 1951, 962.
13. Kletz and Sumner, *ibid.*, 1948, 1456.
14. Shreve, Heether, Knight and Swern, *Analyt. Chem.*, 1950, 22, 1498.
15. Bernstein and Powling, *J. Amer. Chem. Soc.*, 1951, 73, 1843.
16. Rasmussen and Brattain, *J. Chem. Phys.*, 1947, 15, 120.
17. Bradacs and Kahovec, *Z. physikal. Chem.*, 1940, B48, 63.
18. Rasmussen, Tunnicliff and Brattain, *J. Chem. Phys.*, 1943, 11, 432.
19. Blout, Fields and Karplus, *J. Amer. Chem. Soc.*, 1948, 70, 194.
20. Rasmussen, Tunnicliff and Brattain, *ibid.*, 1949, 71, 1068.
21. Kohlrausch and Pongratz, *Z. Chem. Phys.*, 1934, B27, 176.
22. Furchgott, Rosenkrantz, Harris and Shorr, *J. Biol. Chem.*, 1946, 163, 375.
23. Torkington and Thompson, *Trans. Faraday Soc.*, 1945, 41, 236.
24. Edgell, *J. Amer. Chem. Soc.*, 1948, 70, 2816.
25. Park, Lycan and Lacher, *ibid.*, 1951, 73, 711.
26. Fox and Martin, *Proc. Roy. Soc.*, 1940, A175, 208.
27. Saier and Coggeshall, *Analyt. Chem.*, 1948, 20, 812.
28. Jones, Williams, Whalen and Dobriner, *J. Amer. Chem. Soc.*, 1948, 70, 2024.
29. Rose, *J. Res. Nat. Bur. Stand.*, 1938, 20, 129.
30. Lambert and Lecomte, *Compt. rend. Acad. Sci. Paris*, 1938, 206, 1007.
31. *Idem, Ann. Phys.*, 1938, 10, 503.
32. Gore and Johnson, *Phys. Rev.*, 1945, 68, 283.
33. Rasmussen and Brattain, *J. Chem. Phys.*, 1947, 15, 131, 135.
34. Sheppard and Sutherland, *Proc. Roy. Soc.*, 1949, A196, 195.
35. Kilpatrick and Pitzer, *J. Res. Nat. Bur. Stand.*, 1947, 38, 191.
36. Hampton, *Analyt. Chem.*, 1949, 21, 923.
37. Freeman, *J. Amer. Chem. Soc.*, 1953, 75, 1859.
38. Shreve, Heether, Knight, and Swern, *Analyt. Chem.*, 1950, 22, 1261.
39. Thompson, Vago, Corfield and Orr, *J. Chem. Soc.*, 1950, 214.
40. Anderson and Seyfried, *Analyt. Chem.*, 1948, 20, 998.
41. Fred and Putscher, *ibid.*, 1949, 21, 900.
42. Johnston, Appleby and Baker, *ibid.*, 1948, 20, 805.
43. Treumann and Wall, *ibid.*, 1949, 21, 1161.
44. Dinsmore and Smith, *ibid.*, 1948, 20, 11.
45. Torkington, *Proc. Roy. Soc.*, 1951, A206, 17.
46. Bateman, Cunneen, Fabian and Koch, *J. Chem. Soc.*, 1950, 936.
47. Sheppard and Simpson, *Quart. Rev.*, 1952, 6, 1.
48. Jones, *J. Amer. Chem. Soc.*, 1950, 72, 5332.
49. Grummit and Christoph, *ibid.*, 1951, 73, 3479.
50. Marvel and Young, *ibid.*, p. 1066.
51. Crombie and Harper, *J. Chem. Soc.*, 1950, 1707, 1714.

52. Hatch and Nesbitt, *J. Amer. Chem. Soc.*, 1950, **72**, 727.
53. Philpotts and Thain, *Nature*, 1950, **166**, 1028.
54. Clark, *Applied Spectroscopy*, 1968, **22**, 204.
55. Sheppard, *J. Inst. Pet.*, 1951, **37**, 95.
56. Thompson and Torkington, *Proc. Roy. Soc.*, 1945, **A184**, 21.
57. Ruzicka, Jeger *et al.*, *Helv. Chim. Acta*, 1950, **33**, 672, 711, 1050.
58. McMurry and Thornton, *Analyt. Chem.*, 1952, **24**, 318.
59. Sheppard and Sutherland, *Trans. Faraday Soc.*, 1945, **41**, 261.
60. Salomon, Van der Schee, Ketelaar and Van Eyk., *Discuss. Faraday Soc.*, 1950, **9**, 291.
61. Bladon, Fabian, Henbest, Koch and Wood, *J. Chem. Soc.*, 1951, 2402.
62. Bowles, George and Cunliffe Jones, *J. Chem. Soc.* (B), 1970, 1070.
63. Saunders and Smith, *J. Applied Phys.*, 1949, **20**, 953.
64. Hatcher and Yost, *J. Chem. Phys.*, 1937, **5**, 992.
65. Morino, Kuchitsu and Shimanouchi, *ibid.*, 1952, **20**, 726.
66. Nielsen, Liang and Smith, *ibid.*, 1090.
67. Haszeldine, *Nature*, 1951, **168**, 1028.
68. Putscher, *Analyt. Chem.*, 1952, **24**, 1551.
69. Richardson and Sacher, *J. Polymer Sci.*, 1953, **10**, 353.
70. Crombie, *J. Chem. Soc.*, 1952, 2997, 4338.
71. Sinclair, McKay, Myers and Jones, *J. Amer. Chem. Soc.*, 1952, **74**, 2578.
72. Hirschmann, *ibid.*, 5357.
73. Mixer, Heck, Winstein and Young, *ibid.*, 1953, **75**, 4098.
74. Davison and Bates, *J. Chem. Soc.*, 1953, 2607.
75. Lord and Walker, *J. Amer. Chem. Soc.*, 1954, **76**, 2518.
76. Kitson, *Analyt. Chem.*, 1953, **25**, 1470.
77. Henbest, Meakins and Wood, *J. Chem. Soc.*, 1954, 800.
78. Leonard and Gash, *J. Amer. Chem. Soc.*, 1954, **76**, 2781.
79. Oroshnik and Mebane, *ibid.*, 5719.
80. Lecomte and Naves, *J. Chim. Phys.*, 1956, 462.
81. Gruzdev, *Zhur, Fiz, Khim.*, 1954, **28**, 507.
82. O'Connor, *J. Amer. Oil. Chem. Soc.*, 1956, **33**, 1.
83. Werner and Lark, *J. Chem. Soc.*, 1954, 1152.
84. Lord and Miller, *Applied Spectroscopy*, 1956, **10**, 115.
85. Jones and Herling, *J. Org. Chem.*, 1954, **19**, 1252.
86. Jones and Sandorfy, *Chemical Applications of Spectroscopy* (Interscience, 1956), p. 370.
87. Allan, Meakins and Whiting, *J. Chem. Soc.*, 1955, 1874.
88. Brice, Lazerte, Hals and Petersen, *J. Amer. Chem. Soc.*, 1953, **75**, 2698.
89. Meakins, *J. Chem. Soc.*, 1953, 4170.
90. Slowinski, Emil and Claver, *J. Polymer. Sci.*, 1955, **17**, 269.
91. Bellamy, *J. Chem. Soc.*, 1955, 4221.
92. Scrocco and Salvetti, *Bull. Sci. della Fac. Chim. Bologna*, 1954, **12**, 1.
93. Philpotts, *private communication*.
94. Ahlers, Brett and McTaggart, *J. App. Chem.*, 1953, **3**, 433.
95. Francis, *Analyt. Chem.*, 1956, **28**, 1171.
96. Chapman and Taylor, *Nature*, 1954, **174**, 1011.
97. DeTar and Carpino, *J. Amer. Chem. Soc.*, 1956, **78**, 475.
98. Bickford, DuPré, Mack and O'Connor, *J. Amer. Oil. Chem. Soc.*, 1953, **30**, 379.
99. Hall and Mikos, *Analyt. Chem.*, 1949, **21**, 422.
100. Kuhn, Inhoffen, Staab and Otting, *Chem. Ber.*, 1953, **86**, 965.
101. Zechmeister, *Experimentia*, 1954, **10**, 9.

102. McManis, *Applied Spectroscopy*, 1970, **24**, 495.
103. Brown, *J. Amer. Chem. Soc.*, 1955, **77**, 6341.
104. Brügel, *Angew, Chem.*, 1956, **68**, 441.
105. Sheppard and Szasz, *Trans. Faraday Soc.*, 1953, **49**, 358.
106. Traynham and Sehnert, *J. Amer. Chem. Soc.*, 1956, **78**, 4024.
107. Jackson, Paschke, Tolberg, Boyd and Wheeler, *J. Amer. Oil. Chem. Soc.*, 1952, **29**, 229.
108. Henbest, Meakins, Nicholls and Wilson, *J. Chem. Soc.*, 1957, 997.
109. Cole and Thornton, *ibid.*, 1332.
110. Blomquist, Wolinsky, Meinwald and Longone, *J. Amer. Chem. Soc.* 1956, **78**, 6057.
111. Masetti and Zerbi, *Spectrochim. Acta*, 1970, **26A**, 1891.
112. Heilmann, Gaudemaris and Arnaud. *Bull. Soc. Chim. Fr.*, 1957, **112**, 119.
113. Gamboni, Theus and Schinz, *Helv. Chim. Acta*, 1955, **38**, 255.
114. Theus, Surber, Colombi and Schinz, *ibid.*, 239.
115. Sörensen and Sörensen, *Acta Chem. Scand.*, 1954, 8, 1741, 1763.
116. Bohlmann and Mannhardt, *Chem. Ber.*, 1955, **88**, 429, 1336.
117. Gunzler, Kienitz and Neuhaus, *Naturwiss.*, 1956, **43**, 299.
118. Weiblen, *Fluorine Chemistry*. Ed. Simons. Vol. 2 (Academic Press, New York, 1954), p. 449.
119. Orr, *Spectrochim. Acta*, 1956, 8, 218.
120. Jones, Mansfield and Whiting, *J. Chem. Soc.*, 1956, 4073.
121. Johansen, *Bull. Acad. Sci. U.S.S.R.*, 1954, **18**, 708.
122. Dollish, Fateley and Bentley, *Characteristic Raman Frequencies of Organic Compounds* (Wiley, New York, 1974).
123. Breslow, Ryan and Groves, *J. Amer. Chem. Soc.*, 1970, **92**, 988.
124. Mitchell and Merritt, *Spectrochim. Acta*, 1971, **27A**, 1609.
125. Miller, Brown and Rhee, *Spectrochim. Acta*, 1972, **28A**, 1467.
126. Rhee and Miller, *Spectrochim. Acta*, 1971, **27A**, I.
127. Noack, *Spectrochim. Acta*, 1962, **18**, 697, 1625.
128. Bellamy, *Advances in Infrared Group Frequencies* (Methuen, London, 1968).
129. Duncan, *Spectrochim. Acta*, 1970, **26A**, 429.
130. Whetsel, *Applied Spectroscopy Reviews*, 1968, 2, I.

4

Alkynes and Allenes

4.1. Introduction and table

The possibilities of detection of acetylenic links in unknown materials depend largely on the degree of substitution at the triple bond. Mono-substituted acetylenes can be readily identified by the highly characteristic $\equiv$C–H stretching vibration and also by the C$\equiv$C vibration, which gives a strong band when the triple bond is situated at the end of the chain. In di-substituted materials the former band is, of course, absent, whilst the second is very variable in intensity, depending on the position of the triple bond. In symmetrical structures the C$\equiv$C absorption is often too weak to be detectable, and in these circumstances there is no sure way of recognising this type of bond, other than by the corresponding Raman spectrum. On the other hand, when the C$\equiv$C stretching band is present, its position is highly characteristic, whilst the usual shifts arising from conjugation and from aromatic substitution are also very informative.

The only other regions from which information might be sought are, in the case of mono-substituted materials, the $\equiv$CH bending and combination frequencies. Deformation vibrations occur near 650 cm^{-1} and a combination frequency near 1250 cm^{-1} has been described in some cases.

A review of the absorption bands found in simple alkynes has been given by Sheppard and Simpson [20], who have also summarised the observed frequencies for the C$\equiv$C and $\equiv$CH absorptions in these materials. An extensive study of conjugated acetylenes has been made by Allen, Meakin and Whiting [25].

Allenic materials are not common, and relatively few have been

64

studied. They appear to behave as though the bonds were more nearly triple and single, rather than as two double bonds, and characteristic absorption bands arise near 1950 cm^{-1} and 1060 cm^{-1}, due to the stretching modes of the C=C=C structure. Absorption in the 2000 cm^{-1} region also arises from other similar structures [1], such as HN=C=O (2274 cm^{-1}), O=C=O (2349 cm^{-1}), HN=N≡N (2140 cm^{-1}), O=N≡N (2223 cm^{-1}), NH=C=S (1963 cm^{-1}) and O=C=S (2050 cm^{-1}).

The correlations discussed are listed in Table 4.

Table 4

Acetylenic Compounds

≡CH stretching bands		3320–3310 cm^{-1} (m.)
C≡C stretching bands	mono-substituted	2140–2100 cm^{-1} (m.)
	alkyl mono-substituted	2130–2120 cm^{-1} (m.)
	di-substituted	2260–2190 cm^{-1} (w.)
	alkyl di-substituted	2240–2230 cm^{-1} (w.)
≡CH deformations	mono-substituted	680– 610 cm^{-1} (m.)
	alkyl	near 630 cm^{-1}
	aryl	Two bands near 640 and 615 cm^{-1}

Allenes

C=C=C stretching bands	mono-substituted	1980–1945 cm^{-1} (m.)
	di-substituted	1955–1930 cm^{-1} (w.)
=CH$_2$ deformation	mono-substituted	875– 840 cm^{-1} (s.)

4.2. ≡CH Stretching vibrations

The work of Fox and Martin [2] and others on hydrogenic stretching vibrations has already been described in relation to alkanes and alkenes. In the light of these results, an absorption is to be expected somewhere above 3000 cm^{-1} which is due to the stretching of the hydrogen atom attached to a triply-bonded carbon atom in a mono-substituted acetylenic material. Furthermore, this band can be expected to be reasonably strong and to be largely free from interference from the remainder of the molecule, especially as in this case it is not possible for any other substituents to be directly joined to the same carbon atom as the single hydrogen. This expectation is, of course, realised in practice; Randall *et al.* [3] quote data for a number of mathematically analysed molecules in which the ≡C–H stretching absorption occurs in the range 3390–3290 cm^{-1}, whilst Sheppard [4] gives values of 3300 cm^{-1} for

1-butyne and 3305 cm^{-1} for vinyl acetylene (Raman). Wotiz and co-workers [5, 6] have examined about a dozen mono-substituted acetylenes, in all of which this band appears near 3270 cm^{-1}. They note, however, that owing to the poor dispersion of rock-salt in this region their values are 20–40 cm^{-1} lower than known Raman values for a few of these products, and this correction would bring their findings into line with the others. Later Raman studies, and infra-red work at higher resolution [15, 16] has established that the band usually falls between 3320 and 3310 cm^{-1} in alkyl-substituted alkynes, and in derivatives with substituents such as halogen, oxygen or hydroxyl at the α-carbon atom. Substitution of a halogen directly on the triple bond raises the frequency so that the bromo compound absorbs at 3325 cm^{-1} and the fluoro derivative at 3355 cm^{-1}. Conversely, the attachment of elements of low electronegativity, such as phosphorus, arsenic *etc.* give rather lower frequencies [26] down to 3290 cm^{-1}. Conjugation with a carbonyl group or a double bond lowers the frequency to about 3305 cm^{-1}.

In many cases the CH band shows a weaker shoulder on the low frequency side. This has been attributed to a Fermi-resonance effect. Doubling of the CH band can also occur through self association as the CH proton of alkynes is sufficiently acidic to take part in hydrogen bonding. Shifts from association are not large but with the strongest hydrogen bonds they can amount to about 50 cm^{-1}. It is interesting to note that the CH stretching absorption of HCN [7] also occurs at 3311 cm^{-1}, indicating that the nature of the attachment of the triple bond does not exert any marked influence on the absorption frequency.

Mono-substituted acetylenes can therefore be readily differentiated from double-bonded and saturated materials by the examination of this region. The only other groupings likely to absorb in the same region are bonded OH and NH groups. These can be readily differentiated by their broad appearance, and in any case confirmation of the acetylenic structure can be obtained by reference to the C≡C stretching region.

4.3. C≡C Stretching vibrations

4.3(a). Position. The C≡C stretching vibration is known, from fundamental studies on simple molecules [3, 4, 8, 20], to occur near 2100 cm^{-1}. Wotiz and co-workers [5, 6] have examined twenty-eight variously substituted acetylenes and studied the position and intensity of this absorption band. Their findings have been amplified by Sheppard and Simpson [20], Blomquist [21], and, more recently by Nyquist and Potts [33]. There is also a substantial amount of supporting Raman data [8, 16]. There is a frequency difference of about 100 cm^{-1} between mono- and di-substituted alkynes so these can usually be differentiated without difficulty. Alkyl mono-substituted alkynes absorb in the narrow range 2130–2120 cm^{-1}, although the *tert.-* butyl derivative is an exception, absorbing at 2105 cm^{-1}. Halogen substitution at the α carbon atom leads to a small frequency rise [33] so that $CF_3 C{\equiv}CH$ absorbs at 2156 cm^{-1}. The attachment of elements of low electronegativity, with vacant d orbitals, such as sulphur, phosphorus or arsenic, results in a marked fall in frequency into the 2050–2020 cm^{-1} region [26, 40, 41]. The direct attachment of fluorine results in a substantial frequency rise whereas bromine and iodine give lower frequencies. The general trend is that of the strength of the C–X bonds [15], with higher frequencies than alkyl compounds when C–X is stronger than C–C and lower frequencies when the inverse is true.

Dialkyl alkynes absorb between 2230 and 2240 cm^{-1}. The direct attachment of chlorine does not alter this, but bromine and iodine give lower values (2210 and 2200 cm^{-1}), corresponding to the lower force constants of the C–X bonds. CF_3-substitution raises the frequency so that $CH_3 {-} C{\equiv}CCF_3$ absorbs at 2271 cm^{-1}, and the hexafluro compound at 2305 cm^{-1}. In the great majority of cases this band is accompanied by another, weaker band at higher frequencies [10, 32]. This originates in a Fermi-resonance effect.

As with mono-substituted compounds, the direct attachment of elements with vacant d orbitals can lead to a substantial frequency reduction. Sacher *et al.* [42] have examined a number of derivatives with the alkyne link attached to silicon, germanium, arsenic, antimony *etc.* and find the C≡C stretching band in the 2193–2156 cm^{-1} range. A number of references to similar work are given by Bellamy [15].

4.3(b). The influence of conjugation. There has been a good deal of interest in the chemistry of conjugated acetylenes, and a certain amount of infra-red data are available [25, 27–31, 35]. Conjugation with acetylenic or olefinic groups usually causes some intensification of the absorption, together with a small shift towards lower frequencies. The changes are, however, much smaller than those which occur in the olefines. Conjugation with carbonyl groups does not alter the frequency appreciably [25, 37–39] in most cases, but a few α-carbonyl compounds show shifts to $2170\ cm^{-1}$ although others do not.

Similarly, in an extensive series of conjugated acids and esters studied by Allan *et al.* [25] the main $C \equiv C$ frequency occurred in the region $2260–2235\ cm^{-1}$, which is towards the top of the normal di-substituted region. Wotiz and Miller [5] find the $C \equiv C$ stretching absorption of $C_5 H_{11} C \equiv CBr$ at $2120\ cm^{-1}$, which is below the normal range for di-substituted acetylenes, whilst diacetylene [9] and vinyl acetylene [4] absorb at 2024 and $2099\ cm^{-1}$, respectively. This frequency shift is not shown by $\alpha\beta$-$\alpha'\beta'$-conjugated alkynes in which the individual conjugation effects appear to cancel each other. Di-cyanoacetylene [22], for example, absorbs at $2267\ cm^{-1}$. This effect has been observed also in conjugated acids and esters $R(-C \equiv C-)_n COOR$, where N is greater than 1. In some cases three bands are observed although only two triple bonds are present. 1 : 4-Nonadiyne and 5 : 8-tridecadiyne, for example, give three bands in the $2000\ cm^{-1}$ region [34].

4.3(c). Intensity. The intensity of the fundamental $C \equiv C$ stretching vibration is extremely variable and is directly related to the position of the triple bond in the molecule. Wotiz and Miller [5] found the band to be strong in compounds in which the triple bond terminated the chain, and that the intensity diminished progressively in compounds in which the triple bond was situated farther and farther away from the end. This arises from the establishment of a pseudo centre of symmetry about which the stretching of the triple bond is symmetrical and, therefore, inactive in the infra-red. A strictly comparable phenomenon exists with the $C = C$ linkage, and it has been fully discussed in Chapter 3. The consequence is that the $C \equiv C$ fundamental is not detectable in a compound such as $C_3 H_7 - C \equiv C-(CH_2)_2 Cl$, and the absence of an absorption band in the triple-bond region cannot therefore be taken as evidence for the absence of such a bond. In contrast, this band is, of course, always strong in Raman spectra. Conjugation usually increases the intensity a little, but in di-substituted materials it is necessary to have as many as four conjugated $C \equiv C$ groups to bring up a band of medium intensity [27]. In $CH_3 (C \equiv C)_3 CH_3$, for example, the absorption at $2222\ cm^{-1}$ is weak, whilst $CH_3 (C \equiv C)_4 CH_3$ has a medium-strength band at $2237\ cm^{-1}$.

Conjugation with carbonyl groups also increases the intensity, and extinction coefficient data for a number of such materials have been given by Allan *et al.* [25].

4.4. ≡CH Bending vibrations

Mono-substituted alkynes show a characteristic CH bending absorption in the range 680–610 cm^{-1} which is usually quite strong. With alkyl substituents it is near 630 cm^{-1}. The band originates in a degenerate pair and the degeneracy is split when the substituent is planar [33]. In consequence aromatic derivatives show two bands in this region. Phenyl acetylene for example absorbs at 642 and 613 cm^{-1}. Moritz has pointed out that the position of this band varies systematically with substitution in exactly the same way as does the CH_2 out of plane deformation of $C=CH_2$ groups, so that some idea of the probable position of this band can be obtained from the data on alkenes [43]. Like the $=CH_2$ deformations the position of this band is very little affected by conjugation. In a number of cases a strong overtone or combination band in the 1300–1200 cm^{-1} range has been recorded as being due to these bands [4, 11, 12, 20], but so many different types of structure give rise to absorption in this region that this observation is not likely to be of much value for structural diagnosis, unless later work should show a selective narrowing of this range for various structural types.

4.5. Allenic structures

Allenes like other cumulative double bond systems do not show a normal double bond absorption. They have instead, a high frequency antisymmetric absorption which is usually of medium intensity in the infra-red, and a symmetric band at lower frequency which is usually very weak or forbidden in the infra-red and is best seen in the Raman spectra. The antisymmetric band occurs at 1980–1945 cm^{-1} in mono-substituted allenes, and at 1930–1955 cm^{-1} when di-substituted so that there is no terminal $=CH_2$. Asymmetric di-substitution of the latter type gives this band at 1930–1915 cm^{-1}. The symmetric mode appears strongly in the Raman spectra near 1080 cm^{-1}.

Allene itself has been studied by several workers [13, 14] and its substituted derivatives have been examined by Wotiz [19, 44] and by Borisov [45]. Polar groups raise the antisymmetric frequency towards the top of the frequency range, and conjugation lowers it towards the bottom [23, 24, 44]. Doubling of this band is often found in systems in which a polar group is directly attached [19, 20]. The frequency range is very characteristic, and although other X=Y=Z systems absorb in much the same region, it is usually possible to identify an allene by its medium intensity and absence of much band structure.

Allenes can also be identified by their CH deformation absorption, provided they are not fully substituted. A mono-substituted allene, or an asymmetric di-substituted derivative, has a strong band in the $875-840$ cm^{-1} range which is due to the $=CH_2$ out of plane deformation [44, 46]. This is parallel to the corresponding band of vinyl compounds at 900 cm^{-1}. The band position is little influenced by substituent effects as the groups are well removed down the chain. A characteristic overtone is also found at slightly more than twice the fundamental frequency. This is also parallel to the similar vinyl group behaviour. This band appears to absorb in much the same frequency range in allenes having only a single hydrogen atom at each carbon, but the number of compounds studied is not large.

4.6. Bibliography

1. Herzberg and Reid, *Discuss. Faraday Soc.*, 1950, **9**, 92.
2. Fox and Martin, *Proc. Roy. Soc.*, 1940, **A175**, 208.
3. Randall, Fowler, Fuson and Dangl, *Infra-red Determination of Organic Structures* (Van Nostrand, 1949).
4. Sheppard, *J. Chem. Phys.*, 1949, **17**, 74.
5. Wotiz and Miller, *J. Amer. Chem. Soc.*, 1949, **71**, 3441.
6. Wotiz, Miller and Palchak, *ibid.*, 1950, **72**, 5055.
7. Herzberg, *Molecular Spectra and Molecular Structure : Diatomic Molecules* (Van Nostrand, 1950), p. 279.
8. Hibben, *The Raman Effect and its Chemical Application* (Reinhold, 1939), pp. 200 ff.
9. Herzberg, *Infra-red and Raman Spectra of Polyatomic Molecules* (Van Nostrand, 1945) p. 323.
10. Henne and Nager, *J. Amer. Chem. Soc.*, 1951, **73**, 1042.
11. Wu, *Vibrational Spectra and Structure of Polyatomic Molecules* (Edwards, 1946), p. 176.
12. Crawford, *J. Chem. Phys.*, 1940, **8**, 526.
13. Linnett and Avery, *ibid.*, 1938, **6**, 686; cf. also Herzberg, *op. cit.*, p. 339.
14. Thompson and Harris, *Trans. Faraday Soc.*, 1944, **40**, 295.
15. Bellamy, *Advances in Infra-red Group Frequencies* (Methuen, London, 1968).
16. Dollish, Fateley and Bentley, *Characteristic Raman Frequencies of Organic Compounds* (Wiley, New York, 1974).
17. Nyquist and Muelder, *J. Mol. Structure,* 1968, **2**, 465.
18. Nyquist, *Spectrochim. Acta,* 1971, **27A**, 2531.
19. Wotiz and Celmer, *J. Amer. Chem. Soc.,* 1952, **74**, 1860.

20. Sheppard and Simpson, *Quarterly Reviews,* 1952, 6, 1.
21. Blomquist, Burge, Huang Liu, Bohrer, Sucry and Kleis, *J. Amer. Chem. Soc.,* 1951, 73, 5510.
22. Miller and Hannan, *J. Chem. Phys.,* 1953, 21, 110.
23. Oroshnik, Mebane and Karmas, *J. Amer. Chem. Soc.,* 1953, 75, 1050.
24. Celmer and Solomons, *ibid.,* 1372.
25. Allan, Meakins and Whiting, *J. Chem. Soc.,* 1955, 1874.
26. Mathis, Barthelot, Charrier and Mathis, *J. Mol. Structure,* 1968, 1, 481.
27. Cook, Jones and Whiting, *ibid.,* 1952, 2883.
28. Bohlmann, *Chem. Ber.,* 1955, 88, 1755.
29. Bohlmann and Viehe, *ibid.,* 1955, 88, 1017, 1245.
30. Schubach and Trautschold, *Annalen,* 1955, 594, 67.
31. Georgieff, Cave and Blakie, *J. Amer. Chem. Soc.,* 1954, 76, 5494.
32. Henne and Finnegan, *ibid.,* 1949, 71, 298.
33. Nyquist and Potts, *Spectrochim. Acta,* 1960, 16, 419.
34. Gensler, Mahadeven and Casella, *J. Amer. Chem. Soc.,* 1956, 78, 163.
35. Sörensen and Sörensen, *Acta Chem. Scand.,* 1954, 8, 1741, 1763.
36. Bohlmann and Mannhardt, *Chem. Ber.,* 1955, 88, 1330.
37. Theus, Surber, Colombi and Schinz, *Helv. Chim. Acta,* 1955, 38, 239.
38. Gamboni, Theus and Schinz, *ibid.,* 255.
39. Grob and Fischer, *ibid.,* 1794.
40. Sacher, Davidshon and Miller, *Spectrochim. Acta,* 1970, 26A, 1011.
41. Christensen and Engelsen, *Spectrochim. Acta,* 1970, 26A, 1743.
42. Sacher, Park, Miller and Brown, *Spectrochim. Acta,* 1972, 28A, 1361.
43. Moritz, *Spectrochim. Acta,* 1967, 23A, 167.
44. Wotiz and Mancuso, *J. Org. Chem.,* 1957, 22, 207.
45. Borisov and Sverdlov, *Optics and Spectroscopy,* 1963, 15, 14.
46. Petrov, Jakoleva and Kormer, *Optics and Spectroscopy,* 1959, 7, 237.

5

Aromatic Compounds

5.1. Introduction and table

Aromatic-type compounds give rise to a large number of very sharp, characteristic bands, so that their identification by infra-red methods is usually straightforward; furthermore, the changes in certain regions which result from substitution are largely independent of the nature of the substituents, so that it is usually possible to determine, also, the degree and type of substitution present.

In a molecule as complex as benzene a very considerable number of normal vibrational modes are possible, but infra-red studies have contributed largely to the demonstration of the planar nature of the ring [1], and a number of workers have studied the normal vibrations and enabled many of them to be identified [2—8, 15]. The origin of many of the bands used in correlation work is therefore known so that it is often possible to assess the extent of any changes which are likely to result from structural modifications. Studies on the correlation of spectral changes following upon substitution of various types were initiated by Lecomte [9, 10, 11] in a considerable series of papers between 1937 and 1946, and these have been further developed by later workers. Varsanyi's book *The Vibrational Spectra of Benzene Derivatives* [15], gives a very detailed review of the origins and behaviour of all the characteristic bands, both in the infra-red and Raman spectra.

The presence of an aromatic-type structure is best recognised by the presence of the $=C-H$ stretching vibrations near 3030 cm^{-1} and the ring breathing vibrations in the 1600—1500 cm^{-1} region. In these regions the absorption bands are very little affected by the

72

substitution pattern, although a study of the latter region can also throw light on the nature of substitution in certain cases, and on the presence of conjugation with the double bonds of the ring.

Once the presence of an aromatic ring is established, the presence and relative positions of substituents can be studied in the regions $2000-1660 \text{ cm}^{-1}$, $1250-1000 \text{ cm}^{-1}$ and $1000-650 \text{ cm}^{-1}$. Of these regions, the first is generally the most definite, and usually gives a clear indication of the type of substitution. Confirmation of this can then be sought in the low-frequency region and, to a lesser extent, in the $1250-1000 \text{ cm}^{-1}$ region. The presence of groupings absorbing or giving overtones in the $2000-1660 \text{ cm}^{-1}$ region complicates the interpretation in some cases, whilst strong electron attracting or donating substituents, such as nitro-groups, exert a profound influence on all these three regions. The interpretation of the spectrum is then rendered much less precise.

The regions of absorption to be discussed are listed in Table 5.

5.2. =CH Stretching vibrations

Studies on the CH stretching absorptions of saturated and un-saturated compounds have already been discussed fairly fully (Chapter 3), and the same considerations as to stability of position apply in the aromatic series. Fox and Martin [12, 13, 14] have examined many aromatic materials in the 3000 cm^{-1} region using the high dispersion of a grating, including polycyclic materials such as naphthalene and quinoline. They find that the CH aromatic stretching vibrations produce bands close to 3038 cm^{-1} which are generally three in number. Many mono-substituted aromatics give a characteristic triplet of this type about 3058 cm^{-1} in carbon tetrachloride solution, but there are sometimes more bands than this, and in multiple-ring systems very complex patterns are produced. In nearly all cases, however, one of the bands in this region is considerably more intense than the others. This region has been studied in some detail at high resolution by Josien and Lebas [61] using mono-substituted aromatics. They have shown that in many cases the multiple-band systems can be explained very simply in terms of a single fundamental absorption accompanied by a series of weaker absorptions arising from combinations of the various bands in the $1600-1400 \text{ cm}^{-1}$ region. The fundamentals themselves were assigned within the range 3079 cm^{-1}: (nitrobenzene) to 3030 cm^{-1} (toluene). In mono-substituted benzenes two other fundamentals are

Table 5

=C—H *Stretching Modes*

Sharp absorptions near 3030 cm^{-1}, and 3070 cm^{-1} (w)

2000–1660 *cm*$^{-1}$

Absorption patterns typical of the substitution type (see Fig. 9)

Skeletal Ring Breathing Modes

Near 1600 cm^{-1} (v.)
Near 1500 cm^{-1} (v.)
Near 1580 cm^{-1} (w. unless conjugated)
Near 1450 cm^{-1} (m.)

CH *Out-of-plane Deformations*

Five adjacent free hydrogen atoms	770–730 cm^{-1} (v.s.) and 710–690 cm^{-1} (s.)
Four adjacent free hydrogen atoms	770–735 cm^{-1} (v.s.)
Three adjacent free hydrogen atoms	810–750 cm^{-1} (v.s.)
Two adjacent free hydrogen atoms	860–800 cm^{-1} (v.s.)
One free hydrogen atom.	900–860 cm^{-1} (m.)

The Region 1225–950 cm^{-1}

1 : 2-, 1 : 4- and 1 : 2 : 4-substitution	1225–1175 cm^{-1}, 1125–1090 cm^{-1}, 1070–1000 cm^{-1} (2 bands) (all w.)
Mono-, 1 : 3-, 1 : 2 : 3- and 1 : 3 : 5-substitution	1175–1125 cm^{-1}, 1110–1070 cm^{-1} (not 1 : 3 : 5-), 1070–1000 cm^{-1} (1 band) (all w.)
1 : 2-, 1 : 2 : 3- and 1 : 2 : 4-substitution	1000–960 cm^{-1} (w.)
Tropolones	1630–1600 cm^{-1}, 1570–1535 cm^{-1} (v.)

allowed in the infra-red in this region, but the strongest band is in the 3030 cm^{-1} region. In the Raman spectra only a single band appears near 3057 cm^{-1}.

Workers have discussed the relationships which these frequencies may have to the electron withdrawing or donating powers of the substituent [62, 68], and to such properties as dipole moments, electron densities as calculated by molecular diagrams, and reactivities. It is interesting to note, for example, that with the exception of the halogen atoms, which are anomalous, an approximately linear relationship connects the C—H stretching frequency and the percentage of *m*-substituted material obtained on nitration. There is also an interesting intensity relationship. In general the 3030 cm^{-1} band is the strongest, but the intensities of all the CH stretching bands fall off progressively as the Hammett σ, value of the substituent increases. The rate of fall of intensity with increase of σ_1 is much greater for the 3030 cm^{-1} band than for the others [56]. In consequence, whereas the 3030 cm^{-1} band in ethyl benzene is twice as strong as that at 3065 cm^{-1}, the reverse is true in nitrobenzene where the first band is very weak compared with the second, which now falls at 3079 cm^{-1}. This accounts for the abnormally high value usually quoted for the CH stretch of nitrobenzene and illustrates very well the difficulties in defining precise frequency ranges in which a particular band will always be found.

In general, therefore, there is very little difficulty, in practice, in identifying the CH absorption in aromatic compounds and usually, using a rock-salt prism, it appears as a sharp, relatively weak band about 3030 cm^{-1} on the side of the main, much stronger, CH$_2$ and CH$_3$ absorption bands below 3000 cm^{-1}. The intensity is appreciably less than the saturated CH$_2$ vibrations, but increases in proportion to the amount of aromatic residues present. With higher resolution it is usually possible to recognise at least two bands, near 3070 and 3030 cm^{-1}, and sometimes more.

Other factors which can cause absorption above 3000 cm^{-1} are CH stretching absorptions of normal double bonds of various types and of CHX groups, where X is a halogen. The former can usually be identified by the characteristic bands they show elsewhere in the spectrum. The differentiation from halogenated materials is less easy, but confirmation of an aromatic structure is usually obtainable from the 1600—1500 cm^{-1} region. Interferences arising from hydrogen-bonded OH and NH groups are usually eliminated by the use of dilute solutions, but in some cases where intramolecular bonds are involved the =CH aromatic stretching region may be masked.

These stretching vibrations show overtones in the 8000—5000 cm^{-1} region which are also characteristic of the aromatic structure [66], and Rose [16] has employed this region for the analysis of the aromatic components of hydrocarbons.

5.3. Absorptions in the region 2000—1650 cm^{-1}

Studies of the absorption patterns produced by substituted aromatics in the 2000—1650 cm^{-1} range afford one of the best and most selective methods for the identification of benzene-ring substitution. This method is due to Young, Duvall and Wright [17], who have shown that each of the various types of substitution gives rise to a typical absorption pattern which, within limits, is independent of the nature of the substituents.

Aromatic compounds, generally, give a much richer absorption pattern in this region than most other materials, and a series of overtone and combination bands is obtained. They are relatively weak compared with the fundamentals, and for studying this region fairly thick cells (0.1 mm for liquids and 1.0 mm for 10 per cent solutions of solids) are employed. In the latter case double-beam operation, with solvent cancellation of the small carbon tetrachloride absorptions in this region, is desirable.

Whiffen [64] has recently been able to show that all the bands in this region can be assigned with reasonably high precision to summation bands of the CH out-of-plane fundamentals which occur between 1000 and 700 cm^{-1}. This accounts for the variations in characteristic pattern with alterations in the positions of substitution and explains why the general shapes are more important in this case than the absolute frequencies. This also explains the simplification of the bands with more symmetrical structures in which a smaller number of out-of-plane CH fundamentals occur.

Young *et al.* [17] state that they have employed this spectral region for a number of years in the characterisation of benzene ring substitution and that it has stood the test of experience with a large number and variety of compounds. Since their publication we have used this region in our own laboratories, with considerable success, on a wide range of aromatic compounds without encountering any difficulties other than the occasional production of an irregular pattern when certain limited substituents are present. This last fact had been noted by the original authors, and is discussed below.

The essential difference between analysis in this region and normal infra-red techniques lies in the fact that the intensities and numbers of the bands are relatively more significant than the precise wavelengths at which they occur. It is therefore desirable that workers in this field should have a set of reference patterns readily available for comparison. Typical patterns are illustrated by Young *et al.* [17], but it is a relatively simple matter to obtain a typical set by the examination of the various toluenes, xylenes, etc., on the instrument being used, and this is to be preferred.

These patterns are sufficiently distinctive to permit small variations in intensity and position to occur without seriously complicating the recognition of the characteristic type. Thus, mono-substituted materials generally produce a series of four bands of gradually diminishing intensity towards the long wave-length region. In the case of chloro- and bromo-benzenes the intensity of the first of these is actually a little less than that of the second, whilst in phenol the fourth band is stronger than the second and third, and all bands show a shift towards longer wave-lengths. Nevertheless, the patterns are all sufficiently different from those produced by other types of substitution, and sufficiently similar to the standard, to allow the ready identification of these materials as mono-substituted aromatics.

Di-, tri- and tetra-substituted materials show even more regularity with respect to the appropriate typical patterns. Young *et al.* [17] note that occasionally there is a more marked splitting of doublets than is usual, but state that they have had no serious trouble in their identification. Patterns are also given for penta- and hexa-substituted materials, but these must be treated with more caution, partly because the pattern is less characteristic as substitution is increased, and partly because the original authors do not consider they have yet examined enough samples of these types to be statistically significant. Whiffen [64] has pointed out that 1 : 2 : 3 : 4- and 1 : 2 : 3 : 5-tetra-substituted aromatics have only two out-of-plane CH frequencies, and so can give rise to only three summation bands ($a + a$, $a + b$ and $b + b$), of which only two will fall in the $2000-1650$ cm^{-1} range. In the former case they will be well separated, but in the latter they will be close together and may sometimes appear as a single band. In 1 : 2 : 4 : 5- and in penta-substituted aromatics only one fundamental CH is permitted. This normally falls near 870 cm^{-1}, so that a strong single band near 1740 cm^{-1} is to be expected.

Interference with these typical patterns can arise from two causes: the occurrence of a fundamental or some other overtone band in the same region, and the presence of certain substituents which alter appreciably the original fundamental frequencies.

Fundamentals and overtones of other bands in this region are not difficult to recognise, and appear simply as an additional band superimposed upon the original pattern. In styrene, for example, a band appears at 1818 cm^{-1} which is the overtone of the out of plane CH deformation vibration at 910 cm^{-1}, and this is superimposed on the normal mono-substituted pattern which appears on either side.

When carbonyl groups are present, the intensity of the CO fundamental, at the cell thickness required, usually blots out a considerable proportion of the region and reduces the possibilities of the identification.

The presence of fluorine, ether or nitro-substituents in mono-substituted compounds causes considerable alterations in the general pattern. In the first two instances the spectrum is very similar to the normal, but has an additional band above 2000 cm^{-1}, which is a little confusing. This can be shown to be also due to a summation band [64], and it can be safely ignored in making a diagnosis. The pattern derived from nitrobenzene is markedly different from the others, because, as will be indicated later, the fundamental CH out-of-plane frequencies are themselves considerably displaced in this compound. It is a general rule that if the pattern for a compound differs from the normal one for that specific configuration, the pattern so obtained will not resemble any of the others, so that only in occasional rare cases is it not possible to obtain any useful information from this region. With further substitution the patterns become normal in most cases, and the nitrotoluenes and ni-troxylenes, for example, show normal or near normal patterns. However, heavy nitro-group substitution can stil result in atypical patterns in some cases.

Little data are available on the applicability of these absorptions to naphthenic and polycyclic hydrocarbons, but it is to be expected that the patterns resulting from the combined effects of two or more rings, each differently substituted, will invalidate the use of this region for identifications. The A.P.I. series of spectra includes a number of variously substituted naphthalenes which would not be classifiable from the patterns in this region, but, as in earlier cases, the patterns obtained are quite distinct from any standard pattern. This has been confirmed by Fuson and Josien [63] in studies on substituted benzanthracenes. Whilst they do not regard the absorption pattern in this region as being indicative of the substi-tution pattern, they have noted a number of regularities in this region which may be helpful in the subdivision of these compounds into different classes.

5.4. Ring stretching vibrations

The fundamental studies on benzene vibrations, referred to above [2–7], have shown that the characteristic skeletal stretching modes of the semi-unsaturated carbon–carbon bonds lead to the appearance of a group of four bands between 1650 and 1450 cm^{-1};

of these, those near 1600 cm^{-1} and 1500 cm^{-1} are highly characteristic of the aromatic ring itself and, taken in conjunction with the
C—H stretching bands near 3030 cm^{-1}, they afford a ready means of
recognition for this structure. The actual positions of the bands are
influenced to some extent by the nature of substituent groups, but
depend, to a rather greater degree, on the way in which the latter are
arranged round the ring, so that in some cases it is possible to obtain
a certain amount of confirmatory evidence as to the type of
substitution. It will be seen, however, that the amount of overlapping
between the ranges for various kinds of substitution is such that the
number of cases in which such a differentiation is possible is limited.

These bands are now thoroughly well documented in the infra-red
and Raman spectra, and their assignments are based on large numbers
of vibrational analyses of individual molecules as well as on a massive
collection of data. Early workers who recognised the importance of
this region for identification included Barnes *et al.* [18], Randall *et
al.* [19], Colthup [20], Cannon and Sutherland [42], and
Lecomte [10]. Randle and Whiffen [65] systematised the data and
assigned the various group frequencies. Since then, Katritzky and his
co-workers have added a good deal to our knowledge of both
frequencies and intensities [79–82]; Scheerer *et al.* [83–86] has
carried out normal coordinate studies on the complete series of
chlorinated benzenes; and Green *et al.* [87, 88] have reported on
many other derivatives as has Bogmolov [89].

There is general agreement that the four bands in the 1600 to
1500 cm^{-1} region originate in skeletal ring breathing modes. There
are essentially two pairs which are degenerate in benzene itself, so
that only two bands are shown. On substitution in the ring the
degeneracy is split and all four bands are present. However, as will be
seen, there is a very considerable variation in the relative intensities
of these bands with changes in the substituents and in their relative
arrangements, so that in some compounds one or other of these
bands may be virtually undetectable in the infra-red. The splitting of
the 1500 cm^{-1} band produces that at 1450 cm^{-1} which is clearly
seen in halogenated benzenes but is usually overlaid in alkyl benzenes
by the CH$_2$ deformation bands. Neither of the bands of this pair
have much intensity in the Raman spectrum.

5.4(a). The 1600 cm^{-1} band. Colthup [20] shows this band as
occurring within the range 1625–1575 cm^{-1} for most aromatic
materials. With *para*-substitution there is a small shift towards higher

frequencies $(1650-1585 \text{ cm}^{-1})$, and this applies also, to a lesser extent, to unsymmetrical tri-substituted materials. With vicinal tri-substituted materials, on the other hand, the shift is towards lower frequencies $(1610-1560 \text{ cm}^{-1})$.

These limits are fully substantiated by our own findings, by the A.P.I. series of spectra and by the work of Whiffen [65, 67], of Josien [61], and of other workers such as Varsanyi [15], and Katritzky [79]. In the great majority of mono-substituted materials the band lies within $\pm 5 \text{ cm}^{-1}$ of 1600 cm^{-1}, but in compounds such as chlorobenzene, bromobenzene and thiophenol it lies close to the lower limit, whilst the frequency is higher than the average in nitrobenzene. This would suggest that the band position is determined, to some extent, by the electronegativity of the substituent groups, but this is only one of a number of factors operating, and the introduction of further electronegative groups does not result in any further reduction in the frequency. Furthermore, these shifts are considerably less than those which can be produced by alterations in the positions of the substituents. The position of this band therefore affords little or no evidence as to the nature of the substituents, and only in favourable circumstances, in which the absorption is appreciably above or below 1600 cm^{-1}, does it give any reliable indication of the type of substitution. However, corresponding shifts also occur in the position of the 1500 cm^{-1} band (see below), so that the two bands taken together will give an indication, at least, of certain types of substitution which is occasionally useful in confirmation of data obtained from other regions of the spectrum. Cannon and Sutherland [42] have shown this correlation to be applicable to a wide range of polycyclic aromatics as well as to simpler materials.

The second band of the 1600 cm^{-1} pair occurs near 1580 cm^{-1}, with very variable intensity. The reasons for this are given in section 5.4(f). below. The absence of this band in the infra-red is not uncommon, but it is always present in the corresponding Raman Spectrum.

5.4(b). The 1600–1560 cm^{-1} region. Before this band was identified as one of a degenerate pair, there was much confusion over its origins, some workers associated it with the benzoyl group [21] and others with fused ring systems [42]. In most compounds this band is very weak in the infra-red and it appears as a weak shoulder on the side of the 1600 cm^{-1} band. In the Raman spectrum, the opposite is true and this band is then the stronger of the two components. The overall frequency range is 1597–1562 cm^{-1} in mono-substituted benzenes.

However, in the cases of benzoyl derivatives, and other conjugated systems such as styrene or fused rings, the interaction between the π cloud and the ring allows a very considerable intensification of this band, which is then clearly seen as a separate band, often of comparable intensity with the other component of the doublet. A parallel effect can sometimes be observed when the atom attached directly to the ring has a lone pair of electrons, so that intensification of the 1580 cm^{-1} band is not an invariable sign of conjugation. Nevertheless, it does serve to indicate compounds in which this is likely, and confirmation can then be sought from other bands such as the carbonyl or C=C frequency.

5.4(c). The 1525–1475 cm^{-1} band. The overall range is considerable (1530–1460 cm^{-1}) but this allows for a few exceptional molecules. Iodobenzene for example, is responsible for the lower frequency limit. The great majority of substituted benzenes absorb in a narrower range close to 1500 cm^{-1}. Katritzky [79] suggests 1510–1480 cm^{-1} for mono-substituted benzenes, 1510–1460 cm^{-1} for *ortho* di-substituted, 1495–1470 cm^{-1} for *meta* di-substituted, 1520–1480 cm^{-1} for *para* di-substituted and 1510 $\mp$ 8 cm^{-1} for unsymmetrical tri-substitution. As with the 1600 cm^{-1} band, wide variations in intensity occur which can be informative as to the nature and the positions of the substituents. This is discussed in section 5.4(f).

5.4(d). The 1450 cm^{-1} band. The existence of a fourth skeletal C–C frequency in the region of 1450 cm^{-1} was suggested by Lecomte [66], and has been fully substantiated by later workers [61, 65, 15]. In a series of mono-substituted aromatics studied by Josien and Lebas [61] it fell in the range 1470–1439 cm^{-1}, and was usually of moderate to strong intensity. In some cases, however, as with aniline and nitrobenzene, the intensity was relatively weak. This band is frequently overlaid by strong CH$_2$ deformations, and its utility for identification purposes is therefore reduced, but it can be readily identified in compounds, such as the halogenated benzenes and similar materials, as well as in polynuclear compounds [63]. It is weak in the Raman spectrum.

5.4(e). Polycyclic compounds. On theoretical grounds the main 1600 and 1500 cm^{-1} bands of the phenyl ring can be expected to occur also in polycyclic materials although, as bond fixation occurs in such materials and not all the bonds are of equal length [22], it is to be expected that there may be some broadening of the overall ranges in which they can occur. This is reflected in the findings, as there is a shift towards shorter wave-lengths of the 1600 cm^{-1} band (1650– 1600 cm^{-1}) and a broadening of the range of its subsidiary band to 1630–1575 cm^{-1} for α- and β-substituted naphthalenes. The third and fourth bands remain unchanged in the 1525–1450 cm^{-1} range. These findings are confirmed, insofar as the higher frequency bands are concerned by early Raman studies by Luther [91] and Gockel [92] on methyl naphthalenes. The 1600 cm^{-1} band ranged from 1634 cm^{-1} in 1 : 6-dimethyl naphthalene to 1599 cm^{-1} for the 1 : 2-derivative. The corresponding frequencies for the second component of the doublet were 1586 and 1560 cm^{-1}. There is never any difficulty, in practice, in recognising the aromatic character of such compounds, or of substituted quinolines and similar materials. These bands can also be recognised without difficulty in the spectra of a number of polynuclear hydrocarbons illustrated by Orr and Thompson [23], in stilbene derivatives by Thompson, Vago, Corfield and Orr [24], in substituted naphthalenes [60, 69] and in the polynuclear hydrocarbon spectra of Cannon and Sutherland [42] and of Fuson and Josien [63]. Absorptions in the 1640–1600 cm^{-1} range have also been successfully used in the recognition of the phenyl group in alkaloids [25] and in aromatic steroids [26, 71].

5.4(f). Absorption intensities of the aromatic bands 1600– 1450 cm^{-1}.
The absorption bands due to aromatic structures in the 1600–1500 cm^{-1} region are notorious for the very wide fluctuations in intensity which are encountered, but this is not always so. The 1600 cm^{-1} doublet arises from vibrations in which the main dipole moment change is produced by the movements of the substituents on opposite sides of the ring acting in opposition. If these substituents are the same as in benzene or *para* dimethyl benzene, there is no change in the overall dipole moment during the vibration, and the band appears only in the Raman spectrum and not in the infra-red. With mono-substituted benzenes there is of course, always a change of dipole moment during this mode so that the band is allowed in the infra-red. However, if the polarity of the group is low, as for example with an alkyl group, the dipole change is small and the 1600 cm^{-1} band is therefore weak. With a more polar substituent the change is larger and the band intensifies, and this

occurs independently of whether the substituent is an electron acceptor or an electron donor. The intensity is therefore related to the numerical value of the Hammett constant regardless of its sign.

With *para* di-substitution the nature of the substituents becomes more significant. With two strong electron donors or two strong electron acceptors, the 1600 cm^{-1} band will be weak or absent, but with one strong donor and one strong acceptor there is a very large dipole-moment change during the vibration and a very strong band results. In *para* hydroxy acetophenone, for example, the 1600 cm^{-1} band is stronger than the carbonyl band itself. The intensity is therefore related to the algebraic difference between the electronic effects of the substituents as measured by their Hammett σ values. With *meta* substitution these effects no longer cancel in this way and a consideration of the nature of the vibration will show that the two groups now reinforce each other to increase the dipole change so that the intensity is related to the algebraic sum of their effects. With *ortho* substitution the position is considerably more complex and less easy to predict, but in general a result intermediate between the *para* and *meta* cases is found [79].

In almost all cases the 1575 cm^{-1} band behaves in exactly the same way as the band at 1600 cm^{-1}, and for the same reasons. As it is intrinsically weaker than the 1600 cm^{-1} band in the infra-red, it is correspondingly more difficult to identify when non-polar substituents are present. The very considerable intensification which results from conjugation is due to the fact that the electron delocalisation allows a considerable dipole-moment change to occur during the vibration.

The intensity of the 1500 cm^{-1} band also varies widely. This also arises from a mode in which the dipole-moment change is produced by the relative movements of substituents on opposite sides of the ring, but this time they are both moving in the same direction. With *para* substituents the intensity is therefore related to the algebraic sum of the substituent constants. The band therefore appears when both substituents are the same, but will vanish if the substituents are an electron donor and an electron acceptor of equal power. The band near 1450 cm^{-1} is always reasonably strong and is less affected by the nature of the substituents.

5.4(g). Identification of aromatic structures in the presence of other groupings. In the great majority of cases the three $1600-1500 \text{ cm}^{-1}$ bands of an aromatic structure can be recognised without difficulty,

provided sufficient attention is paid to the preparation of films of suitable thickness; and, coupled with data from the 3000 cm^{-1} region, this is usually sufficient for the positive identification of this group. However, heterocyclic aromatics such as pyridine and pyrimidines also give a somewhat similar set of bands in the 1600 cm^{-1} region [27, 93]. *cyclo*Octatetraene also gives similar absorption bands in the 1600 cm^{-1} region, and the spectrum closely resembles that of styrene, in which a normal C=C link is conjugated with the ring [28].

Other factors which can complicate the identification of aromatics in this region are the presence of CO—NH, C=C, NO$_2$ and NH$_2$ groupings. Occasionally these will mask one or other of the main regions. The bands from a normal C=C link near 1640 cm^{-1} are usually just above the 1600 cm^{-1} phenyl band, but when conjugated with the ring the C=C frequency often falls within the same absorption region. The second C=C band arising from resonance splitting of two non-aromatic C=C vibrations is, however, at an appreciably higher frequency than the aromatic band at 1500 cm^{-1}. Difficulties are also encountered in the identification of the 1600 cm^{-1} band in certain aromatic β-diketones and other structures in which the carbonyl band is shifted into the same region of absorption. Certain quinones and similar materials, in which bond fixation occurs, are also somewhat less easy to recognise, as the aromatic bands occur towards the extremes of the normal range.

In general, there is little difficulty in this identification, and the correlations can be applied to the spectra of samples examined in solution or as pastes in paraffin oil, as the shifts produced by changes of state are very small [29] and well within the limits of the ranges quoted.

5.5. Out-of-plane CH deformation vibrations

Strong bands appear in the region 1000—650 cm^{-1} in the spectra of aromatic materials due to the out-of-plane deformation vibrations of the hydrogen atoms remaining on the ring. Their position is determined almost wholly by the positions rather than the nature of the substituents and, with certain limitations, they provide an excellent method for the recognition of the type of substitution. The numbers and positions of these bands in variously substituted aromatics have been discussed by Randle and Whiffen [65], Varsanyi [15], Katritzky [79, 82], and others. However, although they are all relatively stable in position, most of them are weak, and attention has been directed almost exclusively at the one or two strong bands of the series. The very high intensity of these bands makes them particularly well suited for quantitative work, and very many applications have been described for the estimation of the relative proportions of *ortho, meta-* and *para*-substituted isomers and for certain tri-substituted products. Whiffen and Thompson [31], for example, have given a method for the quantitative analysis of cresylic acid based on these bands, and have also worked out the basis for a similar analysis of mixed xylenols. Friedel, Peirce and McGovern [38] have studied this analysis further and described a method for the analysis of phenols, cresols, xylenols and ethyl phenols. Other typical examples are Ferguson and Levant's [39] method for the analysis of the four component mixture of chlorobenzene with *ortho-, meta-* and *para*-chlorobromobenzenes; Hales' [40] method for the analysis of mixed *tert.*-butylphenols, and Perry's [41] estimation of mixtures of five C_{10} aromatics.

The occurrence of these bands and their correlation with *ortho-, meta-* and *para*-substitution was noted by Barnes *et al.* [18], and was developed and amplified by Whiffen, Torkington and Thompson [30] and by Whiffen and Thompson [31], who determined their position in a number of mono- and di-substituted materials and in a lesser number of tri- and tetra-substituted products. Their findings were later amplified by further data by Richards and Thompson [32] working on substituted phenols, by Thompson, Vago, Corfield and Orr [24] working on substituted stilbenes, by Orr and Thompson [23] working on polynuclear hydrocarbons, and by Kamada [52], and by Cannon and Sutherland [42] working on a wide range of aromatic materials.

This correlation has been included in the various charts issued by different workers [20, 33, 34], with a gradual increase in the proposed limiting ranges as greater numbers of compounds were examined. The band positions, in relation to the various types of substitution, havy been neatly summarised by Colthup [20] in terms of the numbers of adjacent hydrogen atoms remaining round the ring, and they will be considered below under this classification. Finally, Cole and Thompson [35] have supplied a mathematical basis for this empirical correlation and have been able to calculate the expected position of the principal CH out-of-plane frequencies for the various types of substitution, using a simple force system with only two force constants. Their results are in good agreement with the experimental values which they have determined for a considerable series of halogenated benzenes. The correlations even hold good for *ortho*-, *meta*- and *para*-dideuterobenzenes, although the characteristic bands appear about 20 cm^{-1} above the normal values [53].

These correlations therefore derive from a very considerable volume of data in the literature, which is fully sufficient to substantiate them in respect of di- and tri-substituted products. With tetra- and penta-substituted materials, naphthalenes [69] and poly-cyclic compounds the correlations appear to apply with slightly greater deviations. Thus in *cyclo*pentenophenanthrenes [72] and in benzanthracenes [63] the correlation with the number of adjacent hydrogen atoms in each ring still holds, but as is to be expected with compounds of this complexity, additional bands can occur in the same regions. In these cases therefore the absence of a particular band is more diagnostic than its presence.

Marked deviations from these correlations can occur when the ring is heavily substituted with highly polar groups [54] such as CO, CF$_3$ or NO$_2$, and these special cases will be considered separately below.

5.5(a). Five adjacent ring hydrogen atoms. Mono-substituted aromatics.

Whiffen and Thompson [31] originally proposed the range 760–740 cm^{-1} for the position of the out-of-plane CH bending absorption of mono-substituted aromatics, and this has been widened slightly by Colthup [20] to 770–730 cm^{-1}. If all variants are included, the range is wider still. Toluene for example, absorbs at 729 cm^{-1}, benzoic acid at 808 cm^{-1} and nitrobenzene at 792 cm^{-1}. These are, however, exceptional and most compounds absorb around 750 cm^{-1} with a deviation of about 15 cm^{-1} on either side. Normally the band is easily recognised as being the strongest in this region of the spectrum, although the position of the band changes considerably within the range, with even small changes in molecular structure. There is, for example, a shift of 15 cm^{-1} on passing from phenylhydrazine to phenylhydrazine hydrochloride. The upper part of the range of this band overlaps the range of *ortho*-di-substituted materials, so that it cannot be used alone to differentiate mono-substitution from *ortho*-di-substitution. However, Whiffen [36, 65] has pointed out that mono-substituted materials also absorb strongly near 690 cm^{-1} where *ortho*-substituted materials do not absorb to any appreciable extent, and the two can be differentiated in this way. This latter correlation has been confirmed by Colthup [10], by Randall *et al.* [19], by Josien and Lebas [61] and by Cannon and Sutherland [42], Varsanyi [16], Katritzky [79], and others. It is not due to a hydrogen deformation but to an out-of-plane skeletal mode. The band in mono-substituted materials is almost always within ± 10 cm^{-1} of 700 cm^{-1}, but even here some deviations occur, so that if all examples are included the range widens to 700–675 cm^{-1}. The band should always be examined in conjunction with he 750 cm^{-1} band, as *meta*-substituted products and certain tri-substituted materials also absorb near 700 cm^{-1}, although usually within a rather wider frequency range. The absence of a band within the range 700 cm^{-1} ± 10 cm^{-1} is strong evidence for the absence of a mono-substituted product, but the converse is not necessarily true, and confirmation must be sought from the 750 cm^{-1} region.

In certain very limited cases the 750 cm^{-1} band shows splitting into a group of weaker bands at the same frequency. This occurs in the *cis*-forms of phenyl propenyl carbinol and styryl methyl carbinol [37], although the *trans* forms have only a single band of normal position and intensity. The reasons for this are not known, but it may be connected with steric interferences in the *cis*-compounds. Benzoyl derivatives also give rise to distortions in this region. The 750 cm^{-1} band is not seen in the Raman spectrum and the 690 cm^{-1} band appears only weakly.

5.5(b). Four adjacent ring hydrogen atoms. *ortho*-di-substituted aromatics.

As indicated above, *ortho*-di-substituted materials absorb in approximately the same range of frequencies as mono-substituted derivatives, and the appropriate strong band is found in the range 770–735 cm^{-1}. The reduction in the number of adjacent free hydrogen atoms from five to four does not significantly affect the frequency. This correlation also holds for cyclic materials such as phthalic anhydride, as well as for naphthalenes, quinolines and larger rings. Colthup indicates, however, that a slightly wider range of frequencies is to be expected in such materials. Further detailed assignments are given by Varsanyi [15]. In line with this correlation, Orr and Thompson [23] have traced a strong band near 750 cm^{-1} in a considerable number of carcinogenic hydrocarbons which contain 1 : 2-di-substituted polynuclear structures, although there were shifts of ± 10 cm^{-1} in many cases, and Cannon and Sutherland [42] identify bands at 750 cm^{-1} in 9 : 10-dihydroanthracene and at 725 cm^{-1} in anthracene itself as being due to this vibration. A large number of substituted naphthalenes have been studied by Cencelj and Hadži [69] and by Werner *et al.* [60, 70]. In all cases in which one ring is unsubstituted a strong band corresponding to *ortho*-substitution is shown, whilst it is usually possible to trace the bands derived from the free hydrogen vibrations of the other ring, as corresponding to the appropriate type of tri- or tetra-substituted aromatic compound. When substituents are present in both rings, however, these bands are more difficult to identify, and the applications of these correlations would lead to errors in some cases. Bands in the $746-738$ cm^{-1} range are shown by appropriately substituted benzanthracenes [63], but as indicated above, the occurrence of additional bands lessens the general utility of this correlation for polycyclic compounds.

5.5(c). Three adjacent ring hydrogen atoms. 1 : 3- and 1 : 2 : 3-substitution. Aromatic rings with three adjacent free hydrogen atoms show absorption in the $810-750$ cm^{-1} range with a second band of medium intensity in the region $725-680$ cm^{-1}. This last is not of course a hydrogen deformation mode but a skeletal frequency similar to that responsible for the 690 cm^{-1} band in mono-substituted benzenes. It does not appear with any great intensity in vicinal compounds. Colthup [20] indicates a slightly narrower range in both cases for vicinal tri-substituted materials than for *meta*-di-substituted ones, and this is reflected in the examples summarised by Varansyi [15]. The vicinal compounds generally absorb towards the lower end of the frequency range, but exceptions do occur. Overlapping of these two bands with the ranges of mono-substituted materials may lead to complications in a few cases, but usually the frequency of the first band is around 780 cm^{-1}, whilst the position of the second is much more variable than with mono-substitution, so that it only occasionally falls within the same narrow range.

The main $810-750$ cm^{-1} band can also be traced without difficulty in the spectra of most substituted naphthalenes which contain three free adjacent hydrogen atoms [69, 70]. In the A.P.I. series, for example, 1 : 5-dimethylnaphthalene has a strong band at 790 cm^{-1}, which is the strongest band in this region of the spectrum. Similarly, mono-substituted naphthalenes with the substituent at the 1 position also show this band, in addition to the $770-735$ cm^{-1} band arising from the four free hydrogen atoms on the other ring. With further substitution, however, this region often becomes too complex for ready identification of the type of substitution.

5.5(d). Two adjacent ring hydrogen atoms. 1 : 4-, 1 : 2 : 4- and 1 : 2 : 3 : 4-substitution. With a further reduction in the number of free adjacent hydrogen atoms on the ring, the absorption frequency of the out-of-plane CH vibrations shows a further shift to higher frequency, and a strong band occurs in the range 860–800 cm^{-1}. Data on this band have been reviewed by Randle and Whiffen [65] and by Bomstein [55], and later work is summarised by Katritzky [79], and Varsanyi [15]. The range quoted includes a number of less usual compounds and in most compounds the band appears in the lower end of the range. Katritzky [79], for example, quotes a range of 817 $\mp$ 13 cm^{-1} for this band, and for dialkyl benzenes it narrows further to 833–810 cm^{-1}. The extremes are represented by *para* xylene (790 cm^{-1}), and *para* dichlorbenzene (880 cm^{-1}). No strong band is shown in the 750–700 cm^{-1} range, although weak ones sometimes occur. Nevertheless, in cases in which the main band falls in the 810–800 cm^{-1} region, differentiation from *meta*-type substitution is usually possible by the absence or presence of a second 650–700 cm^{-1} band of medium or strong intensity. The same type of absorption is shown by mono-substituted naphthalenes [70], which have this band in addition to that arising from the other ring. The best method of differentiation of the three possible types absorbing in the 860–800 cm^{-1} range is through a study of the 2000–1600 cm^{-1} region, but unsymmetrical tri-substitution can also be differentiated by the medium intensity band arising from the single free hydrogen atom situated between two substituents (see below). The number of 1 : 2 : 3 : 4-tetra-substituted materials reported in the literature or available for examination is small [65], but the correlation holds good for the cases that are known. 1 : 2 : 3 : 4-Tetra-methylbenzene, for example, absorbs at 804 cm^{-1}.

5.5(e). One isolated ring hydrogen atom. Situated between two substituents. 1 : 3-, 1 : 2 : 4-, 1 : 3 : 5-, 1 : 2 : 3 : 5-, 1 : 2 : 4 : 5-, and penta-substituted aromatics. As the number of adjacent free hydrogen atoms is reduced to one the expected high-frequency shift again occurs, and the CH absorption appears in the region 900–860 cm^{-1}. The reduction in the number of vibrating groups also affects the intensity, so that this band is of medium strength, in contrast to the very strong bands in the cases cited above. In the case of 1 : 3 : 5-tri-substituted materials a strong band is shown in the range 865–810 cm^{-1} and another between 675 and 730 cm^{-1}. The last of these bands usually enables this particular type to be differentiated from *para*-substituted and similar products absorbing near 830 cm^{-1}. McCaulay *et al.* [57] find this band between 874 and 835 cm^{-1} in *sym.*-trialkylbenzenes, and have discussed the effects of chain branching in the substituents upon the frequency. The number of tetra- and penta-substituted materials available for study, or having spectra reported in the literature, is small, but it would seem that tetra-substituted benzenes which have an isolated hydrogen atom, absorb in the same place as penta-substituted compounds (875–860 cm^{-1}). The 900–800 cm^{-1} band can be traced with difficulty in the spectra of suitably substituted polynuclear hydrocarbons, but the pattern in this region is usually too complex to enable the recognition of these medium strength bands to be reliable.

5.5(f). The influence of substituent groups. As has already been explained, the positions of these CH frequencies are determined almost wholly by the position of the substituents and are independent of their nature. This, however, is not strictly true, for the electron donating or accepting properties of the substituent play an important part in determining the precise position of the absorption within the frequency range. In extreme cases, as in heavily nitrated compounds, the effects introduced in this way can be sufficient to shift the absorption out of the expected range altogether. Nitrogroup substitution moves the band towards higher frequencies and the extent of the shift follows the degree of substitution. With very heavy nitro-group substitution the bands reach positions from which they do not show any further movement with additional substitution of nitro-groups. The CH bending frequency of a single free hydrogen atom, for example, is very near 920 cm^{-1} for compounds containing three or more nitro-groups. Specialised studies of this kind can occasionally allow the recognition of the substitution pattern of such materials, provided some other fact, such as the number of nitro-groups, is known, but otherwise it is difficult to differentiate, for example, a *para*-substituted material with one nitro-group from a *meta*-substituted product with two. The interpretation of this spectral region therefore requires extreme caution when several nitro-groups are present, or when, for example, a strongly electron-attracting group is situated in the *para*-position to a strongly electron-donating one, as in *p*-aminobenzoic acid. In such cases multiple bands often appear, and interpretation is then difficult. Margoshes and Fassel [58] have illustrated a number of cases of this type. However, compounds containing high proportions of interfering substituents are rare, so that the picture of the substitution pattern obtained from this region is normally a relatively clear one and, coupled with a study of the 2000–1600 cm^{-1} region, it allows a positive identification of the type of substitution present. As indicated above, however, the certainty with which these correlations can be applied diminishes with the degree of substitution, so that identification of tetra- and penta-substitution is very much less precise than, for example, that of mono-substitution. In cases of doubt it is usually possible to resolve any uncertainties by reference to the corresponding Raman spectra. These are complementary to the infra-red in that those bands which are weak in the one are strong in the other. Thus the in plane deformations of the hydrogen atoms are strong and highly diagnostic in the Raman but weak in the infra-red. By the use of both spectra it is therefore nearly always possible to define the substitution pattern with precision.

A number of workers have studied the possible relationships which these frequencies may have to the changes in the electron density around the ring following the introduction of substituents. For example, Bellamy [59] has discussed these absorptions in relation to the Hammett σ values of the substituents which measure their electron-donating or withdrawing properties. With the exception of halogen substituents, which appear to behave as though they had zero σ values, reasonably good relationships can be found connecting σ values and the frequencies of 3, 2 or 1 adjacent ring-hydrogen atoms, and this is occasionally useful in predicting the frequencies of unknown compounds. The anomaly of the halogens may be explained in terms of the findings of nuclear magnetic-resonance studies, which suggest that the electron densities of the ring carbon atoms are not appreciably altered by halogen substitution [73].

The additivity of Hammett σ values is reflected in an approximate additivity in frequency shifts, and it has been noted, for example, that the frequency displacement from the mean, of a *para*-di-substituted compound, is approximately the sum of the displacements of the two separate mono-substituted materials [58]. Kross *et al.* [74] have also considered this question and put forward an explanation for the major shifts in nitro-compounds, etc., in terms of an orbital-following theory. No treatment is, however, yet entirely satisfactory, as in general both electron-withdrawing and electron-donating substituents lead to δCH frequencies which are higher than those of the methyl compounds.

5.6. Characteristic absorptions between 1225 and 950 cm^{-1}

Aromatic compounds all show a series of relatively weak bands in this region, the positions of which vary with the type of substitution arrangement. Their low intensity renders them less generally useful for analytical purposes than the lower-frequency absorptions already described, but they afford a general confirmation of findings from other parts of the spectrum as to the presence of aromatic substituents and their substitution pattern. Randle and Whiffen [65, 67, 75] have studied these vibrations in some detail and shown them to be largely the C–H in plane-deformation modes, although in some instances frequencies sensitive to the mass of the substituent fall in this region. As noted above, these bands are strong in the Raman spectrum and a good deal of information is now available on their positions [15, 90].

Barnes *et al.* [18] noted the appearance of characteristic bands in this region in the spectra of aromatics and assigned them as follows. Mono-substitution   1075–1065 cm^{-1}, *ortho*-substitution   1125–1085 cm^{-1}, *meta*-substitution   1170–1140 cm^{-1} and *para*-substitution 1120–1090 cm^{-1}. Thompson's [38] table follows this with some increases in the ranges and the observation of an additional *para*-substitution band in the range 1225–1175 cm^{-1}. Colthup [20] has elaborated this considerably, extended it to a study of tri-substituted materials, and shown that up to six characteristic bands can occur in this region with certain types of substitution. There is also a considerable amount of overlapping, so that no one band can be used alone for the recognition of structure. Our own findings are in agreement with Colthup and Whiffen [65], and we would classify the various absorptions in this region as indicated below.

5.6(a). 1 : 2-, 1 : 4- and 1 : 2 : 4-substitution. Compounds substituted thus show weak absorptions in the ranges 1225–1175 cm^{-1} and 1125–1090 cm^{-1}, together with two additional weak bands in the range 1070–1000 cm^{-1}. The bands are always very sharp, although weak, and are usually recognisable without great difficulty. The 1 : 2 : 4-compounds can be differentiated from the others by the presence of an additional band in the range 1175–1125 cm^{-1}. In the Raman spectrum [15, 90] 1 : 2-di-substituted compounds give two well characterised bands in the ranges 1060–1020 cm^{-1} and 1230–1215 cm^{-1}. The 1 : 4-compounds have bands at 1230–

1200 cm^{-1} and $1180-1150 \text{ cm}^{-1}$. $1 : 2 : 4$-substituted compounds have a single band in the wider range $1280-1200 \text{ cm}^{-1}$. These frequencies are strong in the Raman but weak in the infra-red.

5.6(b). Mono-, 1 : 3-, 1 : 2 : 3- and 1 : 3 : 5-substitution. Compounds thus substituted absorb weakly between 1175 cm^{-1} and 1125 cm^{-1}, as do the $1 : 2 : 4$-compounds mentioned above. In addition, the mono-, *meta*- and $1 : 2 : 3$-substituted materials absorb between 1110 cm^{-1} and 1070 cm^{-1}, and all four types show a further single band in the range $1070-1000 \text{ cm}^{-1}$. In the Raman spectrum [90] the most characteristic band of a number of substitution patterns is a ring breathing vibration, sometimes called the Star of David mode. In this each carbon atom of the ring is expanding or contracting in the opposite direction to each of its immediate neighbours. This results in a very strong band in the $1010-990 \text{ cm}^{-1}$ region. It is shown by all mono-substituted benzenes, by all $1 : 3$-derivatives and by the $1 : 3 : 5$-series. It is not shown by any of the others. In the infra-red the corresponding bands are weak in mono-substituted compounds, of medium intensity in the $1 :3$- series and absent from the spectra of $1 : 3 : 5$-compounds when it is forbidden by symmetry. $1 : 2 : 3$-substituted benzenes do not show this band but have a medium intensity band in the $1100-1050 \text{ cm}^{-1}$ region.

5.6(c). 1 : 2-, 1 : 2 : 3- and 1 : 2 : 4-substitution. Compounds substituted thus usually absorb weakly between 1000 cm^{-1} and 960 cm^{-1}. In the Raman spectrum the $1 : 2 : 3$-substitution pattern can be differentiated from the $1 : 2 : 4$- by bands at $1100-1050 \text{ cm}^{-1}$ and at $670-500 \text{ cm}^{-1}$ in the former, and by bands at $1280-1200 \text{ cm}^{-1}$ and $750-650 \text{ cm}^{-1}$ in the latter.

It will be seen, therefore, that differentiation is possible between the main classes indicated above, on the basis of the numbers and positions of the bands in this region, but in the infra-red identifications of single types within the main class will often depend on the recognition of a single weak band. As these bands occur in the general region of the spectrum associated with C–C and C–O stretchings, this is, of course, very unreliable, so that data from this portion of the spectrum cannot be used to supply anything more than a general confirmation of facts ascertained elsewhere.

As regards the positional stability of these bands in relation to the nature of their substituents, it is our experience that they are less affected by the presence of strongly directive groups than those in the low frequency region. The bands from dinitro-compounds, for example, all fall within the expected ranges, but further substitution of nitro-groups results in irregularities. However, Kross and Fassel [76] have shown that at least one band in this region is sensitive to the electronegative nature of the substituent group.

5.7. Other aromatic systems

Tropolones and related materials containing three conjugated double bonds in a seven-membered ring have many of the characteristics of aromatic compounds. The infra-red spectra have been studied by several workers [44–51], and the three regions 1624–1605 cm^{-1}, 1570–1538 cm^{-1} and 1280–1250 cm^{-1} have been suggested as being characteristic; the first two of these are believed to originate in C=C stretching vibrations. However, Nicholls and Tarbell [47] have recently shown the 1570–1538 cm^{-1} band to be absent from a number of benzotropolones, and that the 1275 cm^{-1} band is absent from the spectra of colchicines which have no hydroxyl group [48]. The usefulness of these correlations is therefore questionable, as the first cannot be readily differentiated from a normal phenyl absorption, whilst the last two are apparently not specific.

The aromatic compounds C_5H_5 and C_7H_7 have also been studied and shown to be basically similar to benzene [77]. The partial aromatic character of such compounds as thiopene [78] and similar materials is also reflected in their raised CH-stretching frequencies and general spectral character.

5.8. Bibliography

1. Ingold, *Proc. Roy. Soc.*, 1938, **A169**, 149; Ingold *et al.*, *J. Chem. Soc.*, 1936, 912, 1210, and 1946, 222.
2. Wilson, *Phys. Review*, 1934, **45**, 706.
3. Langseth and Lord, *Danske Vidensk Selskab. Math. Fys.*, 1938, **16**, 6.
4. Angus, Bailey, Hale, Ingold, Leckie, Raisin, Thompson and Wilson, *J. Chem. Soc.*, 1936, 966, 971.
5. Bailey, Hale, Ingold and Thompson, *ibid.*, p. 931.
6. Herzberg, *Infra-red and Raman Spectra of Polyatomic Molecules* (Van Nostrand, 1945), p. 367.
7. Pitzer and Scott, *J. Amer. Chem. Soc.*, 1943, **65**, 803.
8. Plyler, *Discuss. Faraday Soc.*, 1950, **9**, 100.
9. Lecomte, *Compt. rend. Acad. Sci. Paris*, 1937, **204**, 1186; 1938, **206**, 1568; 1939, **208**, 1636.
10. Lecomte, *J. Phys. Radium*, 1937, **8**, 489; 1938, **9**, 13 and 512; 1939, **10**, 423.

11. Delay and Lecomte, *ibid.*, 1946, **7**, 33.
12. Fox and Martin, *J. Chem. Soc.*, 1939, 318.
13. *Idem, Proc. Roy. Soc.*, 1940, **A175**, 208.
14. *Idem, ibid.*, 1938, **A167**, 257.
15. Varsanyi, *Vibrational Spectra of Benzene Derivatives* (Academic Press, New York, 1969).
16. Rose, *J. Res. Nat. Bur. Stand.*, 1938, **20**, 129.
17. Young, Duvall and Wright, *Analyt. Chem.*, 1951, **23**, 709.
18. Barnes, Gore, Liddel and Van Zandt Williams, *Infra-red Spectroscopy* (Reinhold, 1944).
19. Randall, Fowler, Fuson and Dangl, *The Infra-red Determination of Organic Structures* (Van Nostrand, 1949).
20. Colthup, *J. Opt. Soc. Amer.*, 1950, **40**, 397.
21. Rasmussen, Tunnicliff and Brattain, *J. Amer. Chem. Soc.*, 1949, **71**, 1068.
22. Cf. Badger, *Quarterly Reviews*, 1951, **5**, 147.
23. Orr and Thompson, *J. Chem. Soc.*, 1950, 218.
24. Thompson, Vago, Corfield and Orr, *ibid.*, p. 214.
25. Marion, Ramsay and Jones, *J. Amer. Chem. Soc.*, 1951, **73**, 305.
26. Jones, Humphries, Packard and Dobriner, *ibid.*, 1950, **72**, 86.
27. Brownlie, *J. Chem. Soc.*, 1950, 3062.
28. Lippincott, Lord and McDonald, *J. Amer. Chem. Soc.*, 1951, **73**, 3370.
29. Richards and Thompson, *Proc. Roy. Soc.*, 1948, **A195**, 1.
30. Whiffen, Torkington and Thompson, *Trans, Faraday Soc.*, 1945, **41**, 200.
31. Whiffen and Thompson, *J. Chem. Soc.*, 1945, 268.
32. Richards and Thompson, *ibid.*, 1947, 1260.
33. Thompson, *ibid.*, 1948, 328.
34. Van Zandt Williams, *Rev. Sci. Instr.*, 1948, **19**, 135.
35. Cole and Thompson, *Trans. Faraday Soc.*, 1950, **46**, 103.
36. Whiffen, Thesis, Oxford, 1946.
37. Philpotts and Thain, *Nature*, 1950, **166**, 1028.
38. Friedel, Peirce and McGovern, *Analyt. Chem.*, 1950, **22**, 418.
39. Ferguson and Levant, *ibid.*, 1951, **23**, 1510.
40. Hales, *Analyst*, 1950, **75**, 146.
41. Perry, *Analyt. Chem.*, 1951, **23**, 495.
42. Cannon and Sutherland, *Spectrochimica Acta*, 1951, **4**, 373.
43. Laurer and McCaulay, *Analyt. Chem.*, 1951, **23**, 1875.
44. Koch, *J. Chem. Soc.*, 1951, 512.
45. Aulin-Erdtman and Theorell, *Acta Chem. Scand.*, 1950, **4**, 1490.
46. Bartels-Keith and Johnson, *Chem. and Ind.*, 1950, 677.
47. Nicholls and Tarbell, *J. Amer. Chem. Soc.*, 1952, **74**, 4935.
48. *Idem, ibid.*, 1953, **75**, 1104.
49. Doering and Knox, *J. Amer. Chem. Soc.*, 1952, **74**, 5683.
50. *Idem, ibid.*, 1953, **75**, 297.
51. Doering and Hiskey, *ibid.*, 1952, **74**, 5688.
52. Kamada, *Japan Analyst*, 1952, **1**, 141.
53. Tiers and Tiers, *J. Chem. Phys.*, 1952, **20**, 761.
54. Randle and Whiffen, *J. Chem. Soc.*, 1952, 4153.
55. Bomstein, *Analyt. Chem.*, 1953, **25**, 512.
56. Schmid, Z. *Electrochem.*, 1962, **66**, 53.
57. McCaulay, Lien and Launer, *J. Amer. Chem. Soc.*, 1954, **76**, 2364.
58. Margoshes and Fassel, *Spectrochimica Acta*, 1955, **7**, 14.
59. Bellamy, *J. Chem. Soc.*, 1955, 2818.
60. Ferguson and Werner, *ibid.*, 1954, 3645.

61. Josien and Lebas, *Bull. Soc. Chim. Fr.*, 1956, 53, 57, 62.
62. *Idem, Compt. Rend. Acad. Sci. Paris*, 1955, **240**, 181.
63. Fuson and Josien, *J. Amer. Chem. Soc.*, 1956, **78**, 3049.
64. Whiffen, *Spectrochim. Acta*, 1955, 7, 253.
65. Randle and Whiffen, *Molecular Spectroscopy* (Inst. of Petroleum, 1955) p. 111.
66. Depaigne-Delay and Lecomte, *J. Phys. Radium*, 1946, 7, 38.
67. Whiffen, *J. Chem. Soc.*, 1956, 1350.
68. Bellamy, *ibid.*, 1955, 4221.
69. Cencelj and Hadži, *Spectrochim. Acta*, 1955, 7, 274.
70. Werner, Kennard and Rayson, *Austral. J. Chem.*, 1955, **8**, 346.
71. Scheer, Nes and Smeetzer, *J. Amer. Chem. Soc.*, 1955, **77**, 3300.
72. Dannenberg, Schiedt and Steidle, *Z. Naturforsch.*, 1953, **8B**, 269.
73. Corio and Dailey, *J. Amer. Chem. Soc.*, 1956, **78**, 3043.
74. Kross, Fassel and Margoshes, *ibid.*, 1332.
75. Randle and Whiffen, *Trans. Faraday Soc.*, 1956, **52**, 9.
76. Kross and Fassel, *J. Amer. Chem. Soc.*, 1955, **77**, 5858.
77. Nelson, Fateley and Lippincott, *ibid.*, 1956, **78**, 4870.
78. Hildago, *Compt. Rend. Acad. Sci. Paris*, 1954, **239**, 253.
79. Katritzky, *Quart. Reviews*, 1959, **13**, 353.
80. Katritzky and Jones, *J. Chem. Soc.*, 1959, 3670.
81. Katritzky and Lagowski, *J. Chem. Soc.*, 1958, 4155; 1959, 657.
82. Katritzky and Simmons, *J. Chem. Soc.*, 1959, 2051; 2058.
83. Scherer, *Spectrochim. Acta*, 1963, **19**, 601; 1965, **21**, 321; 1966, **22**, 1179; 1967, **23A**, 1489; 1968, **24A**, 747.
84. Scherer, Evans, Muelder and Overend, *Spectrochim. Acta*, 1962, **18**, 57.
85. Scherer, Evans and Muelder, *Spectrochim Acta*, 1962, **18**, 1579.
86. Scherer and Evans, *Spectrochim. Acta*, 1963, **19**, 1739.
87. Green, *Spectrochim. Acta*, 1961, **17**, 607; 1968, **24A**, 1627; 1970, **26A**, 1503; 1523; 1913.
88. Green, Harrison and Kynaston, *Spectrochim. Acta*, 1971, **27A**, 793; 807; 2199; 1972, **28A**, 33.
89. Bogomolov, *Optics and Spectroscopy*, 1960, **9**, 162; 1961, **10**, 162; 1962, **12**, 99; 1962, **13**, 90; 183.
90. Dollish, Fateley and Bentley, *Characteristic Raman Frequencies of Organic Compounds* (Wiley, New York, 1974).
91. Luther, *Z. Electrochem.*, 1948, **52**, 210.
92. Gockel, *Zeit. Phys. Chem.*, 1935, **29B**, 79.
93. Katritzky and Ambler in *Physical Methods in Heterocyclic Chemistry*, Vol. II (Academic Press, New York, 1963) Katritzky and Taylor, *idem*, Vol. 4, 1971.

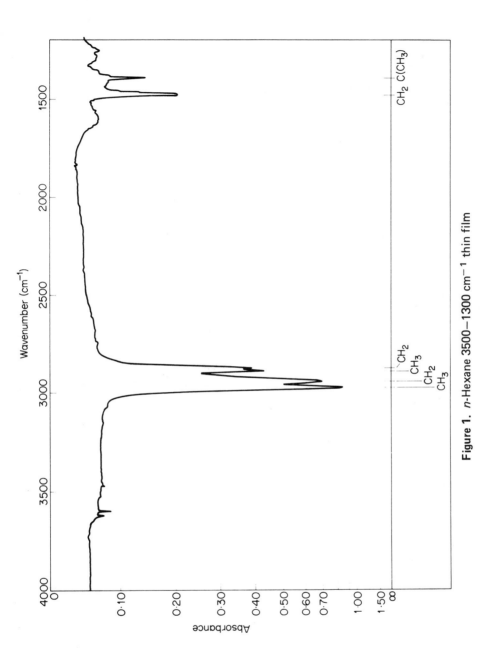

Figure 1. *n*-Hexane 3500–1300 cm^{-1} thin film

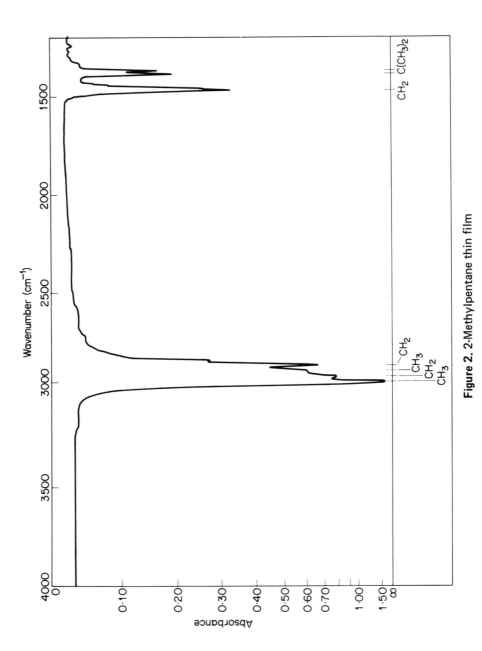

Figure 2. 2-Methylpentane thin film

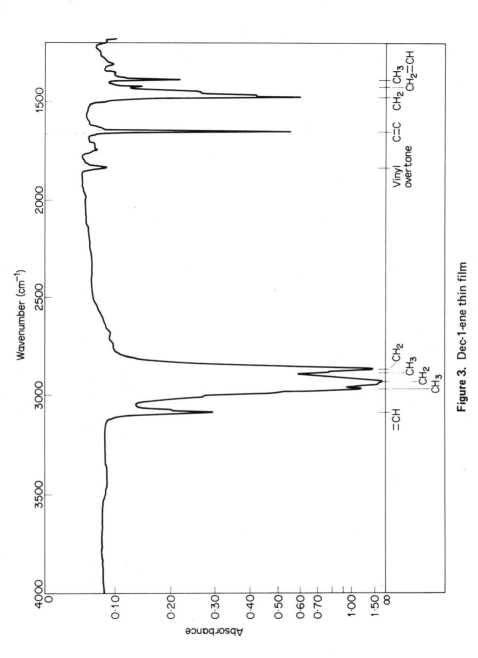

Figure 3. Dec-1-ene thin film

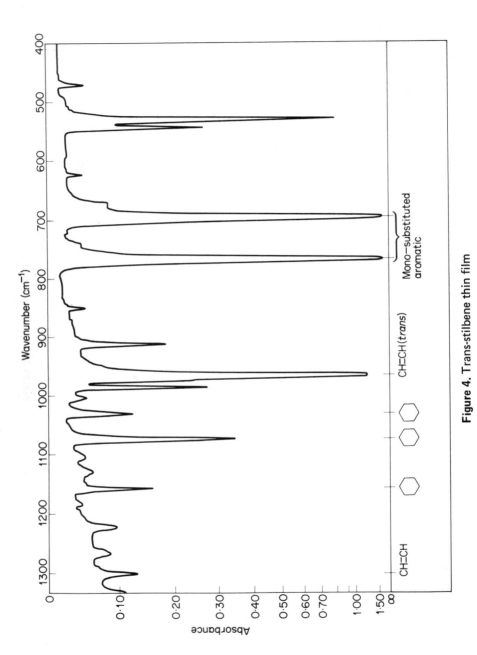

Figure 4. Trans-stilbene thin film

100

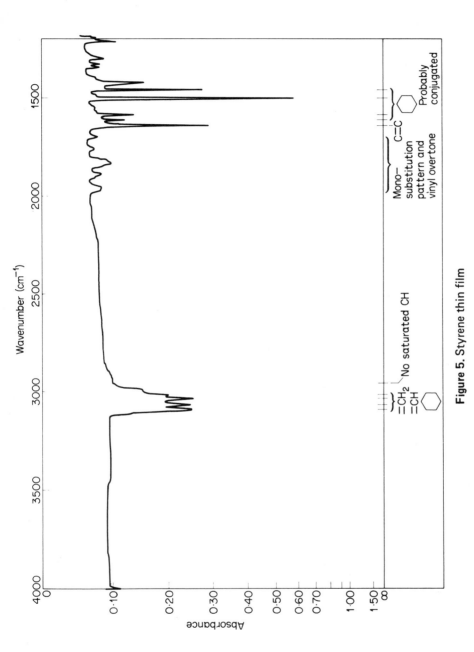

Figure 5. Styrene thin film

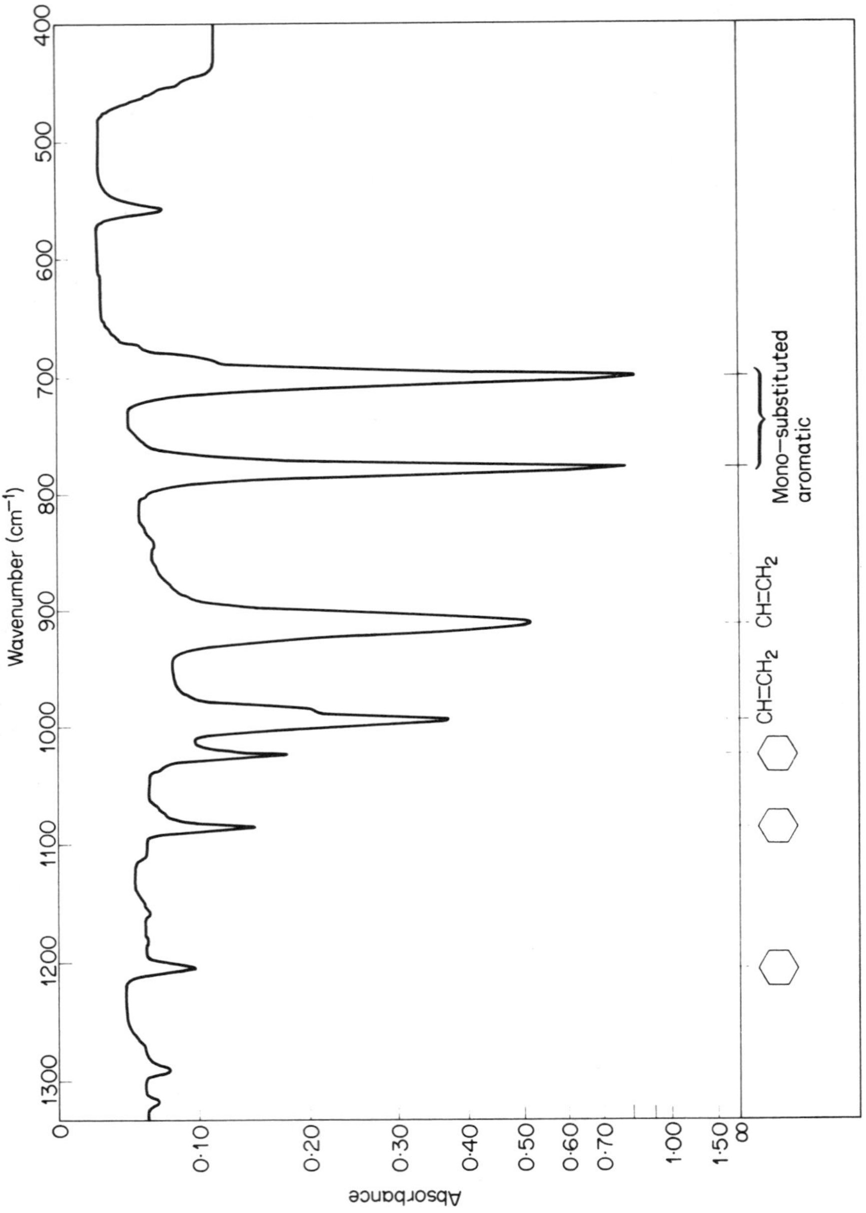

Figure 6. Styrene thin film

102

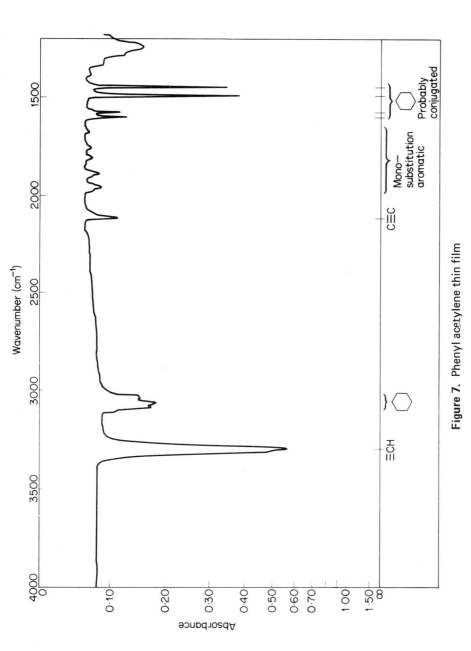

Figure 7. Phenyl acetylene thin film

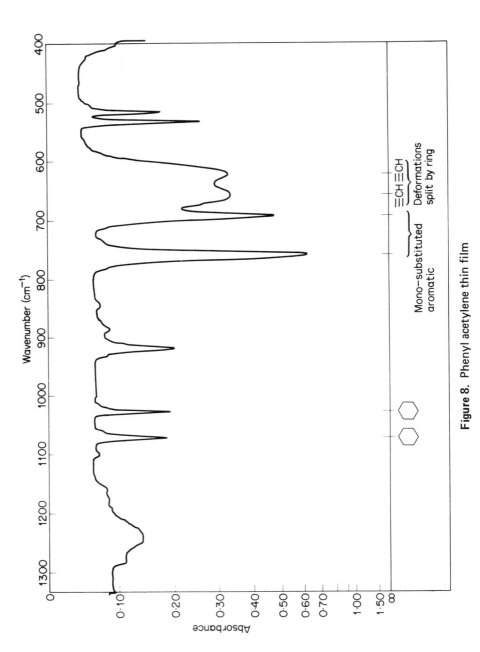

Figure 8. Phenyl acetylene thin film

104

Part Two | Vibrations involving mainly C—O and O—H linkages

6

Alcohols and Phenols

6.1. Introduction and table

The infra-red absorption band arising from the O–H valence vibration is one of the earliest known and most studied of any. It was first noted by Aschkinass [1] and by Ransohoff [2] in 1895 that a band near 3300 cm^{-1} appeared to be associated with the hydroxyl group, and these observations were much extended by the work of Coblentz [3] and of Weniger [4, 5]. Many other early workers also interested themselves in this group absorption, and its study was intensified in 1933, following the observations of Wulf, Hendricks, Hilbert and Liddel [6–9] that infra-red spectrographic studies afforded a simple and convenient method of following the phenomenon of hydrogen bonding, since when this occurs, the OH bond-length is increased and the absorption band shifts to a lower frequency. These authors worked mostly in the overtone regions, but essentially similar results were obtained by Fox and Martin [10, 11] working on the OH fundamentals, whilst Badger and Bauer [12] were able to show that the magnitude of the shift of the OH fundamental, due to hydrogen bonding, could be used as a measure of the strength of the hydrogen bond formed, being about 35 cm^{-1} per kilo calorie.

The spectroscopic study of the hydrogen bond has been very fully reviewed by a number of authors [12–16, 53–58, 94–96], and no attempt will be made here to do more than indicate the general effects of this phenomenon in relation to the interpretation of the infra-red spectra of alcohols. Similarly, the bibliography has been limited by the choice of a restricted number of examples in cases where much work has been done.

The hydroxyl group is very highly polar, and therefore associates with any other molecules having some degree of polar attraction, so that it is only in the vapour state and in dilute solution in non-polar solvents that the absorption of the free OH vibration is observed, although there are certain exceptions among compounds in which steric hindrance can prevent or reduce the amount of bonding. However, even in carbon disulphide in very dilute solution, alcohols show some change in the OH frequency due to solvent association [78], whilst changes of state, such as, for example, the passage from liquids to solid crystals, also give rise to large frequency shifts owing to the increased possibilities for orientation in the crystal state [17, 18]. Similarly, temperature changes will markedly influence the position of OH absorption bands [19, 20]. All these effects make it very difficult to indicate any specific frequency ranges for various types of bonding, but, as will be seen, it is nevertheless possible to differentiate between intermolecular, intramolecular and chelated bonds, and to decide, for example, whether or not intermolecular association is dimeric or polymeric in type. The approximate absorption ranges covering the various types of hydrogen bond are indicated in Table 6, but it should be appreciated that these are only indications, and that in assessing the type of bond involved account will have to be taken of the physical state of the

Table 6

OH *Stretching Frequencies*

Free OH	
Primary alcohols	$3643-3630$ cm^{-1} (m.)
Secondary alcohols	$3635-3620$ cm^{-1} (m.)
Tertiary alcohols	$3620-3600$ cm^{-1} (m.)
Phenols	$3612-3593$ cm^{-1} (m.)
Oximes	Near 3600 cm^{-1} (m.)
Peroxide OH	Near 3550 cm^{-1} (m.)
Bonded OH	
Intermolecular bonds	Single bridge $3550-3450$ cm^{-1} (m.)
	Polymeric $3400-3200$ cm^{-1} (vs.) broad
Intramolecular bonds	Single bridge $3570-3540$ cm^{-1} (m.)
	Chelated $3200-2500$ cm^{-1} broad (s.)

OH *Deformations and* C–O *Stretches*

Primary alcohols	Near 1050 cm^{-1}, (s.), $1350-1260$ cm^{-1} (s.)
Secondary alcohols	Near 1100 cm^{-1} (s.), $1350-1260$ cm^{-1} (s.)
Tertiary alcohols	Near 1150 cm^{-1} (s.), $1410-1310$ cm^{-1} (s.)
Phenols	Near 1200 cm^{-1} (s.), $1410-1310$ cm^{-1} (s.)

sample, and in some cases even of the temperature at which the observations were made.

Absorption bands in the low-frequency regions of the spectrum are also to be expected from the C—O stretching mode and from the OH deformation mode. As will be seen, many alcohols exhibit two absorption bands, each covering a wide frequency range, which may be associated with these vibrations. Both, however, are susceptible to changes following hydrogen bonding, and are of limited value in interpretative work.

The correlations discussed are listed opposite (Table 6).

6.2. The free OH stretching vibration

The absorption range for the OH valence-stretching vibration of an unbonded hydroxyl group is usually quoted as being $3700-3500$ cm^{-1} [22—25]. The upper limit at which the absorption can occur is indicated by water, in which the band occurs at 3760 cm^{-1} [26, 27]. However, this is rather a special case in which both hydrogens are attached to the same oxygen atom, and hydrogen peroxide is probably more typical in absorbing at 3590 cm^{-1} [28]. Due to the difficulty in differentiating free hydroxyl absorptions from those arising from very weak hydrogen bonds with solvents, etc., the lower limit is difficult to define, but single-bridge dimers in which the OH bonding is weak absorb near 3500 cm^{-1} [29], and this point is often taken as being the lower end of the free OH absorption range.

However, this overall range $3700-3500$ cm^{-1} is considerably greater than is found experimentally using unassociated alcohols or phenols in non-polar solvents. The frequency range for each class is very narrow, and the overall spread is no greater than $3650-3590$ cm^{-1}. This is the range quoted in Table 6. Fox and Martin [10], for example, find the free OH absorptions of a considerable series of aliphatic and aromatic alcohols between 3636 cm^{-1} and 3618 cm^{-1}. Kuhn [74] has also carried out accurate measurements of the free OH frequencies on a range of thirty-five alcohols and phenols. In all cases the absorption occurred between 3644 and 3605 cm^{-1}. Within a class of compounds the overall range is smaller still and the free OH band of a large number of 1 : 2-diols [97, 132] occurs at 3630 ± 5 cm^{-1}. Similarly, phenols are grouped together with $\alpha\beta$-unsaturated alcohols and absorb at the bottom of the overall range [74, 98]. The influence of the aromatic ring in lowering the free OH frequency slightly is confirmed by the work of Ingraham *et al.* [75], who find this band at 3616–3588 cm^{-1} in twenty-five substituted phenols and of 3615–3592 cm^{-1} in fourteen substituted catechols. These workers have related the OH stretching frequencies of substituted phenols to the Hammett σ values of the aromatic substituents, and Goulden [98] has shown that they can also be related to Pk$_a$ values. Hunsberger *et al.* [76] also have noted a systematic fall in the OH frequency throughout the series methanol (3645 cm^{-1}), phenol (3628 cm^{-1}), α- or β-naphthol (3618 cm^{-1}), and hydroxyphenanthrene (3605 cm^{-1}). The influence of an adjacent carbonyl group is considerably greater than that of an αβ-double bond, as monomeric acids absorb near 3520 cm^{-1}, but α-fluorine substitution on the other hand has only a very small effect upon the free OH frequency, although it reduces the tendency towards hydrogen bonding [79, 80]. In the Raman spectra the OH stretching frequency is very weak and difficult to detect. It is therefore surprising that it was from Raman rather than infra-red spectra that it was first appreciated by Bateuv and Matveeva [30] that it was possible to differentiate primary, secondary and tertiary alcohols by characteristic absorptions at 3632 cm^{-1}, 3622 cm^{-1} and 3615 cm^{-1}. This has since been fully confirmed in the infra-red, and the origins of the differences are beginning to be understood. Dalton *et al.* [44] first drew attention to the asymmetry of the OH stretching band of many primary and secondary alcohols, which was always broader than the corresponding band of methanol or of tertiary alcohols. They suggested that the band was broadened by the presence of two rotational isomers, and by graphical resolution they were able to show that there was a weaker band on the low frequency side of the main peak. Their results on the origins of this doublet have been confirmed by a number of workers [45, 63, 64, 99] and have been explained by Kreuger *et al.* [145]. It originates in the effect on the OH bond of an interaction between the lone pair electrons on the oxygen with the adjacent CH bond in the *trans* position. This raises ν_{OH}. With no attached hydrogen atoms at the adjacent carbon, as in tertiary alcohols, the frequency is normal and falls very close to 3617 cm^{-1}. At the other extreme, methanol which has hydrogen atoms which are *trans* to each of the oxygen lone pairs, shows a higher OH frequency than any other alcohol, and the band remains narrow. With ethanol and other primary alcohols, two configurations are possible and two bands appear although they need graphical resolution methods to identify them precisely. The higher frequency band at 3637 cm^{-1} corresponds to the form with two hydrogen atoms in the *trans* positions. This gives rise to the main peak. The second band at 3623 cm^{-1} is weaker and is derived from the form with a single *trans* hydrogen atom. With secondary alcohols such as *iso*propanol, the conformers which are possible have either one or no hydrogen atom *trans* to the lone pairs. The bands corresponding to these occur at 3627 and 3617 cm^{-1}. This last frequency is of course the same as in tertiary alcohols. The CH bonds behave in a parallel but inverse way. Cole and Jefferies [132] report unpublished data showing that primary alcohols absorb near 3642 cm^{-1} with extinction coefficients of 70, secondary alcohols near 3629 (ϵ 60–50) and tertiary alcohols near 3618 cm^{-1} (ϵ 45). This has been further studied by Anet and Bavin [133], who find very similar results for a series of simple alcohols. It is very probable that a similar mechanism is responsible for the subtle differences between the OH stretching frequencies of axial and equatorial hydroxyl groups in steroids and triterpenes. Allsop *et al.* [134] have shown that in the former the absorption is at $3629–3630 \text{ cm}^{-1}$, and in the latter at $3637–3639 \text{ cm}^{-1}$.

These values all relate to free OH absorptions in non-polar solvents, and small changes occur with alterations in the polarity of the solvent and with changes of temperature and pressure. The changes with solvent have been shown to arise mainly from association effects so that the magnitude of the shift will depend directly on the basic character of the solvent acceptor site, and on the acidity of the alcohol. Phenol in ether shows a very large shift whereas methanol in a solvent of low polarity will show only a small one. Changes with temperature [100, 162] and with pressure [101] are small, and arise from alterations in the distances separating the individual molecules.

The free OH stretching frequencies of N—OH groups in oximes are similar to alcohols and appear [136–138] between 3650 and 3500 cm^{-1}.

6.3. Hydrogen bonded OH vibrations (general)

Freymann and Heilmann [33], working in the overtone region, were able to differentiate three types of hydroxyl group bonding: intermolecular bonding between two or more molecules, intramolecular bonding, and chelation in which resonance structures are involved. Bonding systems of each of these types have been fully studied by many workers and, as a result, it is usually practicable to decide which of them is involved in any single case. The differentiation between intermolecular and intramolecular bonds is simply made by solution measurements in non-polar solvents. Intramolecular bonds are not broken by dilution as are the intermolecular bonds. With the former there is no change in the proportion of the free and bonded OH bands as dilution proceeds, whereas with the latter the monomer band grows progressively at the expense of the association band. The frequency of the bonded OH absorption is a direct measure of the strength of the hydrogen bond, and, for example, is very much the same in the structure C—OH . . . O—, as in C—OH . . . O=C—, unless resonance conditions are possible, in which case resonance stabilisation gives rise to a much stronger hydrogen bond, with a consequent big low-frequency shift. Diacetone alcohol, for example, absorbs at 3484 cm^{-1}, which is very much the same as the normal OH vibration of dimeric alcohols, whereas acetyl acetone absorbs at 2700 cm^{-1} [34]. Both these are instances of OH groups bonded to a carbonyl group, but the latter is stabilised by resonance, and the oxygen atoms are thereby brought closer together. In chelate compounds of this type it can be shown that the carbonyl frequency (which is an alternative measure of the strength of the hydrogen bond) is a linear function of the double-bond character of the $\alpha\beta$-double bond [76, 102], as this is the primary factor determining O . . . O distances.

Several workers [77, 84, 103, 104, 135] have studied the connection between OH bonded frequencies and the distance apart of the two atoms linked by the hydrogen bond. Lord and Merrifield [84] showed that an approximately linear relationship existed between the OH frequency shift and the O . . . O distance in hydrogen-bonded acids and similar compounds. This treatment has been further elaborated by Nakamoto *et al.* [103], by Pimentel [104, 135], and by Bellamy and Owen [147]. Different relationships exist for any one type of X—H . . . Y bond, but each separate series obeys a simple linear relation, provided the three atoms involved are co-linear. This has led to the suggestion that deviations from linearity might be used as a measure of the angle of distortion of bent hydrogen bonds.

Measurements of the strengths of hydrogen (or deuterium) bonds can also be used to determine the relative proton-accepting powers of bases. Thus the OD stretching frequency of deuteromethanol in substituted pyridines is a direct function of their basic strength [105].

6.4. Intermolecular hydrogen bonds

Intermolecular hydrogen bonding in alcohols or phenols results in a downward frequency shift and a very considerable broadening and intensification of the absorption band. Intramolecular bonds show a similar frequency shift, and the bands are broadened, but in many cases there is a much smaller intensity increase. It is impossible to quote a useful frequency range for the OH stretch of associated alcohols, as the shifts range from the very small, as when the alcohol is associated with an aromatic ring, to the very large when it is associated with a strong base. The origins of the very considerable band widths found in hydrogen-bonded systems are not yet fully understood, but in alcohols and phenols one contributory factor is the variety of different polymeric species which can occur.

Alcohols and phenols linked as single bridge dimers have only weak hydrogen bonds and absorb in the 3500 cm^{-1} region. This has been well established by studies on sterically hindered systems [29, 35, 36] and by measurements at such high dilutions that only monomer and dimer forms are present. At normal concentrations the dimer bands are not seen but are replaced by stronger broader polymer bands at lower frequencies. Thus, in a limited series of compounds studied by Kuhn [74] the absorption of dimeric alcohols was in the range 3525–3472 cm^{-1} and that from polymeric materials in the range 3341–3338 cm^{-1}. Marrinan and Mann [107, 108] quote 3347–3324 cm^{-1} for polymeric association in cellulose and 3404 cm^{-1} for single bridge dimeric bonds. It is interesting to note that hydrogen bonds reinforce each other when joined in chains or rings, so that the original hydrogen bond of a dimer becomes significantly stronger when a third alcohol unit is attached at either end. This is due to the fact that the polarity of the unbonded hydrogen at the one end, and of the unbonded oxygen at the other is increased by the association. Conversely, the association of two different OH groups to the same carbonyl oxygen atom (2 : 1 complexes) results in weaker individual hydrogen bonds than those from the association of only one hydroxyl group. This same effect is found in water where the association of both hydrogen atoms to a solvent in a 2 : 1 complex gives smaller frequency shifts than those of the corresponding 1 : 1 complex [148].

6.4(a). Single hydrogen-bridge complexes. In very dilute solutions in which dimerisation is precluded the strength of hydrogen bonds between alcohols and solvents can be studied. Work of this type has been done by many workers, such as Mecke [38], who has examined the OH absorption in phenol in eighteen different solvents. The results are in agreement with those of earlier workers in that, on changing from indifferent solvents, such as *cyclo*-hexane and CCl_4, to solvents with polar properties, a broadening of the band accompanied by an increase in intensity is observed. The relative influence of solvents in effecting this change is given by *cyclo*-hexane < carbon disulphide < chlorobenzene < benzene. It is interesting to note that at higher concentrations, where polymerisation of the alcohol is the predominating effect, this order changes, and is then conditioned also by such factors as dielectric constant and molecular size of the solvent. At such concentrations the association, as evidenced by the OH absorption frequency, increases in the order benzene < chloroform < carbon disulphide [35]. In single-bridge complexes of this type both the C—O—H . . . O— bridge and the C—O—H . . . O=C— bridge absorb at similar frequencies near 3510 cm^{-1}. Coggeshall and Saier [31] have found a band at this point in dilute solutions of ethyl alcohol and acetone in carbon tetrachloride and also of ethyl alcohol and dioxan in the same solvent. Kuhn [74] gives values of $3492-3482 \text{ cm}^{-1}$ for the OH band in the system ROH . . . OEt_2, where R is an alkyl group. Interactions with other solvents such as amines also occur in a similar way [39]. There has over the past few years been a very extensive study of the frequency shifts of OH groups associated with solvents, with π clouds, and with bases, and of the relationships between these and equilibrium constants and ΔH values. This topic is discussed at some length in reference 58 and will not be repeated here. It can be briefly summarised by the statement that the major part of the frequency shifts of alcohols and phenols in solvents arise from association rather than from dielectric constant effects. Whereas the Badger-Bauer relationship shows alarming discrepancies when applied universally, there do appear to be reliable individual relationships between $\Delta \nu$ and ΔH for similar systems. (*i.e.* OH . . . O, OH . . . N, OH . . . π *etc.*).

Apart from interactions between dissimilar molecules, single-bridge complexes are formed only in special cases in which steric effects prevent full polymeric association. A common case of such steric effects is that of *ortho*-substituted phenols, which have been studied by many workers [18, 31, 35, 36, 40, 41, 74, 139]. It has long been known on chemical grounds that *ortho*-substituted phenols show differences from other types, whilst 2 : 6-substituted phenols with large substituents give no phenolic reactions and are insoluble in water. This is explicable on the basis of steric hindrance, and it can be shown that hydrogen bonding does not occur in 2 : 6-di-substituted phenols with large substituent groups, whilst it is much reduced in *ortho*-substituted phenols with groups larger than the methyl group. Sears and Kitchen [40] have published data on an extensive list of phenols of various types, which they have examined in the solid, liquid and solution states, and they have been able to show, for example, that the OH absorption in 2 : 6-di-*tert*.-butyl-phenol occurs in the normal position for free OH absorption, and that there is no significant shift on going from solutions to the liquid or solid states. With partly hindered phenols the wave-length shift from the free OH position is very much less than in the case of unhindered phenols, and Coggeshall [41] has shown that this is due to the fact that in such cases bonding proceeds by dimeric association rather than by polymeric association. However, Hunter [95] has pointed out that an alternative explanation of the limited degree of association would be that the inability of the OH group to achieve full co-planarity with the ring reduces its acidity and proton donating powers.

Another case of single-bridge dimerisation is that of branched-chain aliphatic alcohols in which the steric effects of the branched chains prevent polymeric association. Smith and Creitz [29] have observed this effect in a number of such alcohols, and they assign a frequency of 3500 cm^{-1} to the hydrogen bridge so formed. They note also that in these cases the dimer contributes towards the free OH absorption as well as the monomer. A similar case is that of *tert.*-amyl alcohol, which, in the liquid state, has long been known to exhibit some free OH absorption due to steric effects, whilst the bonded OH absorption is also at a higher frequency than in the straight-chain alcohols. Other cases occur in carbohydrates, aluminium soaps and similar materials [46, 47, 81].

6.4(b). Polymeric association. This represents the commonest case of hydrogen bonding in alcohols studied in the solid or liquid state or at any significant concentration when long chains of molecules are linked by relatively strong hydrogen bonds. The usual absorption region in the absence of chelation effects is between 3400 cm^{-1} and

$) \, cm^{-1}$, the actual position depending on the nature of any solvent, and on whether or not the material is examined in the solid or liquid state. The OH absorption bands arising from rapidly cooled melted films are generally much closer to the normal position for strong solutions than to the position in the crystalline state. This is in accordance with the view that the crystalline state represents that in which the molecules have oriented themselves in such a way as to give the maximum strength of hydrogen bonds, so that the frequencies are invariably lower in the crystal than in solution. A sharpening of the band accompanied by a shift to yet lower frequencies usually results from exposure of the sample to very low temperatures.

The frequency range for this type of OH absorption is too much a matter of common experience to need detailed references, but a typical series of such absorptions is given by the spectra of the various simple alcohols included in the A.P.I. series of spectra. In the series methanol to decanol the bonded OH absorption in concentrated solution in non-polar solvents occurs in the range $3370-3344 \, cm^{-1}$. Kuhn [74] also gives a list of a considerable number of alcohols and diols, the OH frequencies of which, in concentrated solutions, all fall within the ranges given in Table 6.

Studies on the relative intensities of the free and bonded OH absorptions at different concentrations provide a convenient method for measuring the dissociation constants either of polymeric alcohols [140] or of alcohols associated with carbonyl groups [141].

6.5. Intramolecular hydrogen bonds

The majority of cases of intramolecular hydrogen bonding are those in which the bond is resonance stabilised, and this gives rise to a particularly strong bond, with a consequent large OH frequency shift. These cases will be dealt with under chelate compounds in the following section. Intramolecular bonding without resonance stabilisation results in the formation of single-bridge hydrogen bonds similar to those of the dimeric associations discussed above. The hydrogen bond is accordingly weak, and a relatively weak absorption occurs in the range $3570-3450 \, cm^{-1}$. The essential difference between the two types lies in the fact that intermolecular hydrogen bonds are broken by dilution with non-polar solvents, whereas intramolecular bonds are independent of the concentration.

6.5(a). Single-bridge complexes. Typical cases of single-bridge complexes are the internal bonds in ring compounds, as in 2-hydroxy-*cyclo*hexanones [109], phenols [7, 42], nitro-alcohols [110], 1 : 2-chlorhydrins [142] and cyclic 1 : 2-diols [97, 132]. In the last of these Kuhn [97] has shown that hydrogen-bonding studies can be used in the determination of the probable configurations of this series. The most stable bridged structures of this type are, as is to be expected, those involving six-membered rings [111]. With *ortho*-substituted phenols in which hydrogen bonding of the OH on to the other substituent group is possible an interesting case arises. Pauling [43] has pointed out that the C—O bond of phenols has some double-bond character which will cause the hydrogen atom to lie in the plane of the ring. Phenol can therefore have the two equivalent configurations I and II.

With *ortho*-substitution of a halogen or other proton-accepting group, the *cis*- and *trans*-configurations III and IV become possible, which are not equivalent, whilst in 2 : 6-dichlorophenol equivalence is again obtained with the structure V and VI. Phenol and 2 : 6-dichlorophenol are therefore to be expected to show one OH absorption, whereas *ortho*-chlorophenol should show two, as the *cis*-form will be able to form a hydrogen bond, whilst the *trans* will not. This is, in fact, the case [112, 113]. Similar results are also obtained with *ortho*-methoxy-, phenyl- and cyano-compounds [20], whilst certain substituted vinyl alcohols show a similar pair of bands corresponding to two configurations [48]. In catechols also [75], a parallel effect is found, and in this case the single-bridge intra-molecular bond absorbs between 3575 cm^{-1} and 3535 cm^{-1}.

The strengths of hydrogen bonds formed internally in this way may not always be accurately reflected in the OH frequency shifts. The end atoms of the hydrogen bond are not able to take up that particular distance which is conducive to the formation of the strongest bond, but are forced into a specific separation dictated by the internal geometry. It is therefore unwise to conclude that $\Delta\nu$ is related to either ΔH or to the internuclear distance. Thus, the frequency shifts of the *ortho* halogenated phenols would appear to suggest that the strongest hydrogen bonds are to iodine and the weakest to fluorine. This is contrary to all other expectation, and ΔH measurements and equilibrium studies of competitive bonding in 2 : 6-dihalogenated phenols have shown that the bonds to fluorine are in fact the stronger of the two [150].

In all situations of this kind it is usual to see some indication of the monomeric frequency of the *trans* form, but the intensity with which this is shown will depend on the relative strength of the hydrogen bond, and in many cases such bands are difficult to detect. In situations in which there are a number of alternative bonding sites available to the OH group multiple bands appear. A particularly good example is the compound shown below which has been studied by Campbell *et al.* [151]. This shows four different OH frequencies in

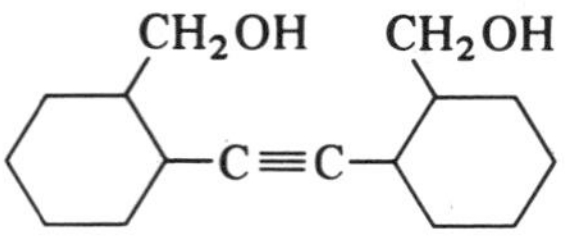

$$CH_2OH \qquad CH_2OH$$

dilute solution. These correspond to, free OH (25%), OH to π clouds (40%), OH to π C≡C clouds (15%) and OH to OH (20%). The π cloud associations are predominant over the OH . . . OH interactions because the latter involve fixed distances which only allow the formation of weak bonds. Multiple bands also appear in α-hydroxy acids in dilute solution. These show bands due to alcohol to acid, and acid to alcohol, as well as the two distinct monomeric OH bands [149].

An effect similar to the *ortho*-effect in intermolecular bonding of phenols can also be observed in the intramolecular bonding of substituted bisphenol alkanes [41, 82].Intramolecular bonding between the two phenolic hydroxyls occurs in one configuration, but a second '*trans*' configuration is also stabilised because of the restriction of free rotation by the presence of a large *ortho* substituent group. The *cis* form therefore exhibits hydrogen bonding, whereas the *trans* form absorbs at the free OH position unless the conditions are favourable for intermolecular bonding.

6.5(b). Chelate compounds. In the earlier studies of Hilbert, Wulf, Hendricks and Liddel [6—9] and other workers, it was found that *ortho*-phenols in which carbonyl or nitro-groups were substituted at the *ortho* position exhibited very much stronger hydrogen bonds than the corresponding *meta* or *para* materials or differently substituted phenols. The same effect was observed for acetyl acetone, and this compound, together with other enolic β-diketones, was re-examined in the fundamental region by Rasmussen *et al.* [34]. They found that the OH absorption in these compounds occurred near 2700 cm^{-1} and that the carbonyl frequency also shifted to a much greater extent than was usual in hydrogen bonding phenomena. These authors pointed out the similarity between the OH frequencies of these compounds and those of fatty acid dimers, which also exhibit unusual characteristics in this respect, Pauling [13] has accounted for the strong bonding in the fatty acid dimers on the basis of a large contribution of an ionic resonance structure along with the normal covalent structure; the OH bond is therefore subject to additional loosening owing to the charges, with the consequent formation of a stronger hydrogen bond.

Rasmussen *et al.* [34] suggest that a somewhat similar explanation can be applied to the enolic β-diketones which can have contributions from the resonance structures VIII and IX.

$$ \underset{\text{VIII.}}{\overset{\displaystyle O-H\cdots\cdots O}{R'C=CR''-CR'''}} \quad \text{and} \quad \underset{\text{IX.}}{\overset{\displaystyle {}^{+}O-H\cdots\cdots O}{R'-C-CR''=CR'''}} $$

The term 'conjugate chelation' has been suggested by Rasmussen *et al.* to describe this phenomenon. Similar findings in respect of the OH frequency have been reported by Martin [49], who showed that salicaldehyde and *ortho*-hydroxyacetophenone exhibit broad weak OH absorption bands extending from 3500 cm^{-1} to beyond 2900 cm^{-1}, comparable with, but much weaker than those of salicylic acid. *ortho*-Hydroxy benzyl alcohol, however, in which the possibility of an ionic structure is not present, showed a sharp band at 3600 cm^{-1} (free OH) and another at 3436 cm^{-1} (single-bridge intramolecular bond). Hunsberger [50] has demonstrated similar effects in hydroxynaphthaldehydes and similar materials, and many other examples of this effect have been described, as in acetyl acetones [114, 115, 143], hydroxyanthraquinones [51, 83, 93, 116, 144], hydroxyflavones [117], hydroxyindanones [118] and related compounds [119, 120]. An OH frequency of 2600 cm^{-1} has also been found in compounds involving a chelated hydroxyl group and a heterocyclic nitrogen atom [121]. In many cases of this type the absence of any OH absorption in the fundamental region has been reported, presumably owing to its being a weak band which has been superimposed upon the strong CH stretching absorption near 3000 cm^{-1}. One case in which this conjugate chelation can arise through intermolecular, rather than intramolecular, association is 5 : 5-dimethyl-1 : 3-*cyclo*hexanedione. This compound contains the β-keto-enol structure, and shows an OH stretching absorption at 2632 cm^{-1}, but steric considerations will not allow of a direct intramolecular hydrogen bond, and Rasmussen *et al.* [34] suggest that it must therefore arise through dimerisation into a structure capable of a similar resonance stabilisation.

The lower frequency limits at which OH absorptions of this type occur are extremely difficult to define, as the absorption bands are relatively weak and extremely broad. Reid and Ruby [52] find a shallow OH absorption in 6-phenyl-4-hydroxy-5 : 6-dihydropyrone as low as 2500 cm^{-1}. Most of the near symmetric or symmetric hydrogen bonds which give even lower frequencies are intermolecular in origin. However one of the strongest hydrogen bonds known occurs in nickel dimethylglyoxime where the OH frequency has been identified by deuteration studies [77] as an extremely weak band near 1775 cm^{-1}.

6.6. The intensity and shape of OH stretching absorptions

The shape of the OH absorption band and its relative intensity can sometimes be used to give an indication of the type of structure present. The broad nature of these bands in the spectra of liquids has usually been ascribed to the molecules being involved in hydrogen bonding to different extents, so that the shape of the band is an indication of the distribution of various strengths of hydrogen bonding [10, 57]. However, as Kletz has pointed out [59], this explanation cannot be applied to crystals, in which all hydrogen bonds would be expected to be equivalent. In the spectra of crystals of phenols the OH absorption band is sharper than in the spectrum of the liquid, but still much broader than in dilute solution. An alternative suggestion is that the width of these bands is due to combination with a number of low-frequency vibrations, including the deformation frequencies [12, 60, 61], and perhaps more importantly the low lying O . . . O stretching frequencies [152, 153]. In agreement with this is the observation that the width of the band is generally greater with increased hydrogen bond strength, whilst with deuterium substitution the breadth of the band is much reduced. Theories of the origins of intensity increases on bondir have also been discussed by Tsubomura [122, 123], and intensity changes with temperature have been considered by Hughes *et al.* [100] and by Finch and Lippincott [106].

The intensities of free OH stretching absorptions obey Beer's Law, and the extinction coefficients are similar for any single class of related materials, such as, for example, phenols [51]. The OH intensities of normal alcohols are also reasonably constant, so that substitution at the 2 or 3 carbon atom has no effect unless an intramolecular bond becomes possible. The intensities are all considerably lower than those of the phenols [124, 145]. The intensities of the phenols can be related to the Hammett σ values and are therefore connected with the OH polarity [154, 155].

With single-bridge bonding the shape of the OH absorption band remains essentially sharp [38], and observations on limited groups of materials indicate that within any one class, at least, the extinction coefficients are reasonably constant and that Beer's Law is obeyed [29, 35]. In the case of single-bridge dimers the dimeric forms contribute to both the free and bonded OH absorption.

In solvents in which association effects occur the intensity is greatly increased [124], so that in ether the intensity of the methanol hydroxyl group is approximately eight times as large as in carbon tetrachloride, and in triethylamine this ratio increases to twelve times.

With stronger intermolecular bonding, broad and very intense bands are observed. In general the bands become broader and more intense as the strength of the bonding increases. This may be related to the increasing polarity of the OH bond which might therefore be expected to show a greater rate of change of dipole during the vibration. However the bands are so broad, and so often overlaid with minor peaks due to Fermi-resonance interactions that any measurement of their absolute intensities is a fruitless exercise.

The examination of the appearance or otherwise of OH absorption bands from single crystals examined in polarised radiation has also been employed in the determination of structures — as, for example, in the study of the bonded OH groups in polyvinyl alcohol [65] and in carbohydrates [46]. As in the case of free OH groups, deuteration is often of considerable assistance in identifying OH groups and in studying bonding effects, and it has been employed for this purpose by Rowen and Plyler [66], Sheppard [116], Sutherland [125] and many others.

6.7. C—O Stretching and O—H deformation vibrations

Both the above vibrations might reasonably be expected to give rise to strong absorption bands by which alcohols could be characterised. As will be seen, most alcohols do give rise to two bands in the lower frequency region. Unfortunately, however, due to coupling effects there has been a good deal of confusion as to which of them is which.

The first of these bands, and usually the strongest, was first noted by Weniger [4] in 1910, who found a strong band in the spectra of liquid primary alcohols near 1050 cm^{-1}, in secondary alcohols near 1100 cm^{-1} and in tertiary alcohols near 1150 cm^{-1}. This was confirmed by Lecomte [67, 85], who was able to show that they were associated with the hydroxyl group, as, over a long series of aliphatic alcohols, the intensity was observed to diminish systematically as the chain length was increased and the proportion of OH groups reduced. He also noted the sensitivity of these absorptions to structural changes, and observed a marked reduction in the intensity of absorption at 1100 cm^{-1} in some vinyl carbinols as compared with saturated secondary alcohols. The spectrum of phenol was considered too complex for any useful assignments to be made.

More recently a systematic study of the position of this absorption in relation to the surrounding structure has been made by Zeiss and Tsutsui [86]. In saturated straight-chain alcohols they find that this band occurs within the relatively narrow frequency ranges of $1075-1010 \text{ cm}^{-1}$ for primary alcohols, $1119-1105 \text{ cm}^{-1}$ for secondary alcohols and near 1140 cm^{-1} for tertiary alcohols. These findings have been substantially confirmed by Stuart and Sutherland [125], working on liquid alcohols. However, α-branched, αβ-unsaturated and alicyclic alcohols all show frequency shift away from these ranges, and the extent and directions of these show a number of curious anomalies. Chain branching at the α-carbon atom reduces the frequency in all three cases by $10-15 \text{ cm}^{-1}$, but the effect is reversed when the chain is doubly branched at this point so that di-*tert.*-butyl alcohol absorbs at 1163 cm^{-1} within the normal range. αβ-Unsaturation lowers the frequency considerably, and in secondary alcohols the band falls between 1074 and 1012 cm^{-1}, but the shift is found to be greater for aryl than for alkenyl unsaturation despite the weaker nature of the double bonds involved. Although fewer examples are available a similar shift to low frequencies appears to occur in the corresponding primary and tertiary alcohols, and here again allyl alcohol has been found to absorb at a higher frequency (1030 cm^{-1}) than benzyl alcohol (1010 cm^{-1}). In alicyclic secondary alcohols the frequency gradually falls with increasing ring size from *cyclo*butanol (1090 cm^{-1}) to *cyclo*heptanol (1025 cm^{-1}), so that those showing the greatest frequency shift are those with the lowest ring strain. In sterols the band falls between 1075 and 1000 cm^{-1} in solution, but as a result of a good deal of detailed work, its position within this range has been correlated with the sterochemical arrangement of the C_3 hydroxyl group [87–90], so that valuable structural information can now be obtained in this way. Similarly in the triterpenes, differentiation between axial and equatorial 3-hydroxy groups is possible [134]. A somewhat similar effect has been observed in the decanols [91], although the differences are less clearly defined.

The considerable degree of overlapping of the frequency ranges for different types of alcohols with various functional groups makes it very difficult to identify the type of structure present from observations on this region alone, but when additional evidence is available it can often afford some extremely valuable information.

The second band of alcohols usually arises in the region of 1300 cm^{-1} [69], but its position varies considerably with structural effects. Both these absorptions are sensitive to changes of state, indicating that they are associated with hydrogen-bonding effects, and there is a good deal of disagreement in the literature as to which of them is to be assigned to the OH deformation vibration. In the case of methyl alcohol, some workers assign the OH deformation to a band at 1109 cm^{-1} [70], whilst others prefer the band at 1340 cm^{-1} [71, 26]. In addition, methanol has a strong 1034 cm^{-1} band which is associated with the C–O vibration [72, 73]. A detailed deuteration study by Falk and Whalley [156] on methanol has largely resolved this problem. It now seems clear that the band which occurs at 1346 cm^{-1} in the vapour and rises to 1420 cm^{-1} in the liquid is primarily associated with the in plane OH bending mode, whereas the band at 1034 cm^{-1} is largely due to the C–O stretch. However the OH bending mode is very sensitive to coupling effects. Perchard and Josien [157] have shown that it is displaced to 1244 cm^{-1} in ethanol due to coupling with a torsional mode of the CH_2 group. When the coupling is removed in CH_3CD_2OH the band moves upwards to 1293 cm^{-1} the C–O stretch in this compound appears at 1050 cm^{-1}. There is some doubt as to the assignments in *n*-propanol, as although this compound has a strong band at

1065 cm^{-1}, a normal coordinate study [158] has suggested that the C—O stretch lies at 911 cm^{-1} and in the plane OH bend at 1218 cm^{-1}.

In phenol, despite studies in a series of different states, and on deuteration [32, 68], there was until recently doubt as to whether the OH deformation mode occurs near 1200 cm^{-1} or near 1350 cm^{-1}, as both these absorptions are sensitive to changes in the hydrogen bonding pattern. In this last case the two bands are closer together, and some coupling is to be expected in analogy with carboxylic acids [37]. Kuratani [92] in deuteration studies on pentachlorphenol assigns two bands (at 1280 cm^{-1} and 1195 cm^{-1}) to the OH deformation on account of coupling effects. The identification of these bands in substituted phenol has also been discussed by Mecke and Rossmy [126] and by Davies and Jones [127]. The most recent work supports the assignment of the OH in plane bend to the 1200 cm^{-1} absorption. Evans [159] assigns this band at 1228 cm^{-1} in polymeric phenol, at 1198 cm^{-1} for the terminal molecule of the chain and at 1179 cm^{-1} for the unassociated compound. Green also assigns this band at 1175 cm^{-1} in unassociated phenols. Jakobsen and Brewer [161] put it at 1205 to 1239 cm^{-1} in liquid *para*-substituted phenols. In the sterol series most workers regard the $1075-1000$ cm^{-1} absorption as deriving from the C—O stretching mode [87—89], but others relate it to the OH deformation [90]. The position is even further complicated by the appearance in certain alcohols of additional bands which are sensitive to deuteration, which may arise from rotational isomerism [146].

A good deal of work has been carried out on deuterated alcohols in an attempt to resolve this difficulty, notably by Sutherland and his co-workers [21, 125, 128, 129] by Quinan and Wiberley [130, 131] and by Falk *et al.* [156] and Perchard *et al.* [157] it does appear to be more probable that the band near 1100 cm^{-1} in aliphatic alcohols is associated primarily with the C—O stretching mode, whereas the single or double peak in the $1400-1300$ cm^{-1} region is connected with the in-plane OH deformation frequency. It has been pointed out that the latter would be expected to couple with adjacent CH_2 deformations leading to mixed vibrations which would not follow the expected pattern of change on deuteration. Bratoz *et al.* [114] have made a somewhat similar set of assignments for acetyl acetone derivatives, in which the OH deformation is placed at 1435 cm^{-1}, and the C—O stretching mode, the frequency of which is raised by resonance, at 1284 cm^{-1}. In addition, all alcohols show very broad bands in the liquid state in the range 750–650 cm^{-1}. These arise from out of plane bonded OH deformation frequencies, but they have no useful applications in correlation work.

6.8. Bibliography

1. Aschkinass, *Wied. Ann.*, 1895, **55**, 401.
2. Ransohoff, Dissert Berlin, 1896.
3. Coblentz, *Investigations of Infra-red Spectra*, Part I (Carnegie Institute 1905), p. 35.
4. Weniger, *Phys. Review*, 1910, **31**, 388.
5. *Idem, J. Opt. Soc. Amer.*, 1923, **7**, 517.
6. Liddel and Wulf, *J. Amer. Chem. Soc.*, 1933, **55**, 3574.
7. Wulf and Liddel, *ibid.*, 1935, **57**, 1464.
8. Hilbert, Wulf, Hendricks and Liddel, *Nature*, 1935, **135**, 147.
9. *Idem, J. Amer. Chem. Soc.*, 1936, **58**, 548 and 1991.
10. Fox and Martin, *Proc. Roy. Soc.*, 1937, **A162**, 419; 1941, **A175**, 208.
11. *Idem, Trans. Faraday Soc.*, 1940, **36**, 897.
12. Badger and Bauer, *J. Chem. Phys.*, 1937, **5**, 839.
13. Pauling, *The Nature of the Chemical Bond* (Oxford University Press, 1950), pp. 316 *et seq.*
14. Wright, *Ind. Eng. Chem. (Anal. Ed.)*, 1941, **13**, 1.
15. Hunter, Price and Martin, *Report of a Symposium on the Hydrogen Bond* (Institute of Chemistry Monograph, 1950).
16. Huggins, *J. Org. Chem.*, 1936, **1**, 407.
17. Kletz and Price, *J. Chem. Soc.*, 1947, 644.
18. Richards and Thompson, *ibid.*, 1947, 1260.
19. Hoffmann, *Z. physikal. Chem.*, 1943, **B53**, 179.
20. Lüttke and Mecke, *Z. Elektrochem.*, 1949, **53**, 241.
21. Cf. Ramsay and Sutherland, *Discuss. Faraday Soc.*, 1950, **9**, 274.
22. Thompson, *J. Chem. Soc.*, 1948, 328.
23. Colthup, *J. Opt. Soc. Amer.*, 1950, **40**, 397.
24. Barnes, Gore, Liddel and Van Zandt Williams, *Infra-red Spectroscopy* (Reinhold, 1944).
25. Randall, Fowler, Fuson and Dangl, *Infra-red Determination of Organic Structure* (Van Nostrand, 1949).
26. Herzberg, *The Infra-red and Raman Spectra of Polyatomic Molecules* (Van Nostrand, 1945), p. 280.
27. Cross, Burnham and Leighton, *J. Amer. Chem. Soc.*, 1937, **59**, 1134.
28. Giguere, *J. Chem. Phys.*, 1950, **18**, 88.
29. Smith and Creitz, *J. Res. Nat. Bur. Stand.*, 1951, **46**, 145.
30. Bateuv and Matveeva, *Izvest. Akad. Nauk. S.S.S.R. Otdel Khim Nauk.*, 1951, 448.
31. Coggeshall and Saier, *J. Amer. Chem. Soc.*, 1951, **73**, 5414.
32. Williams, Hofstadter and Herman., *J. Chem. Phys.*, **1939**, 7, 802.
33. Freymann and Heilmann, *Compt rend. Acad. Sci. Paris*, 1944, **219**, 415.
34. Rasmussen, Tunnicliff and Brattain, *J. Amer. Chem. Soc.*, 1949, **71**, 1068.
35. Friedel, *ibid.*, 1951, **73**, 2881.
36. Coggeshall, *ibid.*, 1947, **69**, 1620.
37. Hadži and Sheppard, *Proc. Roy. Soc.*, 1953, **A216**, 247.

38. Mecke, *Discuss. Faraday Soc.*, 1950, **9**, 161.
39. Baker, Davies and Gaunt, *J. Chem. Soc.*, 1949, 24.
40. Sears and Kitchen, *J. Amer. Chem. Soc.*, 1949, **71**, 4110.
41. Coggeshall, *ibid.*, 1950, **72**, 2836.
42. Lüttke and Mecke, *Z. physikal. Chem.*, 1950, **56**, 196.
43. Pauling, *J. Amer. Chem. Soc.*, 1936, **58**, 94.
44. Dalton, Meakins, Robinson and Zaharia, *J. Chem. Soc.*, 1962, 1566.
45. Eddy, Showell and Basilo, *J. Amer. Oil Chemists Soc.*, 1963, **40**, 92.
46. Thompson, Nicholson and Short, *Discuss. Faraday Soc.*, 1950, **9**, 222.
47. Brown, Holliday and Trotter, *J. Chem. Soc.*, 1951, 1532.
48. Buswell, Rodebush and Whitney, *J. Amer. Chem. Soc.*, 1947, **89**, 770.
49. Martin, *Nature*, 1950, **166**, 474.
50. Hunsberger, *J. Amer. Chem. Soc.*, 1950, **72**, 5626.
51. Flett, *J. Chem. Soc.*, 1948, 1441.
52. Reid and Ruby, *J. Amer. Chem. Soc.*, 1951, **73**, 1054.
53. Pimentel and McClellan, *The Hydrogen Bond* (Freeman, San Francisco, 1960).
54. Coulson, *Research*, 1957, **10**, 149.
55. Hadzi and Thompson, *Hydrogen Bonding* (Pergamon Press, 1959).
56. Murthy and Rao, *Applied Spectroscopy Reviews*, 1969, 2, 69.
57. Sutherland, *Trans. Faraday Soc.*, 1940, **36**, 889.
58. Bellamy, *Advances in Infrared Group Frequencies* (Methuen, London, 1968).
59. Kletz, *Discuss. Faraday Soc.*, 1950, **9**, 211.
60. Davies and Sutherland, *J. Chem. Phys.*, 1938, **6**, 755, 762.
61. Davies, *J. Chem. Phys.*, 1940, 8, 577; *Discuss. Faraday Soc.*, 1950, **9**, 212.
62. *Ibid.*, p. 213.
63. Saier, Cousins and Basilo, *J. Chem. Phys.*, 1964, **41**, 40.
64. Kreuger and Mettee, *Can. J. Chem.*, 1964, **42**, 347.
65. Glatt, Webber, Seaman and Ellis, *J. Chem. Phys.*, 1950, **18**, 413.
66. Rowen and Plyler, *J. Res. Nat. Bur. Stand.*, 1950, **44**, 313.
67. Lecomte, *Traité de Chimie Organique* (Masson et Cie, Paris, 1936), 2, 143.
68. Brattain, *J. Chem. Phys.*, 1938, **6**, 298.
69. Lecomte, *Compt. rend. Acad. Sci. Paris*, 1925, **180**, 825.
70. Davies, *J. Chem. Phys.*, 1948, **16**, 267.
71. Noether, *ibid.*, 1942, **10**, 693.
72. Borden and Barker, *ibid.*, 1938, **6**, 553.
73. Barker and Bosschieter, *ibid.*, p. 563.
74. Kuhn, *J. Amer. Chem. Soc.*, 1952, **74**, 2492.
75. Ingraham, Corse, Bailey and Stitt, *ibid.*, p. 2297.
76. Hunsberger, Ketcham and Gutowsky, *ibid.*, p. 4839.
77. Rundle and Parasol, *J. Chem. Phys.*, 1952, **20**, 1487.
78. Stuart, *ibid.*, 1953, **21**, 1115.
79. Henne and Francis, *J. Amer. Chem. Soc.*, 1953, **75**, 991.
80. Haszeldine, *J. Chem. Soc.*, 1953, 1757.
81. Harple, Wiberley and Bauer, *Analyt. Chem.*, 1952, **24**, 635.
82. Amberlang and Binder, *J. Amer. Chem. Soc.*, 1953, **75**, 947.
83. Josien, Fuson, Lebas and Gregory, *J. Chem. Phys.*, 1953, **10**, 331.
84. Lord and Merrifield, *ibid.*, 166.
85. Tuot and Lecomte, *Bull. Soc. Chim. France*, 1943, **10**, 542.
86. Zeiss and Tsutsui, *J. Amer. Chem. Soc.*, 1953, **75**, 897.
87. Rosenkrantz, Milhorat and Farber, *J. Biol. Chem.*, 1952, **195**, 509.
88. Cole, Jones and Dobriner, *J. Amer. Chem. Soc.*, 1952, **74**, 5571.

89. Rosenkrantz and Zablow, *ibid.*, 1953, **75**, 903.
90. Furst, Kuhn, Scotoni and Gunthard, *Helv. Chim. Acta*, 1952, **35**, 951.
91. Dauben, Hoerger and Freeman, *J. Amer. Chem. Soc.*, 1952, **74**, 5206.
92. Kuratani, *J. Chem. Soc. Japan*, 1952, **73**, 758.
93. Hergert and Kurth, *J. Amer. Chem. Soc.*, 1953, **75**, 1622.
94. Wheland, *Resonance in Organic Chemistry* (Wiley, New York, 1955), pp. 47 *et seq.*
95. Hunter, *Progress in Stereochemistry*, Vol. 1 (Butterworth, London, 1954), pp. 224, *et seq.*
96. Ferguson, *Electronic Theories of Organic Chemistry* (Prentice Hall, New York, 1952).
97. Kuhn, *J. Amer. Chem. Soc.*, 1954, **76**, 4323.
98. Goulden, *Spectrochim. Acta*, 1954, **6**, 129.
99. Khairetdinna and Perelygin, *Dokl. Akad. Nauk. SSSR.*, 1967, **174**, 1111.
100. Hughes, Martin and Coggeshall, *J. Chem. Phys.*, 1956, **24**, 489.
101. Fishman and Drickamer, *ibid.*, 548.
102. Bellamy and Beecher, *J. Chem. Soc.*, 1954, 4487.
103. Nakamoto, Margoshes and Rundle, *J. Amer. Chem. Soc.*, 1955, **77**, 6480.
104. Pimentel and Sederholm, *J. Chem. Phys.*, 1956, **24**, 639.
105. Tamres, Searles, Leighly and Mohrman, *J. Amer. Chem. Soc.*, 1954, **76**, 3983.
106. Finch and Lippincott, *J. Chem. Phys.*, 1956, **24**, 908.
107. Marrinan and Mann, *J. App. Chem.*, 1954, **4**, 204.
108. *Idem, Trans. Faraday Soc.*, 1956, **52**, 481, 487, 492.
109. Huitric and Kumler, *J. Amer. Chem. Soc.*, 1956, **78**, 1147.
110. Urbanski, *Bull. Acad. Polon. Sci. Cl. III*, 1956, **4**, 87.
111. Hoyer, *Chem. Ber.*, 1956, **89**, 146.
112. Rossmy, Lüttke and Mecke, *J. Chem. Phys.*, 1953, **21**, 1606.
113. Keussler and Rossmy, *Z. Electrochem.*, 1956, **60**, 136.
114. Bratoz, Hadži and Rossmy, *Trans. Faraday Soc.*, 1956, **52**, 464.
115. Eistert and Reiss, *Chem. Ber.*, 1955, **88**, 92.
116. Hadži and Sheppard, *Trans. Faraday Soc.*, 1954, **50**, 911.
117. Shaw and Simpson, *J. Chem. Soc.*, 1955, 655.
118. Farmer, Hayes and Thomson, *ibid.*, 1956, 3600.
119. Musso, *Chem. Ber.*, 1955, **88**, 1915.
120. Duncanson, *J. Chem. Soc.*, 1953, 1207.
121. Branch, *Nature*, 1956, **177**, 671.
122. Tsubomura, *J. Chem. Phys.*, 1956, **24**, 927.
123. *Idem, ibid.*, 1955, **23**, 2130.
124. Barrow, *J. Phys. Chem.*, 1955, **59**, 1129.
125. Stuart and Sutherland, *J. Chem. Phys.*, 1956, **24**, 559.
126. Mecke and Rossmy, *Z. Electrochem.*, 1955, **59**, 866.
127. Davies and Jones, *J. Chem. Soc.*, 1954, 120.
128. Stuart and Sutherland, *J. Phys. Radium*, 1954, **15**, 321.
129. Krimm, Liang and Sutherland, *J. Chem. Phys.*, 1956, **24**, 778.
130. Quinan and Wiberley, *ibid.*, 1953, **21**, 1896.
131. *Idem, Analyt. Chem.*, 1954, **26**, 1762.
132. Cole and Jefferies, *J. Chem. Soc.*, 1956, 4391.
133. Anet and Bavin, *Can. J. Chem.*, 1956, **34**, 1756.
134. Allsop, Cole, White and Willix, *J. Chem. Soc.*, 1956, 4868.
135. Huggins and Pimentel, *J. Phys. Chem.*, 1956, **60**, 1615.
136. Palm and Werbin, *Can. J. Chem.*, 1953, **31**, 1004.
137. Califano and Lüttke, *Z. Phys. Chem.*, 1955, **5**, 240.

138. *Idem, ibid.*, 1956, **6**, 83.
139. Bowman, Stevens and Baldwin, *J. Amer. Chem. Soc.*, 1957, **79**, 87.
140. Ens and Murray, *Can. J. Chem.*, 1957, **35**, 170.
141. Widom, Philippe and Hobbs, *J. Amer. Chem. Soc.*, 1957, **79**, 1383.
142. Nickson, *ibid.*, 243.
143. Mecke and Funck, *Z. Electrochem.*, 1956, **60**, 1124.
144. Shigorin and Dokunichiv, *Doklady Akad. Nauk. S.S.S.R.*, 1955, **100**, 323.
145. Kreuger, Jan and Wieser, *J. Mol. Structure*, 1970, **5**, 375.
146. Maclou and Henry, *Compt. Rend. Acad. Sci. Paris*, 1957, **244**, 1494.
147. Bellamy and Owen, *Spectrochim. Acta*, 1969, **25A**, 329.
148. Bellamy, Blandamer, Symonds and Waddington, *Trans. Faraday Soc.*, 1971, **67**, 3435.
149. George, Green and Pailthorpe, *J. Mol. Structure*, 1971, **10**, 297.
150. Baker and Shulgin, *Can. J. Chem.*, 1965, **43**, 650.
151. Campbell, Eglington and Raphael, *J. Chem. Soc.* (B), 1968, 338.
152. Sheppard *in reference 55.*
153. Rosenberg and Iogansen, *Optics and Spectroscopy*, 1971, **31**, 380.
154. Stone and Thompson, *Spectrochim. Acta*, 1957, **10**, 17.
155. Brown, *J. Phys. Chem.*, 1957, **61**, 820.
156. Falk and Whalley, *J. Chem. Phys.*, 1961, **34**, 1554.
157. Perchard and Josien, *J. Chim. Phys.*, 1961, **65**, 1834; 1856.
158. Fukushima and Zwolinski, *J. Mol. Spectroscopy*, 1968, **26**, 368.
159. Evans, *Spectrochim. Acta*, 1960, **16**, 1382.
160. Green, *Chem. and Ind.*, 1962, 1575.
161. Jakobsen and Brewer, *J. App. Spectroscopy*, 1962, **16**, 32.
162. Robinson, Schrieber and Spencer, *J. Phys. Chem.*, 1971, **75**, 2219.

7

Ethers, Peroxides and Ozonides

7.1. Introduction and table

The only characteristic feature of ethers by which they can be identified from their infra-red spectra is the very intense absorption arising from the single bond $C-O-$ stretching vibration. It has been seen that in alcohols this occurs near 1050 cm^{-1} in structures containing the CH_2OH group, but that this frequency is very sensitive to environmental changes, so that even chain branching, as in secondary alcohols, causes a shift towards 1100 cm^{-1}. In the case of ethers, therefore, it is not to be expected that this band position will be constant, particularly as in this case interactions will occur between the $C-O-$ and $O-R$ vibrations, so that the nature of the substituents on either side of the link will be liable to alter the frequency. However, correlations are possible for limited types of structure, as in the case of the alcohols, and the group CH_2-O-CH_2, for example, can be expected to absorb within a considerably narrower range than the $1150-1000 \text{ cm}^{-1}$ which was originally given by Lecomte [1] for saturated ethers as a whole. However, although the band arising from this structure can usually be recognised from its very high intensity, this correlation cannot be used to give more than a possible indication of the presence of this group. This is due to the influence of structural changes on the frequency.

One instance in which structural changes result in a very considerable frequency shift of the $C-O$ vibration is in compounds containing the group $=C-O-$. In the case of alcohols, such as phenol, a high-frequency shift towards 1250 cm^{-1} is found as a result of this, and esters and acids show the same effect. It is therefore not surprising that aryl ethers and similarly unsaturated ethers also show their strong C–O absorption near 1250 cm^{-1}. The fact that so many related materials also show this band is another illustration of the difficulty in using this region for structural identifications. A strong band near 1250 cm^{-1} is, at best, only a reasonably good indication of the presence of a C–O link, the carbon atom of which is unsaturated, and a study of the rest of the spectrum must be made to ascertain whether or not this is likely to arise from esters, alcohols or any of the other alternative structures.

A second complication can arise from the fact that the position and nature of substituents on the aromatic ring also have an effect upon the C–O– bond-length. Colthup [2] finds that the C–O stretching frequency of *meta*-substituted aromatic ethers is higher than normal, and his chart shows values up to 1300 cm^{-1} for this vibration. He also indicates the presence of a medium intensity band between 1070 cm^{-1} and 1000 cm^{-1} in mono-aryl ethers, CH_2–O–Ph. We have observed such a band in a number of cases, as have others [30], and it may possibly be connected with the vibration of the CH_2–O– residue of the molecule. However, it would be an oversimplification to assign the 1250 and 1050 cm^{-1} bands to the different individual bonds. The vibrations are very strongly coupled, so that in alkyl ethers for example the 1050 cm^{-1} is the antisymmetric component of a pair, the second of which occurs at 820 cm^{-1}. As the symmetric mode the latter is appreciably weaker in the infra-red. The coupling may well be reduced in aryl/alkyl ethers but it is not clear whether the difference in bond character is sufficient to make this the dominant factor in determining the band positions. The position is further complicated by the occurence of extensive rotational isomerism in the liquid and solution states. This gives rise to multiple bands in many instances.

Cyclic ethers are in rather a different position from the open-chain products, in that the ring itself may have certain modes of its own which, although not C–O vibrations as such, are nevertheless sufficiently characteristic to enable the ring to be identified. Work along these lines on the epoxy-group and on cyclic structures in the sugars has given useful results. It is, for example, possible in this way to differentiate α and β anomers, and furanose derivatives from characteristic frequencies related to tetrahydropyrane and tetrahydrofuran rings [32–35].

A correlation for the –O–O– peroxide linkage has been suggested which associates this group with absorption at 890–830 cm^{-1}, and this has been supported by Raman data [21]. Peroxides containing the –CO–O–O– group may also be identified by the characteristic carbonyl vibration (Chapter 9). No satisfactory correlations for ozonides have yet been developed.

The correlations discussed are listed in Table 7.

Table 7

Alkyl ethers —CH$_2$–O–CH$_2$—	1150–1060 cm^{-1} (v.s.)
Aryl ethers and others with the group =C–O—	1270–1230 cm^{-1} (v.s.)
Cyclic ethers (1) Epoxy compounds	Near 1250 cm^{-1} (s.)
	Near 890 cm^{-1} (*trans*)
	Near 830 cm^{-1} (*cis*) (m.)
(2) Larger ring compounds	1140–1070 cm^{-1} (v.s.)
Alkyl peroxides	890–820 cm^{-1} (v.-w.)

7.2. Aliphatic ethers

The spectra of a number of simple aliphatic ethers have been given Barnes *et al.* [3]. The exact positions of the C–O–C bands are not given, but from the spectra it can be seen that they all absorb in the approximate range 1150–1060 cm^{-1}. They have also given the spectra of allyl ethyl ether and of diallyl ether, in both of which the C–O–C band is in approximately the same position as in the saturated materials. This last observation would appear to indicate that the structure C=C–C–O–C does not have any marked effect on the C–O frequency, and this agrees with Lecomte's [1] observation that the C–O stretching frequencies of vinyl alcohols are essentially normal.

Again, by analogy with alcohols, it is likely that chain branching at the carbon atom carrying the ether link will seriously modify the C–O frequency. Little or no information is available on this, but it is worth noting that the spectrum of di*iso*propyl ether given by Barnes *et al.* [3] shows a triplet structure in the 1100–1070 region, whereas di*iso*amyl ether [4] in which the chain branching is removed farther from the C–O band has only a single strong band at 1111 cm^{-1}. Tschamler and Leutner [30, 36] quote 1150–1080 cm^{-1} for saturated ethers and have noted that ethers containing the group C–O–C–O–C a strong doublet is shown in this range. This observation has been extended by Bergmann and Pinchas [37], who have studied eighteen ketals and acetals, many of which are cyclic dioxolanes, and identified a series of four bands in the 1200–

1000 cm^{-1} region which they regard as specific for the C–O–C–O–C grouping and which they assign to various coupled C–O vibrations. However, Lagrange and Mastagli [38] have discussed this correlation and pointed out that many of these bands appear in dioxane derivatives also and that they are not always all present in dioxolanes. More recent work on the acetals [39, 56–58] suggests that the antisymmetric C–O–C–O–C band appears in the 1145 to 1129 cm^{-1} range, with the symmetric mode between 900 and 800 cm^{-1}.

Studies on alkyl ethers which include successive deuteration in methyl ethyl ether [59], have already established that the antisymmetric C–O–C band is at 1120 cm^{-1} for the *trans* configuration, and at 1068 cm^{-1} for the *gauche* form. Weisner *et al.* [60] have made similar studies on diethyl ether and also find the antisymmetric stretch of the *trans* form at 1120 cm^{-1}.

A useful aid to the identification of the C–O stretching bands, apart from their intensities, is the movement to lower frequencies which takes place on complex formation with aluminium trichloride or boron trifluoride [61, 62]. In diethyl ether for example the band shifts to 1070 cm^{-1}. Derovault [62] has also obtained useful data by this method. The results, together with those of Synder and Zerbi [63] and by Mashiko *et al.* [39] fully confirm that the main C–O band of alkyl ethers in the infra-red is to be expected in the 1120 cm^{-1} region with some variations on either side due to structural and conformational effects.

7.3. Aryl and aralkyl ethers

As in the case of aliphatic ethers, the data available on aryl ethers are largely contained in the fairly extensive series of spectra of simple ethers of this type given by Barnes *et al.* [3]. In all of these a strong band is shown in the approximate region $1270–1230 \text{ cm}^{-1}$ and the comparison of *ortho-*, *meta-* and *para-*isomerides indicates that, in agreement with Colthup [2], the *meta-*substituted materials show this band at slightly higher frequencies than the others.

The work of Lecomte [1] also supports this assignment. In addition to a number of normal aryl ethers he has examined compounds of the series $C_6H_5O(CH_2)_nOC_6H_5$, in which n is varied from 1 to 10. In all of these the strong bands near 1250 cm^{-1} remained stationary. In these spectra a second band occurs in the $1150–1060 \text{ cm}^{-1}$ region which is probably due to the CH_2–O–

vibration. Tschamler and Leutner [30] have also found bands in both regions in aralkyl ethers, and diphenyl ethers have been studied by Kimoto [43]. Our own observations on a number of aryl ethers agree with these assignments. Much of the limited information available concerns aryl ethers, and the application of this correlation to vinyl ethers is less certain. For example, the *cis-* and *trans*-butylenyl butyl ethers show considerable differences in this region, and the strongest bands occur in the $1200-1100$ cm^{-1} range [11]. From the limited number of such materials examined [30, 44] it would appear that the C–O vibration occurs somewhere between 1240 cm^{-1} and 1150 cm^{-1}. There is also some evidence that conjugation of the double bonds does not affect this frequency. Anisole has been studied by a number of workers [64–68] and has been the subject of much controversy. It shows strong bands at 1035 and 1176 cm^{-1}. Some workers assign the former band to the C–O stretch whilst others prefer the latter. However, this is to assume complete coupling and the presence of only one antisymmetric stretching mode. The position is not resolved, but all alkyl aryl ethers show two very strong bands at these points. Their intensities imply that each of them has some connection with the C–O stretch.

7.4. Cyclic ethers

7.4(a). Epoxy compounds. Ethylene oxide has been studied by a number of workers [5, 6, 7, 45], who have associated bands at 865 cm^{-1}, 1165 cm^{-1} and 1265 cm^{-1} with the epoxy group. This group is of considerable interest in connection with the oxidation of vegetable oils and hydrocarbons, and a number of workers have therefore interested themselves in the possibility of finding a characteristic absorption frequency for it. Field, Cole and Wood-ford [8] studied eight epoxy compounds of some complexity, such as *iso*butylene oxide, styrene oxide, etc., and examined the spectra for characteristic bands in the regions given above. Although they identified bands in each of these regions in all these compounds, they consider that only the 1250 cm^{-1} band could be identified with any reasonable certainty as being due to the epoxy group, and in the compounds examined it fell constantly in the range $1260-1240$ cm^{-1}. The identification of the remainder they regarded as being much less certain on account of the increased number of C–C and C–H rocking frequencies in these molecules.

Shreve *et al.* [9] have also studied this problem, and have given the spectra of a few simple epoxy compounds, and also of the epoxy derivatives of some 1 : 2-olefines of long-chain length. Their results agree with those of Field *et al.* [8] in so far as they find a band close to 1260–1250 cm^{-1} in all cases, and it is noticeable that, although this band is relatively strong in the simple compounds, it is of only medium to weak intensity in, for example, 1 : 2 epoxytetradecane, in which the molar concentration of epoxy group is much lower. In α-methyl styrene oxide this band is absent, although Patterson [46] has detected it in twenty-five other epoxy compounds. Günthard *et al.* have also noted the absence of this band in steroid epoxides [47]. However, Shreve *et al.* [9] do not consider this band, which, with others at 1430 cm^{-1} and 1136 cm^{-1}, they believe to be associated with C–O stretching modes, to be the most characteristic for epoxy structures in all cases. In all the compounds referred to they found bands close to 910 cm^{-1} and 830 cm^{-1}, although the second of these was shifted to 818 cm^{-1} in 3 : 4-epoxy-1-butene, possibly under the influence of the conjugated double bond. Extension of this work to the epoxy derivatives of mono-unsaturated fatty acids, esters and alcohols showed that spectra of the *trans*-epoxy materials were qualitatively similar to the saturated compounds, except for a new band near 890 cm^{-1}, whilst the *cis*-epoxy materials all showed a band near 830 cm^{-1}. They have therefore tentatively ascribed these bands to ring vibrations of the epoxy ring in *cis*- and *trans*-compounds. These lower-frequency absorptions have also been studied by Patterson [46] and by Günthard *et al.* [47] In the compounds studied by the former, two bands were found in the overall range 950–863 cm^{-1} and 864–786 cm^{-1}, with the majority falling near 910 and 830 cm^{-1} in good agreement with the earlier workers. On the other hand, in steroid epoxides Günthard *et al.* find bands corresponding to the epoxy group between 900 and 800 cm^{-1} in some compounds and between 1050 and 1035 cm^{-1} in others.

The application of these findings will need considerable caution until further spectra are available and, in view of the many different groups which can cause absorption near 830 cm^{-1} and 890 cm^{-1}, it would be unwise to apply these correlations, other than very tentatively, to compounds not closely similar to those described.

In addition, Tuot and Barchewitz [12] have suggested that in the overtone region, absorption at 7042 cm^{-1} may be characteristic of epoxides, but the number of materials examined is small, and this possibility requires further study.

7.4(b). Four-, five- and six-membered ring compounds. With larger rings in which the strain is reduced, the C–O–C stretching frequency should approach nearer to the normal value. Shreve *et al.* [9] give the spectra of tetrahydropyran and dioxan, in both of which a series of bands in the range 1124–1030 cm⁻¹ are shown. The band at 1100 cm⁻¹ in tetrahydropyrane has been ascribed to the C–O vibration [31]. Tschamler *et al.* [30, 48] quote a few similar cases, and again find that the group C–O–C–O–C gives rise to a doublet in the same frequency range. Studies on dioxanes and dioxolanes have confirmed that multiple-band structures are shown, but the precise number of characteristic frequencies produced in the two cases are not clearly defined [37, 38]. In unsaturated cyclic compounds with the structure =C–O–C= the band moves to higher frequencies. Tetrahydrofuran, with a five-membered ring, absorbs at 1076 cm⁻¹, and Barrow and Searles [31] have found five other related compounds absorbing between 1100 and 1075 cm⁻¹. With other heterocyclic systems, such as azoximes and oxadiazoles, a strong 1030 cm⁻¹ band is found which is absent from the furazans [10] and is probably associated with the C–O bond. The influence of ring-strain becomes more pronounced with four-membered rings and the C–O absorption of trimethylene oxides occurs in the 980–970 cm⁻¹ region [31]. This supports indirectly the 890–830 cm⁻¹ range suggested for epoxy compounds which would be expected to absorb at lower frequencies.

However, too little information is yet available for any detailed correlations to be made, beyond the generalisation that the ether link in large ring compounds of this type probably absorbs at something like the same frequency as it does in open-chain products. Jones and Sandorfy have compiled a useful bibliographical table of a number of cyclic ethers [49].

7.5. Peroxides

The possibility of detecting peroxide and hydroperoxide structures by infra-red methods is also of practical interest in connection with the mechanism of oxidative processes in hydrocarbons. The theoretical aspects of such a correlation have been discussed by Sheppard [13], who points out that it is unlikely that peroxides of the type R–O–O–R′ will have any very characteristic bands for the O–O vibration, as this mode would be relatively symmetrical, and not associated with any great change in dipole moment, so that the corresponding frequency would be expected to be weak in the infra-red. Furthermore, the masses and force constants of the O—O group are so similar to those of C—O and C—C groups that it is unlikely that any very characteristic frequency will result. With hydroperoxides, where the O—O group is at the end of a chain, the possibilities are a little greater but, even so, the O—O vibration will contribute to the skeletal modes of the molecule as a whole, and the frequency will therefore be subject to changes with the nature of the substituent groups. Sheppard suggested, however, that some correlations might be found, covering limited groups of substituents in the same way as correlations have been found, for example, for primary, secondary and tertiary alcohols, although this will involve a study of a considerable number of compounds of a variety of different types.

The region in which any O—O skeletal vibration is to be expected is readily obtainable from the spectrum of hydrogen peroxide which has been studied by several workers [14, 15]. Giguère [14] has shown that the 877 cm^{-1} band of H_2O_2 is essentially unchanged in position in D_2O_2, so that it is reasonable to assign it to the O—O stretching mode. Furthermore, this assignment corresponds to a bond length nearly equal to that derived from electron diffraction measurements.

This observation has been followed up by Leadbeater [16], who obtained the infra-red and Raman spectrum of the peroxide of ethyl ether, and the Raman spectrum of the corresponding α-mono- and α-di-hydroxyperoxides, and of benzoyl peroxide. In all cases he found a fairly strong band in the range 883—881 cm^{-1}, and, in contrast to Sheppard's comments, he notes that on passing from the O—O vibration of H_2O_2 to higher molecules there are no frequency changes such as occur with the C—C links of hydrocarbons. The range of materials studied is, however, very limited, and one of them, at least, must be considered as being suspect in regard to its purity, as Shreve et al. [17] have found no band at 883 cm^{-1} in the infra-red spectrum of benzoyl peroxide, and we have confirmed this observation.

Nevertheless, it is fairly clear from the work of others that aliphatic materials of this type do absorb in the 870 cm^{-1} region, although, as is to be expected, the frequency range is relatively wide. Minkoff [18, 50] has studied some thirty pure peroxides, hydroperoxides and peracids, and in all cases has found a band in the range 890—830 cm^{-1}. Whilst assigning this provisionally to the O—O vibration, he points out that the intensity of the band is not high and that it may be hidden by, or confused with, several skeletal frequencies of comparable intensities in large molecules. All the compounds examined were aliphatic, and the majority were simple methyl, ethyl and butyl derivatives. However, despite the low intensity of this band, absorption at 813 cm^{-1} has been used by Raley *et al.* [20] for estimating di-*tert.*-butyl peroxide in *tert.*-butyl alcohol.

Confirmation of this assignment is forthcoming from other studies on groups of compounds containing the O–O link [17, 19, 51–53], and from a number of scattered observations on isolated molecules [54–56]. However, as before, the band is weak and difficult to identify. This is particularly apparent, for example, in the work of Williams and Mosher [51], who compared the spectra of a number of alkyl-hydroperoxides with those of the corresponding alcohols. Although marked differences were found in this region, many of the alcohols themselves absorbed in the same spectral region. As mentioned above, benzoyl peroxide does not follow this correlation, but shows a strong absorption near 1000 cm^{-1}. Shreve *et al.* [17] point out that phthaloyl peroxide, *p*-chlorobenzoyl peroxide and benzaldehyde peroxide also absorb near this point, and suggest tentatively that this may represent the characteristic frequency for the O–O absorption in aryl peroxides. We have examined perbenzoic acid and find absorption near 1030 cm^{-1} and 880 cm^{-1}.

The O–O link is however not well characterised in the infra-red spectrum. In the Raman spectrum on the other hand, the emmission is very strong and the bond is more readily identified. Dollish *et al.* [21] quote many examples of specific frequencies for this mode. These range from 943 cm^{-1} in CF_3OOCl to 771 cm^{-1} in di-*tert.*-butyl peroxide. Most alkyl peroxides have this band in the 770 to 800 cm^{-1} range.

In hydroperoxides a number of attempts have been made to assign OH deformation and C–O stretching frequencies [19, 52, 53], but the direct comparisons with alcohols [51] indicates that these are less likely to be valuable. In the case of peracids the OH deformation frequency occurs near 950 cm^{-1}, as in carboxylic acids, but is noticeably weaker [53]. A few other tentative correlations for hydroperoxides [51] and for narrow classes, such as percarbonates and peresters [50], have been suggested, but require more detailed study before they can be fully substantiated.

7.6. Ozonides

The possibility of using infra-red methods for the study of the formation, stability and decomposition of ozonides was first explored by Briner and his co-workers [22–24]. In preliminary studies they have passed ozone into carbon tetrachloride solutions of unsaturated compounds and observed the spectral changes following varying degrees of ozonolysis. This at first led them to believe that ozonides absorbed in the carbonyl region, but subsequent work by their own group [25, 26] and by Criege *et al.* [27] has shown that these bands arise from decomposition products. The latter workers studied some eighteen ozonides of various types and noted that in the greater proportion of them a new medium strong absorption band appeared after ozonolysis in the range $1064-1042 \text{ cm}^{-1}$. These probably correspond to C–O stretching frequencies of the ozonide system. Briner and Dallwigk [25, 26] have confirmed this finding, and Garvin and Schubert [28] have also found ozonide absorptions in this region. The correlation is, however, of very limited value in view of the many other types of C–O stretching absorptions found hereabouts.

Some studies have also been made on the action of ozone on elastomers such as natural rubber, neoprene, etc. [8]. In these cases, however, there is no suggestion of the detection of any ozonides, but only of decomposition products characterised by OH and carbonyl absorptions.

7.7. Bibliography

1. Lecomte, *Traité de Chimie Organique* (ed. Grignard et Baud, Masson et Cie, Paris, 1936), 2, 143.
2. Colthup, *J. Opt. Soc. Amer.*, 1950, **40**, 397.
3. Barnes, Gore, Liddel and Van Zandt Williams, *Infra-red Spectroscopy* (Reinhold, 1944).
4. Randall, Fowler, Fuson and Dangl, *Infra-red Determination of Organic Structures* (Van Nostrand, 1949).
5. Herzberg, *Infra-red and Raman Spectra of Polyatomic Molecules* (Van Nostrand, 1945), p. 340.
6. Bonner, *J. Chem. Phys.*, 1937, **5**, 704.
7. Thompson and Cave, *Trans. Faraday Soc.*, 1951, **47**, 946.
8. Field, Cole and Woodford, *J. Chem. Phys.*, 1950, **18**, 1298.
9. Shreve, Heether, Knight and Swern, *Analyt. Chem.*, 1951, **23**, 277.
10. Milone and Borello, *Gazz. Chim. Italia*, 1951, **81**, 368.
11. Hall, Philpotts, Stern and Thain, *J. Chem. Soc.*, 1951, 3341.
12. Tuot and Barchewitz, *Bull. Soc. chim. (France)*, 1950, 851.
13. Sheppard, *Discuss. Faraday Soc.*, 1950, **9**, 322.

14. Giguère, *J. Chem. Phys.,* 1950, **18**, 88.
15. Taylor, *ibid.,* 898.
16. Leadbeater, *Compt. rend. Acad. Sci. Paris,* 1950, **230**, 829.
17. Shreve, Heether, Knight and Swern, *Analyt. Chem.,* 1951, **23**, 282.
18. Minkoff, *Discuss. Faraday Soc.,* 1950, **9**, 320.
19. Whitcomb, Moorhead and Sharp, *Symposium on Molecular Structure and Spectrography* (Ohio, 1950).
20. Raley, Rust and Vaughan, *J. Amer. Chem. Soc.,* 1948, **70**, 1336.
21. Dollish, Fateley and Bentley, *Characteristic Raman Frequencies of Organic Compounds* (Wiley, New York, 1974).
22. Briner, Susz and Dallwigk, *Arch. Sci. Phys. Nat.,* 1951, **4**, 199.
23. Susz, Dallwigk and Briner, *ibid.,* 202.
24. Briner, Susz and Dallwigk, *Compt. rend. Acad. Sci. Paris,* 1952, **234**, 1932.
25. Briner and Dallwigk, *Helv. Chim. Acta,* 1956, **39**, 1446, 1826.
26. *Idem, Compt. Rend. Acad. Sci. Paris,* 1956, **243**, 630.
27. Criegee, Kerckow and Zinke, *Chem. Ber.,* 1955, **88**, 1878.
28. Garvin and Schubert, *J. Phys. Chem.,* 1956, **60**, 807.
29. Allison and Stanley, *Analyt. Chem.,* 1952, **29**, 630.
30. Tschamler and Leutner, *Monatsh.,* 1952, **83**, 1502.
31. Barrow and Searles, *J. Amer. Chem. Soc.,* 1953, **75**, 1175.
32. Barker, Bourne, Stacey and Whiffen, *J. Chem. Soc.,* 1954, 171.
33. Barker, Bourne, Stephens and Whiffen, *ibid.,* 1954, 3468, 4211.
34. Barker and Stephens, *ibid.,* 1954, 4550.
35. Whistler and House, *Analyt. Chem.,* 1953, **25**, 1463.
36. Tschamler, *Spectrochim. Acta.,* 1953, **6**, 95.
37. Bergmann and Pinchas, *Rec. trav. chim.,* 1952, **71**, 161.
38. Lagrance and Mastagli, *Compt. Rend. Acad. Sci. Paris,* 1955, **241**, 1947.
39. Mashiko, Saeki, Nukada, Kanazawa and Sazuki, *Proc. International Symp.,* Tokyo, 1962, **A**, 220.
40. Chakhovskoy, Martin and Van Nechel, *Bull. Soc. Chim. Belges,* 1956, **65**, 453.
41. Philpotts, Evans and Sheppard, *Trans. Faraday Soc.,* 1955, **51**, 1051.
42. Price and Osgan, *J. Amer. Chem. Soc.,* 1956, **78**, 4787.
43. Kimoto, *J. Pharm. Soc. Japan,* 1955, **75**, 763.
44. Meakins, *J. Chem. Soc.,* 1953, 4170.
45. Lord and Nolin, *J. Chem. Phys.,* 1956, **24**, 656.
46. Patterson, *Analyt. Chem.,* 1954, **26**, 823.
47. *Günthard, Heusser and Fürst, Helv. Chim. Acta,* 1953, **36**, 1900.
48. Tschamler and Voetter, *Monatsh.,* 1952, **83**, 303, 1228.
49. Jones and Sandorfy, *Chemical Applications of Spectroscopy* (Interscience, New York, 1956), p. 438.
50. Minkoff, *Proc. Roy. Soc.,* 1954, **A.224**, 176.
51. Williams and Mosher, *Analyt. Chem.,* 1955, **27**, 517.
52. Karyakin, Nikitin and Ivanov, *Zhur. Fiz. Khim.,* 1953, **27**, 1856.
53. Swern, Witnauer, Eddy and Parker, *J. Amer. Chem. Soc.,* 1955, **77**, 5537.
54. Criegee and Paulig, *Chem. Ber.,* 1955, **88**, 712.
55. Milos and Mageli, *J. Amer. Chem. Soc.,* 1953, **75**, 5970.
56. Nukada, *Spectrochim. Acta,* 1962, **18**, 745.
57. Nukada, *Bull. Chem. Soc. Japan,* 1961, **34**, 1615, 1624; 1962, **35**, 3.
58. Little and Martell, *J. Phys. Colloid Chem.,* 1949, **53**, 472.
59. Perchard, *Spectrochim. Acta,* 1970, **26A**, 707.
60. Weisner, Laidlaw, Kreuger and Fuhrer, *Spectrochim. Acta,* 1968, **27A**, 1055.

61. Weisner and Kreuger, *Spectrochim. Acta,* 1970, **26A**, 1349.
62. Derovault and Forel, *Spectrochim. Acta,* 1969, **25A**, 67.
63. Snyder and Zerbi, *Spectrochim. Acta,* 1967, **23A**, 391.
64. Stephenson, Coburn and Wilkox, *Spectrochim. Acta,* 1961, **17**, 933.
65. Mooney, *Spectrochim. Acta,* 1963, **19**, 877.
66. Green, *Spectrochim. Acta,* 1962, **18**, 39.
67. Garrigou Legranfe, Lebas and Josien, *Spectrochim. Acta,* 1958, **12**, 305.
68. Rai and Upadhya, *Spectrochim. Acta,* 1966, **22**, 1427.

8

Acid Halides, Carbonates, Anhydrides and Metallic Carbonyls

8.1. Introduction and table

The correlations discussed are shown in Table 8.

Table 8

C=O *Vibrations* (all strong bands)

Acid halides	$1815-1770$ cm^{-1}. (Conjugated materials in lower part of the range.)
Carbonates	Dialkyl $1745-1735$ cm^{-1}, Diaryl $1819-1775$ cm^{-1}
Anhydrides (open-chain)	$1850-1800$ cm^{-1} and $1790-1740$ cm^{-1} ⎫ Conjugation lowers by
Anhydrides Cyclic (5-ring)	$1870-1820$ cm^{-1} and $1800-1750$ cm^{-1} ⎬ 20 cm^{-1} in each case
Peroxides (alkyl)	$1820-1810$ cm^{-1} and $1800-1780$ cm^{-1}
Peroxides (aryl)	$1805-1780$ cm^{-1} and $1785-1775$ cm^{-1}

C—O *Vibrations*

Anhydrides (open-chain)	$1170-1050$ cm^{-1} (s.)
Anhydrides (cyclic)	$1300-1200$ cm^{-1} (s.)

C≡O *Vibrations*   Near 2050 and 2040 cm^{-1}

8.2. Acid halides

In esters, acids and ketones the substitution of a chlorine or bromine atom on an α-carbon atom to the carbonyl group results in a shift in the C=O frequency to a higher value. With a direct attachment of the halogen to the carbonyl group, a considerably bigger shift in this direction is to be expected, and this is what is found in practice.

141

Phosgene gas absorbs at 1828 cm^{-1}, and acetyl chloride [1, 2] and acetyl bromide at 1802 cm^{-1} and 1812 cm^{-1} respectively. An increase in the size of the remainder of the molecule does not appear to affect this frequency significantly, and phenylacetyl chloride absorbs at 1802 cm^{-1} [1, 2], whilst we have found a value of 1790 cm^{-1} for *n*-octoyl chloride. Similar results are given in Raman spectra for oxalyl chloride (1776 cm^{-1}) and propionyl chloride (1786 cm^{-1}).

Conjugation of the acid halide with an $\alpha\beta$ double bond or an aryl group would be expected to lower the CO frequency again, and this is found in benzoyl chloride, which absorbs at 1773 cm^{-1} [1, 2]. This compound, however, shows a second band at 1736 cm^{-1}, due to Fermi resonance with a low lying fundamental. Very few acid fluorides have been reported in the infra-red literature, and owing to the very high electronegativity of fluorine, these absorb at considerably higher frequencies. The compound COF_2, for example [18], absorbs at 1928 cm^{-1}, which is one of the highest carbonyl frequencies known whilst acetyl fluoride absorbs at 1872 cm^{-1} in the vapour state. α-Halogen substitution also raises the carbonyl frequency, and trifluoroacetyl fluoride absorbs at 1901 cm^{-1}. With less halogen substitution the possiblity of rotational isomerism leads to two carbonyl frequencies, in chloroacetyl chloride and similar compounds studied by Mizushima *et al.* [19]. In general, this leads to a pair of carbonyl frequencies, one of which is higher than acetyl chloride and the other lower [20]. The higher frequency is associated with the isomer in which the chlorine atom is near in space to the oxygen atom, whilst the lower corresponds to the form in which the two halogen atoms approach each other. In the latter case the internal field effects probably operate to lower the effective electro-negativity of the directly attached halogen atom, with a consequent fall in the frequency. The frequencies of a number of alkyl acid halides and their halogen-substituted derivatives have been discussed by Bellamy and Williams [20], and fluorinated halides have also been studied by Haszeldine [17], Husted and Ahlbrecht [16] and Wieblen [21]. The rotational barrier in trifluoroacetyl fluoride has been measured by Lord [3].

Recently the carbonyl frequencies of some forty aromatic acid halides (benzoyl, naphthoyl and phthaloyl) have been given by Al Jallo *et al.* [31]. Their values are in general agreement with the ranges listed above.

8.3. Carbonates and chloroformates

It is to be expected that the influence of two oxygen atoms in carbonates will raise the frequency of the carbonyl absorption above that of esters, and this will be true also in chloroformates. This is realised in practice. Diacyl carbonates ROCOOR absorb [41] close to 1740 cm^{-1}. Nyquist and Potts [8] quote the very narrow range of 1741 to 1739 cm^{-1} but other workers suggest a somewhat wider spread [8–11, 41]. Nevertheless the band is usually very close to 1740 cm^{-1} and appears with considerably greater intensity than a normal ester group [9]. In aryl carbonates the environment of the carbonyl group is similar to that in vinyl esters and the frequency rises in just the same way. In most aryl/alkyl carbonates this band appears between 1787 and 1757 cm^{-1}. In diaryl carbonates the effect is increased and the frequency range is between 1819 and 1775 cm^{-1}. Halogen substitution on the R substituents raises the carbonyl frequencies further, and $CCl_3OCOOCCl_3$, for example, absorbs at 1832 cm^{-1}. In cyclic carbonates the influence of ring strain is superimposed on other effects, and in the five-membered ring systems studied [41, 42, 12] the carbonyl frequencies fall in the 1830–1800 cm^{-1} range. There is some variation in these values with changes of state which is illustrated by ethylene carbonate [42], which absorbs at 1870 cm^{-1} in the vapour, 1830–1820 cm^{-1} in non-polar solvents and 1805 cm^{-1} in the liquid. In this compound a second absorption is shown near 1780 cm^{-1}, which is due to an overtone frequency.

Both cyclic and open-chain carbonates also show strong bands due to the C–O stretching frequencies [41]. In open-chain compounds these appear to be somewhat similar to the corresponding bands in esters, but in five-membered ring systems some shifts occur under the influence of ring strain.

Chloroformates absorb at somewhat higher frequencies than carbonates, and the C=O band is usually found between 1780 and 1770 cm^{-1}. This is again raised by α-halogen substitution, the CCl_3 derivative absorbing at 1806 cm^{-1}. The bulk of the available data relates to alkyl chloroformates studied by Ory [13], Nyquist and Potts [8], and Katritzky [11]. The effect of a double bond at the ester oxygen atom has not been much studied, but it would seem that ν_{CO} is raised as in the aryl carbonates. Thus allyl chloroformate absorbs at 1799 cm^{-1}.

8.4. Anhydrides

Anhydrides all show two carbonyl absorption bands. The position, separation and intensity of these depend on whether or not the carbonyls are conjugated, and whether they are part of a strained five-membered ring, as in cyclic anhydrides. Randall *et al.* [1] directed attention to the fact that the two bands are always approximately the same distance apart (*c.* 60 cm^{-1}), and, although a number of exceptions are known, this is a very useful pointer in their identification. Typical values for unstrained materials are as follows:

Acetic anhydride [1]	1824 cm^{-1} and 1748 cm^{-1}
Phenyl acetic anhydride	1808 cm^{-1} and 1745 cm^{-1}
Caproic anhydride [4]	1825 cm^{-1} and 1760 cm^{-1}
Glutaric anhydride [6]	1802 cm^{-1} and 1761 cm^{-1}

$\alpha\beta$-conjugation results in a lowering of these values by 20–40 cm^{-1}. Crotonic anhydride [4] absorbs near 1780 cm^{-1} and 1725 cm^{-1}, and we have found these bands in benzoylbenzoate to occur at 1789 cm^{-1} and 1727 cm^{-1}. It will be seen from these values that the distance apart of the two bands is reasonably consistent, despite the variations in individual frequencies. The factors which determine the separation of these bands have been discussed by Cooke [25], and values for the carbonyl frequencies of a substantial number of anhydrides have been given by Bellamy *et al.* [14] and by Dauben and Epstein [29].

With cyclic materials in which the CO groups form part of a five-membered ring, the influence of ring strain induces a shift to higher frequencies, similar to that found in cyclic ketones and lactones with five-membered rings. Succinic anhydride [1], for example, absorbs at 1865 cm^{-1} and 1782 cm^{-1}. In cyclic materials also, conjugation lowers these values to some extent, so that differentiation between an $\alpha\beta$-unsaturated cyclic anhydride and a saturated open-chain anhydride is not always possible on the basis of the carbonyl absorptions alone. Typical values for unsaturated cyclic products are 1845 cm^{-1} and 1775 cm^{-1} for phthalic anhydride [4], and 1848 cm^{-1} and 1790 cm^{-1} for maleic anhydride [4]. Naphthalene-1 : 2-dicarboxylic anhydride and some substituted derivatives [7] absorb at 1848–1845 cm^{-1} and 1783–1779 cm^{-1}. In the case of cyclic six-membered ring anhydrides condensed to aromatic systems a much smaller separation of the two frequencies (34 cm^{-1}) has been recorded [22] with rather lower frequencies (1770 cm^{-1}, 1736 cm^{-1}) than is usual. Other isolated observations on anhydrides are included in papers by Walker [24], Koo [25], Stork and Breslow [26], and Hochstein *et al.* [27], the last including 3-methoxy-6-methyl pyromellitic acid anhydride with two anhydride systems attached to the same aromatic ring, which leads to four carbonyl frequencies. α-Fluorine substitution, as is usual with carbonyl frequencies, results in an upward shift of both bands, and perfluoroacetic anhydride absorbs [28] at 1884 cm^{-1} and 1818 cm^{-1}.

The doubling of the carbonyl frequency in anhydrides and peroxides (see below) presents an interesting puzzle and this has been discussed by Bellamy *et al.* [14] and by Dauben [29]. The resonance within the CO–O–CO system causes it to be coplanar, and this enhances the coupling between the carbonyl groups to an extent which is not possible in the non-planar diketo form of acetyl acetone. The two carbonyl frequencies therefore correspond to the symmetric and antisymmetric vibrations of the carbonyl groups. The higher of the two frequencies has been shown to arise from the symmetric mode. This is unusual as in most coupled systems of this kind the antisymmetric frequency is the higher. However a number of other cases have been recognised and it appears to a general fact that the symmetric frequency is the higher when the coupled groups are separated by a single atom.

In open chain anhydrides the higher frequency band is always the more intense of the two, but in cyclic systems the dipole-moment change of the symmetric mode which gives rise to this band is reduced. This is readily seen by a consideration of the theoretical four-membered ring anhydride in which the carbonyl groups are in direct opposition. There is then no dipole-moment change in the symmetric vibration and the band is therefore forbidden in the infra-red. Accordingly, as the ring size diminishes there is a progressive fall in the intensity of the symmetric band in relation to the antisymmetric band. Even in six-membered ring anhydrides, the higher frequency band is the weaker of the two, and in five-membered rings the difference is very clearly marked. The relative intensities of these two bands therefore provide a very clear indication of the structure.

Konarski [30] has suggested a somewhat different explanation for the splitting and intensity relationships based on a pulsed vibration between the carbonyl groups and the lone pair electrons of the central oxygen atom. This would result in the same kinds of changes with bond angles as those given above.

A limited number of α-halogenated anhydrides have been described by Rogstad *et al.* [32] and by Redington and Lin [33].

In addition to the carbonyl absorptions, anhydrides also show strong bands due to the C−O−C stretching vibration. These can usually be identified by their high intensity but, as in the case of ethers, the frequency range in which they occur is large. Colthup [5] quotes a range of $1175-1045$ cm^{-1} for this vibration in open-chain materials, and $1310-1210$ cm^{-1} in cyclic materials in which ring strain is involved. These bands are useful in giving negative evidence, and the absence of an anhydride can reasonably be deduced if there are no very strong bands in either of these regions. On the other hand, the occurrence of a band here is not evidence for the presence of an anhydride, as many other types of compound can give rise to strong bands in this region.

8.5. Peroxides

The CO absorptions of peroxides containing the structures CO−O−O−CO are very closely similar to those of anhydrides. The actual bands occur at comparable frequencies in each case, but in the case of the peroxides the splitting of these two in peroxides is rarely more than 30 cm^{-1}, whereas the corresponding value for anhydrides is usually about 60 cm^{-1}.

Most of the data in this field are due to Davison [15], who examined a series of acyl, aryl and unsymmetrical peroxides and compared the frequencies obtained. Seven dialkyl peroxides absorbed between 1820 cm^{-1} and 1811 cm^{-1} and between 1796 cm^{-1} and 1785 cm^{-1}, with an average distance apart of 25 cm^{-1}. The introduction of two aryl groups lowers both frequencies, as is to be expected, although the distance apart remains about the same.

Eleven diaryl peroxides absorbed in the ranges $1805-1780$ cm^{-1} and $1783-1758$ cm^{-1}. With unsymmetrical products in which one aryl and one alkyl group are involved, the mean splitting of the CO frequencies is greater than normal, due to the effect of the aryl conjugation on the nearest carbonyl group. However, the effect is not sufficient to raise the band separation to anything like the level of anhydrides.

Peracids and esters have also been examined in small numbers, by

Davison [15] and by Swern *et al.* whilst Minkoff reproduces a few additional spectra without comment on the carbonyl frequencies. These compounds show only a single carbonyl absorption in the overall range $1790-1740^{-1}$. Davison finds a rather wider frequency spread than do Swern *et al.*, who found a considerable number of peracids absorbing at $1747-1748$ cm^{-1}. This frequency did not change appreciably on passing from the solid phase into solution, confirming that these acids exist in intramolecularly bonded forms.

8.6. Metallic carbonyls

Metallic carbonyls were included in earlier editions of this book at a time when they were relatively rare and exotic compounds. Since that time they have been exhaustively studied by the coordination chemists and a very considerable volume of specialised information has been built up. Adams [34], for example, discusses these compounds at great length and gives a very comprehensive bibliography. It would be of little value to attempt a summary of this extensive but highly specialised topic when a very full treatment is already available in the literature, and it would also be inappropriate to treat this topic at length in a text which is primarily concerned with the spectra of organic compounds. This subject will not therefore be discussed other than to note that metallic carbonyls absorb in the 2000 cm^{-1} region unless they occur in bridged systems when the frequencies fall to 1800 cm^{-1}.

8.7. Bibliography

1. Randall, Fowler, Fuson and Dangl, *Infra-red Determination of Organic Structures* (Van Nostrand, 1949).
2. Rasmussen and Brattain, *J. Amer. Chem. Soc.*, 1949, **71**, 1073.
3. Loos and Lord, *Spectrochim. Acta*, 1965, **21**, 119.
4. Barnes, Gore, Liddel and Van Zandt Williams, *Infra-red Spectroscopy* (Reinhold, 1944).
5. Colthup, *J. Opt. Soc. Amer.*, 1950, **40**, 397.
6. Wasserman and Zimmerman, *J. Amer. Chem. Soc.*, 1950, **72**, 5787.
7. Modest and Szmuszkovicz, *ibid.*, p. 577.
8. Nyquist and Potts, *Spectrochim. Acta*, 1961, **17**, 679.
9. Thompson and Jameson, *Spectrochim. Acta*, 1958, **13**, 236.
10. Gatehouse, Livingstone, and Nyholm, *J. Chem. Soc.*, 1958, 3137.
11. Katritzky, Lagowski and Beard, *Spectrochim. Acta*, 1960, **16**, 964.
12. Sarel, Potoryles and Ben Shosham, *J. Org. Chem.*, 1958, **24**, 1873.
13. Ory, *Spectrochim. Acta*, 1960, **16**, 1488.

148   The Infra-red Spectra of Complex Molecules

14. Bellamy, Connelly, Philpotts and Williams, *Zeit. Electrochem.*, 1960, **64**, 563.
15. Davison, *J. Chem. Soc.,* 1951, 2456.
16. Husted and Ahlbrecht, *J. Amer. Chem. Soc.,* 1953, **75**, 1605.
17. Haszeldine, *Nature,* 1951, **168**, 1028.
18. Nielsen, Burke, Woltz and Jones, *J. Chem. Phys.,* 1952, **20**, 596.
19. Nakagawa, Ichishima, Kurtani, Miyazawa, Shimanouchi and Mizushima, *ibid.,* 1952, **20**, 1720.
20. Bellamy and Williams, *J. Chem. Soc.,* 1968, 3465.
21. Wieblen, *Fluorine Chemistry,* Ed. Simons. Vol. 2 (Academic Press, New York, 1954), p. 449.
22. Brown and Todd, *J. Chem. Soc.,* 1954, 1280.
23. Cooke, *Chem. and Ind.,* 1955, 142.
24. Walker, *J. Amer. Chem. Soc.,* 1953, **75**, 3387, 3390.
25. Koo, *ibid.,* 1953, **75**, 720.
26. Stork and Breslow, *ibid.,* 1953, **75**, 3291.
27. Hochstein, Conover, Regna, Pasternack, Gordon and Woodward, *ibid.,* 1953, **75**, 5455.
28. Fuson, Josien, Jones and Lawson, *J. Chem. Phys.,* 1952, **20**, 1627.
29. Dauben and Epstein, *J. Org. Chem.,* 1959, **24**, 1595.
30. Konarski, *J. Mol. Structure,* 1972, **13**, 45.
31. Al Jallo and Jahloom, *Spectrochim. Acta,* 1972, **28A**, 1655.
32. Rogstad, Klahoe, Cyvin and Christansen, *Spectrochim. Acta,* 1972, **28A**, 111.
33. Redington and Lin, *Spectrochim. Acta,* 1971, **22A**, 2445.
34. Adams, *Metal Ligand and Related Vibrations* (Arnold, London, 1966).

9

Aldehydes and Ketones

9.1. Introduction and tables

The infra-red absorption band arising from the C=O stretching vibration has probably been more extensively studied than any other, and a good deal is now known of the factors which influence its frequency and its intensity. In both cases the frequency of the carbonyl absorption is determined almost wholly by the nature of its immediate environment, and the structure of the rest of the molecule is of little importance unless it is such as to give rise to chelation or some similar effect. Thus the carbonyl frequency shifts away from the normal position in $\alpha\beta$-unsaturated materials and in carbonyl compounds with strongly electronegative substituents on the α-carbon atom, whilst in cyclic ketones the frequency shift and its direction are related to the degree of strain of the ring. Frequency shifts due to chelation and to mutual interference effects can also be considerable in some cases. However, in each of these cases the extent of the frequency shift to be expected is known, and the new range of frequencies falls within comparatively narrow limits. It is therefore often possible to identify the presence of a carbonyl group, and at the same time obtain a considerable amount of data as to the nature of its environment, by the study of the position and intensity of this one band.

This subject has been reviewed by Lecomte [82] and some very useful tables of typical carbonyl frequencies have been compiled by Jones and Sandorfy [83], Jones and Herling [84], Rao [178], Bellamy [179], and Renema [180] amongst others. The factors responsible for group frequency shifts have been considered by a

Table 9

*† CHARACTERISTIC FREQUENCIES OF KETONIC
   CARBONYL VIBRATIONS

Saturated open-chain ketones	$-CH_2-CO-CH_2$	$1725-1705$ cm^{-1}
$\alpha\beta$-unsaturated ketones	$-CH=CH-CO-$	$1690-1675$ cm^{-1}
$\alpha\beta-\alpha'\beta'$-unsaturated ketones	$-CH=CH-CO-CH=CH-$	$1670-1660$ cm^{-1}
‡ Aryl ketones	$Ph-CO-$	$1690-1680$ cm^{-1}
‡ Diaryl ketones	$Ph-CO-Ph$	$1670-1660$ cm^{-1}
α-halogen-substituted ketones	$-CBr-CO$ etc. 2 bands	$1745-1725$ cm^{-1}
$\alpha\alpha'$-dihalogen-substituted ketones	$-CBr-CO-CBr$ etc. 3 bands	$1765-1745$ cm^{-1}
α-diketones	$-CO-CO-$	$1730-1710$ cm^{-1}
β-diketones (enolic)	$-CO-CH_2-CO-$	$1640-1540$ cm^{-1}
$\alpha\beta$-unsaturated-β-hydroxy- or amino-ketones	$-CO-C=C-$ $\quad\quad\ \ |\ \ |$ $\quad\quad OH$ or NH_2	$1640-1540$ cm^{-1}
1-keto-2-hydroxy- or amino-, aryl ketones	(benzene ring)$-CO-$ OH or NH_2	$1655-1635$ cm^{-1} ‡
γ-diketones	$-CO-CH_2-CH_2-CO-$ $-CO-CH_2-O-CO-$	$1725-1705$ cm^{-1} $1745-1725$ cm^{-1}
Six- and seven-membered-ring ketones	–	$1725-1705$ cm^{-1}
Five-membered-ring ketones	–	$1750-1740$ cm^{-1}
Four-membered-ring ketones	–	Near 1775 cm^{-1}
Quinones	2 CO's in 1 ring 2 CO's in 2 rings	$1690-1660$ cm^{-1} $1655-1635$ cm^{-1}
Tropolones	–	Near 1600 cm^{-1}
Other vibrations	Alkyl ketones Aryl ketones	$1325-1215$ cm^{-1} (s.) $1225-1075$ cm^{-1} (s.)

number of workers [83, 85–87, 179, 180], and a number of useful relationships have been suggested connecting carbonyl frequency shifts with changes in such physical properties as bond lengths [89], half-wave potentials [90], electron densities [91], the electronegativities [88] and mesomeric effects of the substituents [87], and, in selected cases, with reactivities [90, 92] and with the C–O bond dipole [93].

There is, of course, a certain amount of overlapping between some of these frequency ranges and those arising from other types of carbonyl absorption, but in some cases this difficulty can be resolved by intensity studies. In addition to the major factors causing a frequency shift which have been mentioned above, small shifts also arise from changes of state and from hydrogen bonding effects. These will be discussed in the section on normal simple ketones, but the findings are, of course, equally applicable to other types of ketones.

Aldehydes have been less extensively studied, but there is a considerable body of evidence to show that the frequency shifts caused by environmental changes are closely parallel to those of ketones. The identification of aldehydes can usually be confirmed by a study of the C—H stretching frequency of the aldehydic group. By virtue of the strong influence of the carbonyl oxygen, this C—H frequency is virtually independent of the rest of the molecule, and is therefore highly characteristic.

The ranges of characteristic frequencies of various types of aldehydes and ketones are shown in Tables 9 and 9a.

Table 9a

*† ALDEHYDES

CO *Vibrations*

Saturated aliphatic aldehydes	$1740-1730$ cm^{-1}
$\alpha\beta$-unsaturated aldehydes	$1705-1680$ cm^{-1}
$\alpha\beta$-$\gamma\delta$-unsaturated aldehydes	$1680-1660$ cm^{-1}
‡ Aryl aldehydes	$1715-1695$ cm^{-1}
$\alpha\beta$-unsaturated β-hydroxyaldehydes	$1670-1645$ cm^{-1}

C—H *Stretching Vibration*

All types	$2900-2700$ cm^{-1} (w.). Usually two bands with one near 2720 cm^{-1}

C—H *Deformation Vibrations*

All types	$975-780$ cm^{-1} (m.)

Other Vibrations

Aliphatic aldehydes	$1440-1325$ cm^{-1} (s.)
Aromatic aldehydes	$1415-1350$ cm^{-1} (s.), $1320-1260$ cm^{-1} (s.), $1230-1160$ cm^{-1} (s.)

* The characteristic shifts indicated are usually additive, so that for example the CO-frequency of an α-bromo-$\alpha\beta$-unsaturated aliphatic ketone will be shifted from the normal by the resultant of the raising of the frequency by the bromine and its lowering by the conjugation.
† All the values quoted refer to dilute solutions. Considerable frequency shifts occur in the condensed phase, in solvents of differing polarity. All carbonyl absorptions are strong.
‡ Modified by the nature and position of ring substituents.

9.2. C=O Stretching vibrations in open-chain ketones

9.2(a). Position. A carbonyl group situated between two methylene groups represents the simplest case of an undisturbed C=O stretching vibration. Weniger [1] was the first to point out in 1910 that aldehydes and ketones always exhibit a strong band near 1680 cm^{-1} which could be associated with the carbonyl grouping, and Barnes *et al.* [2] in 1944 listed a series of ten ketones for which the carbonyl absorptions fell within the range 1725—1690 cm^{-1}, and noted that conjugation induced a low-frequency shift. Since then studies by many workers have shown that in solution the frequency of the carbonyl absorption of simple ketones of this type always lies within the narrow range 1725—1706 cm^{-1}, provided that no hydrogen bonding or other interference effects occur.

Thus, Thompson and Torkington [3] have found a series of eight such materials to absorb uniformly close to 1710 cm^{-1}, and further data in support of this proposed range have been given by Rasmussen, Tunnicliff and Brattain [4]; Bonino and Scrocco [5], Cromwell *et al.* [6], Josien *et al.* [90, 94], and Jones and his co-workers [7—11, 62, 84]. More recent studies on homologous series using carefully controlled conditions, show that normal alkyl ketones absorb within about 3 cm^{-1} of 1720 cm^{-1} in carbon tetrachloride solution. The overall range from acetone to *n*-dihexyl ketone is 1725 to 1717 cm^{-1}. In many cases the band is broad and asymmetric. Hirota *et al.* [95] have shown that this is the result of rotational isomerism with contributions from two forms. In one of these (H.H) in which the carbonyl oxygen is eclipsed by both hydrogen atoms, the frequency is at 1720 cm^{-1}, in the second form (C.H) in which one carbon atom is eclipsed, the frequency falls to 1714 to 1717 cm^{-1}. Corresponding values for the possible (C.C) conformation are less easy to establish as steric factors usually alter the bond angle and so change the frequency independently. This work has been extended by Renema [180] using graphical resolution techniques to separate the two peaks. However, apart from its intrinsic interest in explaining the asymmetry of the bands the effect does not materially affect the observed frequencies as the H.H conformation appears to predominate in most instances.

Branching in the alkyl chain has very little effect upon the frequency unless the side chains are large and so placed as to widen out the angle made by the two carbon-carbon bonds to the carbonyl group. If this occurs the frequency falls as the coupling of these bonds is reduced. Thus, whilst methyl *iso*propyl ketone absorbs at 1718 cm^{-1}, methyl *tert.*-butyl ketone absorbs at 1709 cm^{-1} and di-*tert.*-butyl ketone at 1697 cm^{-1}. Dubois [96] has quoted examples of even more sterically hindered ketones such as 5 : 5-diethyl-2 : 2 : 3 : 3-tetramethyl-4-heptanone in which the carbonyl frequency falls as low as 1674 cm^{-1}. A related effect is found in some large ring cyclic ketones.

The substitution of a polar group is usually effective in altering ν_{CO} only if it is substituted at the α-carbon atom, although an OH group capable of internal hydrogen bonding may, of course, be effective elsewhere. Polar atoms at the α-carbon atom usually give rise to two carbonyl bands, one of which is at a higher frequency than normal. These effects will be discussed later.

The above correlations relate only to spectra obtained in non-polar solvents, and departures from them are common when samples are examined in the solid or liquid state. Hartwell, Richards and Thompson [12] carried out a general study of the influence of physical state on a series of carbonyl compounds and found wide variations in the liquid state. Similarly, in the vapour phase the carbonyl frequencies were found to be appreciably different from those obtained in solution. Acetone, for example, absorbed at 1742 cm^{-1} in the vapour phase, whereas in solution the frequency lay between 1728 cm^{-1} and 1718 cm^{-1}, depending on the solvent. Similarly, didecyl ketone absorbed at 1740 cm^{-1} in the vapour state, and between 1724 cm^{-1} and 1717 cm^{-1} in solution. Dubois *et al.* [97] have recently given the vapour phase frequencies for a number of ketones, where the same effect is shown. It is probable that some form of dipolar association is occurring in the condensed phase [100, 101], resulting in a low-frequency shift of the order of 20 cm^{-1}. As far as possible, therefore, frequency measurements on ketones should be carried out in solution.

The minor variations shown by carbonyl frequencies in different solvents have been studied in some detail and the results have been reviewed at some length by Bellamy [179]. It seems probable that dielectric-constant effects play some considerable part in the frequency shifts which follow a change from the vapour phase to solution in a non-polar solvent. However, the shifts which result thereafter from a change of solvent are very largely due to association effects, and the extents of the shifts parallel the proton donating abilities of the solvents. In a typical ketone a shift of about 15 cm^{-1} results from a solvent change from hexane to chloroform.

Association with an alcohol shifts the band by about 20 cm^{-1} which is comparable with normal values for hydrogen bonding [4, 13].

9.2(b). The influence of olefinic conjugation.

Conjugation of a carbonyl group with a C=C linkage results in a lowering of the frequency by an amount depending on the nature of the double bond. This general finding was known earlier from Raman spectra [14], and has been fully confirmed in the infra-red.

An aliphatic C=C bond in conjugation with a carbonyl group reduces its frequency by about 40 cm^{-1} to about 1680 cm^{-1}. This is well shown by the comparison of compounds with $\alpha\beta$- and $\beta\gamma$-double bonds [15, 102, 90]. The nature of the conjugating group is unimportant, so that acetylenic bonds produce comparable shifts to ethylenic bonds [103−106, 180].

In most $\alpha\beta$-unsaturated ketones two conformations are possible and two carbonyl bands can be seen. These often differ very widely in intensity. The two bands correspond to *s cis* and *s trans* forms, although it is not uncommon for the *s cis* form to be replaced by a skew configuration. Barlet *et al.* [99] have described a method for the determination of the configuration of unsaturated ketones. Typical values for methyl vinyl ketone are 1716 and 1686 cm^{-1}, and for ethyl vinyl ketone [98] 1707 and 1690 cm^{-1}. In all cases it is assumed that the lower frequency band is derived from the *trans* form in which the delocalisation is expected to be the greater. In a few instances steric hindrance allows only one form, as in $CH_3 CH = C(CH_3)_2$ which has only one band at 1693 cm^{-1}.

Further conjugation by a $\gamma\delta$ double bond has only a small influence in shifting this absorption towards the bottom of the frequency range. This is well shown in the sterol series [9], and in examples quoted by Cromwell [6]. This is not unexpected, as Blout, Fields and Karplus [16] have shown that the frequency of an $\alpha\beta$-unsaturated aldehydic carbonyl group is not lowered by more than a few wave numbers even by long conjugated chains of up to seven double bonds.

Conjugation on both sides of the double bond of the carbonyl group does give an appreciable additional shift [98] and drives the frequency down to the 1660 cm^{-1} region. Isomeric forms are again found giving rise to multiple bands. Phenyl styryl ketone for example absorbs at 1669 and 1648 cm^{-1}.

These frequency shifts continue to occur to about the same extent in ketones which have already undergone a shift from some other cause. Thus 2-bromo-3-keto-steroids absorb near 1735 cm^{-1} due to the influence of the halogen, and Δ^1-2-bromo-3-keto-steroids [7] at 1697 cm^{-1}, a fall of 40 cm^{-1}, which is of the same order as in normal $\alpha\beta$-conjugation. Similar effects are also found with carbonyl absorptions the frequency of which has been raised by inclusion in a five-membered ring. It will therefore be appreciated that the correlation given for this type of structure is applicable only when no other factors are present which are also capable of affecting the frequency. However, in general, the individual frequency shifts produced by each factor are additive, so that it is still possible to make a reasonable assessment of the likely position of a carbonyl absorption.

This correlation is not applicable to $\alpha\beta$-unsaturated ketones, aryl or alkyl, which have also a hydroxyl group or an amino-group at the β-carbon atom. In these compounds chelation occurs which has a very big effect on the carbonyl frequencies. These, therefore, represent a special case, and they will be considered separately in the section following that on β-diketones, to which they are closely related.

9.2(c). The influence of aryl conjugation. It was thought at one time [17] that aryl ketones absorbed at somewhat higher frequencies than enones, but the bulk of recent data suggests that there is in fact little difference. Acetophenone absorbs at 1692 cm^{-1} but this is an exception to the general run of aryl ketones, most of which absorb between 1680 and 1690 cm^{-1}. The bands are usually symmetric but asymmetry is found in a few cases such as propiophenone and this is attributed to (C.H) and (H.H) conformations of the acyl group. In benzophenones the carbonyl band shifts into the 1660 cm^{-1} region. Absorption at the same point is also shown by aryl/alkenyl ketones.

It has been noted above that substitution of OH or NH_2 groups on the ring *ortho* to the carbonyl group present a special case of chelation which will be dealt with later. However, other substituents on the ring are also able to affect the frequency to some extent. The substitution of alkyl groups in the *ortho* positions, for example, results in an upward frequency shift due to the reduction in the degree of coplanarity which the carbonyl group can achieve with the ring [107]. Acetomesitylene absorbs at 1701 cm^{-1} in the liquid state, which is 9 cm^{-1} higher than the corresponding value for acetophenone. Similar results have been found in some benzocyclanones. In addition, the carbonyl frequencies of substituted acetophenones vary over a considerable range, depending upon the electron attracting or repelling properties of the substituent groups. Soloway and Freiss [18] showed that there was a considerable variation in the solid state, and more detailed solution studies on acetophenones and benzophenones [90, 169] have shown that the observed carbonyl frequencies can be correlated directly with the Hammett σ values of the substituents. These σ values are derived from kinetic studies, and are a direct measure of the electron donation or withdrawal produced by the substituent group. In the case of benzophenones these frequency shifts have also been related to the polarographic half-wave potentials [90], whilst in aceto-phenones they are also a direct function of the calculated carbonyl bond order [108].

As in the earlier cases, the influence of an α aryl group will be additive with that of any other structure which is capable of influencing the C=O frequency. Thus Gutsche [19] has given data on a number of tricyclic ketones containing one aromatic ring. With a six-membered ring C=O with an α-aryl group, the frequency was found to be 1695–1686 cm^{-1}, which is the same as with similar open-chain materials. With a five-membered ring C=O, however, the frequency rose to 1715–1706 cm^{-1}, the strain of the five-membered ring being offset to some extent by the aromatic conjugation.

9.2(d). The influence of *cyclo*propane conjugation. It is well known that the high electron density at the centre of the *cyclo*propane ring results in its behaving in a manner similar to a olefinic double bond. Not very many data are available on the effects of *cyclo*propane rings when attached to carbonyl groups, but a small number of alkyl *cyclo*propyl ketones examined by Wiberley and Bunce [109] and by Cromwell [170], showed lowered carbonyl frequencies in the liquid state (1704–1686 cm^{-1}). Other examples of this effect have been described by Fuson *et al.* [90, 110, 111], and these include some interesting examples of *i*-sterols, such as *i*-cholestane-6-one. More recently, the conformation of *cyclo*propyl ketones has been studied [127, 145, 181]. *Cyclo*propyl methyl ketone has been shown to exist predominately in the *s cis* form (80%), and it is interesting to find that in this case it is the *cis* form which allows the greater delocalisation. The *cis* form absorbs at 1698 cm^{-1} and the *trans* form at 1708 cm^{-1}. Di*cyclo*propyl ketone absorbs at 1698 cm^{-1} and *cyclo*propyl phenyl ketone at 1675 cm^{-1}.

The influence of other types of strained ring systems is less certain. 'Conjugation' with an epoxy ring in a small number of cases produces no change in the carbonyl frequency, but Cromwell and Hudson [112] have observed a lowering in certain ethyleneimine compounds with an α-carbonyl group. They have discussed the steric requirements for conjugation effects to occur in such systems, but it has been pointed out [90] that the comparison of frequencies obtained in the solid state can be misleading, and further confirmation of their findings is desirable.

9.2(e). The influence of α-halogen groups. It has been known for many years that halogen substitution in the immediate vicinity of a carbonyl group results in a high-frequency shift of the carbonyl absorption. This is particularly marked in the acid chlorides, where a chlorine is directly attached to the carbonyl group, but there is still an appreciable effect when the halogen is situated on the α-carbon atom.

This well-known effect has been extensively investigated in the Raman field [20]. It is applicable to all types of carbonyl absorption, and Hampton [21] and Hartwell, Richards and Thompson [12] have observed a similar effect for esters. Studies were first concentrated on the sterol series by Woodward [22], Djerassi and Scholz [23], Dickson and Page [113], and especially by Jones *et al.* [9, 10, 63, 83, 119]. In these systems the individual rings are rigidly locked in chair or boat forms, so they are particularly well suited to the study of the stereochemical effects of different halogen configurations.

In general, they found that substitution of an α-bromo-group results in a rise in the C=O frequency of about 20 cm^{-1}. Thus non-conjugated 3-keto-steroids absorb at 1719–1715 cm^{-1}, whereas 2- or 4-bromo-3-keto-steroids absorb near 1735 cm^{-1}. Similarly, Δ^1-3-keto-steroids absorb at 1684–1680 cm^{-1} due to the conjugation, and this is raised to 1697 cm^{-1} in Δ^1-2- or -4-bromo-3-keto-steroids.

This displacement of the CO band is increased by halogen substitution on both sides of the carbonyl group, so that 2 : 4-dibromo-3-ketones absorb near 1760 cm^{-1}, the second frequency increment being the same as the first. However, the substitution of two bromines on the same α-carbon atom has very little more effect than one, and 2-bromocholestanone-3, and 2 : 2-dibromocholestanone-3 both absorb at 1739–1735 cm^{-1}. This is due to the fact that this effect is exerted only if the C–Br link is coplanar with the ketonic group so that the proportion of the polar structure II is increased [63].

Thus bromine substituted at an α-equatorial position in a ring with a chair configuration will raise the frequency of an adjacent CO group by 20 cm^{-1}. On the other hand, substitution in a polar position does not result in any alteration in the frequency [63]. Similarly a C_{20} carbonyl group has its frequency increased by 20 cm^{-1} by bromine substitution at C_{21}, but the frequency remains unchanged when bromine is substituted at $C_{17}\alpha$.

Very extensive studies have been carried out by Corey and his co-workers [115–118], by Sandris and Ourisson [119] and by Inayama [120] on the simpler systems of α-halogenated *cyclo*hexanones and pentanones. Their results are closely parallel to those found in the sterol series and provide a measure of additional support for them, but the relative ease of interconversion of the chair and boat forms in such cases must make their findings less decisive. In *cyclo*pentanones the principle that carbonyl halogen interaction occurs in equatorial and not in polar configurations has also been used in an attempt to assess the degree of puckering of the *cyclo*pentanone ring [122]. Parallel effects have also been noted in the *cyclo*heptanone series [25, 70]. Bellamy [179] gives a review of recent work in this series.

The steric specificity of this effect led Jones and Sandorfy [83] and Bellamy and Williams [93, 123] to suggest independently that the mechanism was primarily one of field effects operating across space rather than inductive effects operating along the bonds. In such a case the possibility of rotation about C–C bonds of α-halogen alkyl ketones should lead to the existence of rotational isomers showing two carbonyl frequencies. In the cases of chloroacetone [124] and 1 : 3-dichloroacetone [125] rotational isomerism had already been established, and further work on these, and related materials, and on ω-chloroacetophenones has shown clearly that two carbonyl frequencies are found in mono- and di-halogenated alkyl ketones in the liquid or solution state and that the higher of these is associated with the more polar configuration (I), in which the halogen atom is near in space to the carbonyl oxygen. The lower frequency, which is usually very little different from that of the original ketone, corresponds to form (II) below. As is usual with rotational isomers, the relative intensities of these bands change very considerably with the polarity of the solvent, so that in highly polar solvents or in the liquid itself the lower-frequency band is much reduced in intensity.

I.   II.

On the other hand, only one state exists in the solid, and so only one frequency is observed. In some instances also only one form is stable in the vapour. The steric dependence of this effect affords a ready explanation of the observation that the substitution of a second halogen atom at the same α-carbon does not significantly raise the carbonyl frequency, as only one of these can be close to the oxygen atom. As is to be expected for what is largely an electrostatic interaction, the effect is greatest with fluorine substitution and diminishes progressively throughout the halogen series. ω-chloro acetophenone absorbs at 1713 and 1693 cm^{-1}, the bromo derivative at 1709 and 1688 cm^{-1} and the iodo compound at 1701 and 1683 cm^{-1}. 1 : 1 : 1-Trifluoroacetone [176] absorbs at 1780 cm^{-1}, 40 cm^{-1} higher than acetone in the vapour phase, and Haszeldine reports similar values for α-fluoroketones [68]. Parallel, but smaller effects occur in *ortho* halogenated acetophenones in which the carbonyl oxygen is again close to the halogen atom. Stereospecific effects are found also in α-halogenated esters, acid halides and amides [93].

The bulk of the data on field effects relates to halogen substitution, but more limited studies have shown that very similar effects occur with other polar atoms at the α-carbon atom. Oxygen atoms at this point behave almost exactly as do chlorine atoms. With α nitrogen substitution the upward shift of the higher frequency band is somewhat smaller, but two bands can be distinguished [206].

9.3 C=O Stretching vibration in diketones

9.3(a). α-Diketones. Raman studies on α-diketones indicate that there is very little interaction between adjacent carbonyl groups and that any increase in the frequencies produced in this way is not greater than $5-15 \text{ cm}^{-1}$. Kohlrausch and Pongratz [27] found a high-frequency Raman shift of only 5 cm^{-1} for benzil above acetophenone, whilst Thompson [28] quotes glyoxal as absorbing at 1730 cm^{-1}, as against formaldehyde at 1745 cm^{-1}, and this is a shift in the opposite direction. The single infra-red bands given by benzil and by diacetyl in the carbonyl region are explained by Rasmussen *et al.* [4] by the suggestion that these compounds exist in the *trans* form, so that only one band is active in the infra-red region in each case. The Raman frequencies are only very slightly different, and this affords further evidence that any interaction is slight.

In the steroid field Jones *et al.* [9] have examined a number of 11 : 12-diketones in solution, and find the carbonyl frequencies at 1726 cm^{-1}. This represents only a small frequency rise above the range $1716-1706 \text{ cm}^{-1}$ in which the individual 11- and 12-ketones absorb. Barnes and Pinkney [69], and Leonard *et al.* [70, 127] have also examined numbers of α-diketones in which only very slight interaction effects are observed.

In general, therefore, only a small frequency rise is to be expected for α-diketones, but an exception has been noted for the α-keto ester methyl pyruvate [29], in which both the ester and ketonic carbonyl absorb at 1748 cm^{-1}, and for pyruvic acid, in which they both absorb at 1745 cm^{-1}. Exceptions may also be expected in some cyclic systems in which it is not possible for the carbonyl groups to take up a *trans* configuration. Interactions, which may be dipolar and similar to those of α-halogenated ketones, can then occur. Alder *et al.* [128] have observed elevated carbonyl frequencies in some such cases. In most cases however the effect is small. Thus 1 : 2-diketo-*cyclo*decane has two carbonyl bands due to coupling, but they are separated by only 10 cm^{-1}.

9.3(b). β-Diketones. Rasmussen *et al.* [4] have described a chelation effect in certain β-diketones which gives rise to a very considerable shift of the carbonyl frequency. Both acetyl acetone and dibenzoyl-methane are known from chemical and Raman evidence to exist largely in the mono-enol form, but neither show any CO absorption band corresponding to a normal conjugated ketone. Instead, a very broad band, estimated to be more than a hundred times as strong as the normal carbonyl vibration, is observed in the range $1639-1538 \text{ cm}^{-1}$. These workers believe this absorption to arise from a carbonyl group which has had its double-bond character reduced by resonance between the forms

$$
\begin{array}{c}
\text{OH}\cdots\cdots\text{O} \\
| \qquad\ \| \\
\text{R}-\text{C}=\text{CR}'-\text{C}-\text{R}''
\end{array}
\quad \text{and} \quad
\begin{array}{c}
\text{OH}\cdots\cdots\text{O}^{-} \\
\| \qquad\ | \\
\text{R}-\text{C}-\text{CR}'=\text{C}-\text{R}''
\end{array}
$$

This resonance effect is described as 'conjugate chelation', to differentiate it from a normal hydrogen bonding effect such as that given by diacetone alcohol, and it can arise from any structure

$$
\begin{array}{c}
\text{X}-\text{H} \quad \text{Y} \\
| \qquad\quad \| \\
\text{C}=\text{C}-\text{C}
\end{array}
$$
, in which X and Y can act as electron donors or acceptors,

and a number of examples have been studied. In all these cases very large changes in the OH region of the spectrum lend support to the view that the bonds produced are very much stronger than normal hydrogen bridges. Similar chelation effects in esters have also been described by Rasmussen and Brattain [30]. The extent of the carbonyl frequency shift has been shown to be a direct function of the double-bond character of the $\alpha\beta$ link [64, 129]. In enolic diketones this is normally a full olefinic bond, and frequencies close to 1600 cm^{-1} are found, but in cases in which the $\alpha\beta$ unsaturation is provided by an aromatic system the shifts are somewhat reduced. Other instances of enolic β-diketones showing chelation have been discussed by Shigorin [130], and Park, Brown and Lacher [131], who were able to show that disubstitution of the central methylene group prevented enolisation and gave rise to normal carbonyl frequencies. This phenomenon has also been observed in chalkones [67, 71], in substituted aromatics containing the group −CO−CH$_2$−CO− [69] and even in tetronic acids [143] and in β-triketones in which a single hydrogen atom is retained at the central carbon atom [72].

Resonance systems of this type can also be produced by dimerisation. 5 : 5-Dimethyl*cyclo*hexane-1 : 3-dione, for example [4], absorbs at 1700 cm^{-1} and at 1605 cm^{-1}. The first band is attributed to the normal carbonyl and the second to a conjugated chelated type produced by dimerisation. In this case, however, the intermolecular bonds are broken on dilution in non-polar solvents, and reversion to unchelated frequencies is observed [129, 132]. Corresponding changes in the OH stretching region are also found.

The low carbonyl frequency of the diketo form of this compound appears unusual, but it is paralleled by the data on other β-diketones which have a large group substituted at the central carbon atom. The keto form is stabilised under these conditions. Thus the presence of a secondary or tertiary butyl group in this position in acetyl acetone results in a carbonyl frequency of 1702 cm^{-1}. If this is due to some kind of field effect it is one which operates in the opposite direction to that normally found.

A special situation occurs in β-diketones linked through an aromatic ring or a double bond. If the geometry is such that they become coplanar, as for example in cyclic malonates the coupling is increased and the bands are separated by 30 to 40 cm^{-1}.

9.3(c). $\alpha\beta$-Unsaturated β-hydroxy ketones.

With aromatic compounds it is possible for one of the ring double bonds to function as the $\alpha\beta$-unsaturated unit, so that 1-keto-2-hydroxy-compounds also show abnormal carbonyl frequencies. Gordy [31] has reported the carbonyl absorption of ortho-hydroxyacetophenone at $1639-1613$ cm^{-1}, and Hunsberger [17] has examined a number of naphthalene derivatives in which he has found frequency shifts of $50-60$ cm^{-1} due to this effect. α- and β-Acetonaphthone, for example, absorb at 1685 cm^{-1}, corresponding to a normal $\alpha\beta$-$\alpha'\beta'$-unsaturated aromatic ketone. However, both 1-hydroxy-2-acetonaphthone and 2-hydroxy-1-acetonaphthone absorb at 1625 cm^{-1} due to chelation. In this connection it is particularly interesting to note that 3-hydroxy-2-acetonaphthone absorbs at 1657 cm^{-1}. This represents a considerably smaller shift, and it is attributed to the weaker double-bond character of the 2 : 3-double bond due to bond fixation. Cases of aldehydes showing parallel behaviour are also given. In general, the shift produced in this way is not as great as with true β-diketones, but it is nevertheless very well marked. Similar results have been obtained from phenanthrene derivatives which contain an hydroxyl group at C_{10} and a COOCH$_3$ group at C_9 [64]. In this case the shift is of the order 75 cm^{-1} compared with the methylated phenanthryl ester or ketone, which indicates that the 9 : 10-double bond of phenanthrene has rather more olefinic character than the 1 : 2 double bond of naphthalene. The fact that the lowering is not simply due to the increased aryl conjugation is clearly shown by the fact that the CO frequencies of CHO, COCH$_3$ and COOCH$_3$ groups substituted at C_9 in phenanthrene are only 2 cm^{-1} lower than the corresponding naphthalene derivatives, when no OH groups are present. The study of the frequency shifts of carbonyl groups with an *ortho*-hydroxy substituent has been used by Hunsberger *et al.* [133] to assess the relative double-bond characters of the various cyclic bonds of indane.

One further way in which this phenomenon can arise, is when the carbonyl group is part of a ring system but maintains the $\alpha\beta$-unsaturation and the OH substitution on the β carbon atom. Such cases are 1-hydroxyanthraquinones, anthrones and oxanthrones, which have been extensively studied by Flett [32] and Hadži and Sheppard [134], and the hydroxy quinones studied by Josien *et al.* [73].

Both anthraquinone and 2-hydroxyanthraquinone absorb between 1673 cm^{-1} and 1676 cm^{-1}, whereas 1-hydroxyanthraquinone shows two CO bands at 1680–1675 and 1630–1622, corresponding to free and bonded carbonyl groups. With two hydroxyl groups in each of the β positions only one band is shown at 1639–1623 cm^{-1}, whilst in the extreme case of 1 : 4 : 5 : 8-tetrahydroxyanthraquinone the carbonyl frequency has fallen to 1595 cm^{-1}.

Similarly, 1-hydroxyanthrone shows a fall of 20 cm^{-1} in the carbonyl frequency and 4-hydroxyanthrone a fall of 10 cm^{-1}. The reason for this second case is not clear, but, as will be seen below, numerous other cases are known in which 4-amino-substitution has a similar effect.

Although conjugate chelation would appear to be a generally satisfactory explanation of these effects, there are still a number of factors which are not fully understood and which require further study. Apart from the anomaly of the influence of 4-substitution in the anthraquinones, Woodward *et al.* [33] have found marked carbonyl shifts in β diketo-enol ethers in which substitution precludes either normal hydrogen bonding or chelation. They quote work on dimedone methyl ether, on 3- and 4-methoxytoluoquinones and on adducts of these with certain diones, as showing that strong absorption at 1660 cm^{-1} and 1625 cm^{-1} is characteristic of such systems. Some interesting anomalies also occur in flavones and flavanones [71, 115]. 5-Hydroxy-flavanone is essentially normal in showing a reduction of the carbonyl frequency of 40 cm^{-1} due to the formation of a six-membered chelate ring (I). In 5-hydroxyflavone, however, no frequency shift occurs despite the similarity of structure (II). On the other hand, substitution of a hydroxy group in the 3 position of flavones does result in a frequency fall of some 30 cm^{-1} (III), even though the five-ring chelate produced might be expected to be less stable than a six-membered system. In 3 : 5-dihydroxyflavones there is again no frequency shift, which is even more remarkable, but no satisfactory explanation for this is yet available.

I.   II.   III.

9.3(d). $\alpha\beta$-Unsaturated β-amino ketones. Compounds of this type would be expected to chelate in the same way as the hydroxy-materials dealt with above. However, it is not clear how far this analogy can be taken with safety, as they can also be regarded as being vinylogs of amides, which in some ways they resemble, so that the nitrogen atom may be exerting an influence on the carbonyl frequencies independently of any chelation effects. This is certainly the case, for example, in the γ-quinolones.

In confirmation of this expectation, Cromwell *et al.* [6, 35] have found carbonyl shifts of the order of 20–80 cm^{-1} in β-amino-unsaturated ketones, although α-amino $\alpha\beta$-unsaturated ketones are normal. They find, however, that there is still a considerable frequency shift in fully substituted amines, although it is not so great as that occurring with materials which can form hydrogen bonds. They conclude that in these substituted amines a resonance is occurring

between the structures
$$\begin{array}{c} -N- \\ | \ | \\ R-CO-C=C- \end{array} \rightleftharpoons \begin{array}{c} O^- \ \ -N^+- \\ | \ \ \ \ \ \| \\ R-C=C-C- \end{array}$$

Flett [32] has found similar effects with amino-anthraquinones and he has shown that in these structures hydrogen bonding can play only a small part, even with unsubstituted amines. Thus the NH stretching frequency of 1-amino-anthraquinone is not very different from that of 2-naphthylamine, and the NH frequency of 1-methyl-amino-anthraquinone is only 100 cm^{-1} lower than is usual for secondary amines. This contrasts sharply with the behaviour of the corresponding hydroxy-compounds, in which no OH stretching frequencies are shown in the normal frequency range. Furthermore, Flett finds that 2-amino-anthraquinones show nearly as much movement of the carbonyl absorption as the 1-amino-materials, and he concludes that resonant structures of the types I and II may be involved.

O⁻ NH₂⁺ ... O ... NH₂⁺ ... O ... O₋

I.   II.

Similar changes occur with 4-amino-substituted materials which could not chelate in the same way as the 1-amino-compounds.

It is evident, therefore, that whilst this system shows many analogies to that of $\alpha\beta$-unsaturated ketones with β-hydroxyl groups, it also has characteristic features of its own, so that considerable caution is required in the interpretation of ketonic carbonyl frequencies of $\alpha\beta$-unsaturated materials with β- or even γ-amino-substitution.

9.3(e). Metallic chelates of β-diketones. β-Diketones chelate readily with a number of metallic atoms in the production of compounds such as the metallic acetylacetonates. These have been studied by Lecomte [36, 66] and by Morgan [37]. Lecomte finds a strong band between 1562 cm^{-1} and 1550 cm^{-1} in a series of eleven acetyl acetonates which he attributes to the carbonyl group weakened by resonance between the C—O—M and C=O . . . M links. This is a shift parallel to that given by ionised fatty acids in which a rather similar state of affairs exists [38, 39]. In all these substances there is a second band near 1515 cm^{-1}, which is attributed to the C=C link. These findings have been confirmed by Bellamy and Branch [136], who have examined a series of copper chelates with different β-diketones and also chelates with the same ligands and different metals. In general, the observed frequencies fall in the ranges $1608-1524 \text{ cm}^{-1}$ and $1390-1309 \text{ cm}^{-1}$, but salicaldehyde shows considerably smaller shifts due to the reduced double-bond character, and in this series the carbonyl frequency is a direct function of the stability of the chelate. Bender and Figueras [137] have also noted a frequency shift attributed to the enolate ion in acetylacetone to which sodium ethoxide is added. The carbonyl frequency then occurs at 1604 cm^{-1}. A very similar value is given by zinc glycinate [138]. When a ketonic carbonyl is complexed with aluminium chloride the frequency usually falls by 70 to 150 cm^{-1}. This is due to the complex XCOXAlCl$_3$ [139–141, 182]. Frequency rises such as were earlier suggested by Susz [177] do not occur with ketones, but have been found in compounds such as acid chlorides when the formation of the CH$_3$CO$^+$ ion becomes possible.

9.3(f). γ- and δ-Diketones and related materials. The carbonyl groups of γ- and δ-diketones are too far removed from each other for any direct interaction. However, it should be remembered that lactonisation can occur when one of the carbonyl groups is part of an acid or ester group. The ketonic carbonyl frequency is then lost, and the acid or ester carbonyl frequency is replaced by that of the lactone. In many instances these two forms co-exist, so that all three frequencies can be found. In other cases either form can be produced at will by slight variations in the external conditions. Munday, for example, has suggested that the keto-form of penicillic acid can be wholly converted into the lactone form by grinding in nujol [40]. A number of other cases of lactonisation have been examined by Grove and Willis [13].

One other case of γ-diketone interaction has been found. This concerns the structure CO–C–O–CO–, in which there appears to be some true interaction between the carbonyl groups, as distinct from the chemical re-arrangements of lactonisation. Jones *et al.* [9, 62] have listed a considerable number of 21-acetoxy-20-keto-steroids in which the C=O frequencies of both the keto and ester carbonyl absorptions are raised by about 20 cm^{-1} and 10 cm^{-1}, respectively, due to an interaction effect. The origin of this effect is not understood, but it has been observed for both polar and equatorial configurations of the acetoxy group of 12-acetoxy-11-keto steroids [113]. On the other hand, it does not seem to occur in 17-acetoxy-16-keto compounds in which the keto group is part of a five-membered ring system. This has led Bellamy and Williams [142] to suggest that a dipolar mechanism is involved and that the frequency shift is due to the near approach of the two carbonyl oxygen atoms.

9.4. C=O Stretching vibrations in aliphatic cyclic systems

9.4(a). Seven-membered rings. Carbonyl groups included in aliphatic seven-membered rings show frequencies which are rather lower than those of open chain ketones. Scott and Tarbell [25] gave a value of 1699 cm^{-1} for *cyclo*heptanone, with 2-chloro*cyclo*heptanone absorbing at 1715 cm^{-1}, due to the influence of the halogen atom. Tetrahydrocolchicine, which is believed to be a seven-ring ketone with α-electron-attracting substituents, also absorbs at 1710 cm^{-1}. More recent studies have established the *cyclo*heptanone frequency at 1704 cm^{-1}. This is almost certainly the result of bond angle widening to accommodate steric strain [180, 183].

The behaviour of this carbonyl absorption under the influence of conjugation etc. is closely parallel to that of open-chain materials. Gutsche [43] has shown that 2- and 3-phenyl*cyclo*heptanones absorb at 1690 cm^{-1} in the liquid state, whilst Scott and Tarbell [25] quote 1683 cm^{-1} for 2 : 3-benzo*cyclo*heptanone and 1661 cm^{-1} for the doubly conjugated 2 : 6 : 6-trimethyl*cyclo*heptadiene-2 : 4-one.

There is also considerable Raman data on *cyclo*heptanones [44, 45, 189].

9.4(b). Six-membered rings. It is to be expected that the carbonyl frequencies of unstrained six-membered rings will be the same as for open chains, and this is so. Whiffen and Thompson [46] have found a value of 1710 cm^{-1} for the carbonyl absorption of *cyclo*hexanone, and the general frequency range has been confirmed by the Shell group [41]. Castinel *et al.* [183] quote a value of 1717 cm^{-1} and this agrees with other recent work.

In the steroid field Jones *et al.* [7—11] have examined a very large number of steroids containing carbonyl groups in six-membered and five-membered rings, and they have found all the former to behave exactly like open-chain materials. There are very small but real differences between the carbonyl frequencies of C=O groups situated in the different six-membered sterol rings, but in no case does the frequency fall outside the overall range 1720—1706 cm^{-1} unless conjugation or other interference effects occur.

All the factors, such as α-halogenation, αβ-conjugation, etc., which influence the frequency of normal carbonyl groups influence the frequencies of six-membered ring carbonyl groups to precisely the same extent. For example, twenty-nine αβ-unsaturated six-membered ring ketones were all found by Jones *et al.* [9] to absorb in the range 1684—1674 cm^{-1}, whilst five materials with αβ-α'β'-conjugation absorbed in the range 1663—1660 cm^{-1}. References have already been given to the considerable volume of work on α-halogenated *cyclo*hexanones [115—121].

The introduction of a hetero atom into the ring can, however, give a marked effect. The normal cases of lactones and lactams are considered later in the appropriate chapters, but pyrones and thiapyrones also give abnormal carbonyl frequencies, and their reactivities are widely different from those of normal ketones. Tarbell and Hoffmann [144] have found the carbonyl frequency of 1 : 4-thiapyrone as a very strong band near 1609 cm^{-1}, with a second strong band at 1574 cm^{-1}. That this is due to the greatly increased resonance arising from the lone pair of electrons of the sulphur atom is neatly demonstrated by conversion to the sulphone. This shows a normal carbonyl frequency for an $\alpha\beta$-$\alpha'\beta'$-unsaturated compound (1657 cm^{-1}).

Fusion of additional rings as in bridged cyclic structures usually leads to an increase in ring strain, so that higher carbonyl frequencies than usual result [171, 172].

9.4(c). Five-membered rings. Hartwell, Richards and Thompson [12] found that *cyclo*pentanone absorbed at 1772 cm^{-1} in the vapour state, and Whiffen and Thompson [46] found a value of 1740 cm^{-1} in solution. In more recent work it has been established that the band occurs at 1750 cm^{-1} with a weaker shoulder at 1724 cm^{-1}. The splitting of this band has given rise to many studies which have concluded that it arises from Fermi-resonance interactions [185–187]. Since then Jones's work on sterols [9] has fully confirmed this indication that the carbonyl frequencies are raised in five-membered ring systems, as he has, for example, listed twenty-two materials containing this type of carbonyl group, all of which absorb in the range 1749–1745 cm^{-1} in solution. On this basis he has been able to differentiate carbonyl groups in five- and six-membered rings. Similar frequency increases are shown by five-membered ring lactones, above the values for the corresponding esters.

This strengthening of the C=O bond with increased ring strain appears to be part of a general phenomenon which has been fully discussed by Bellamy [179]. It is a direct consequence of the angle change at the carbonyl bond.

Five-membered ring ketones show frequency shifts on conjugation which are of the same order as those of unstrained rings. Thus $\alpha\beta$-unsaturated CO groups in five-membered rings absorb [9, 90] at 1716 cm^{-1}, a fall of 38 cm^{-1}, which is the same as for normal $\alpha\beta$-unsaturated ketones. Further conjugation does not, however, have an appreciable effect and tetracyclone [146] absorbs at 1713 cm^{-1}.

There are, however, some differences between endo and exo conjugation [90], and the compounds of the latter type absorb at rather lower frequencies (1690 cm^{-1}). Jones [9] has shown that $\beta\gamma$-unsaturation has no effect on the carbonyl frequency, and two Δ^{14}-17-keto-steroids absorb at 1754 cm^{-1}. α-Halogen substitution raises the carbonyl frequency to a smaller extent than in *cyclo*hexanone, and the degree of shift depends in part on the extent of puckering of the *cyclo*pentanone ring [119, 122].

9.4(d). Four-membered and three-membered rings. Four-membered ring ketones are uncommon. *Cyclo*butanone itself absorbs at 1789 cm^{-1}. This further rise above the value for *cyclo*pentanone is to be expected and is consistent with the further closure of the bond angle. *Cyclo*butane-1 : 3-dione has been studied by Miller *et al.* [188]. This shows two very strongly coupled frequencies at 1754 (antisymmetric) and 1809 cm^{-1} (symmetric), although only the first of these is allowed in the infra-red due to symmetry.

*Cyclo*propanone [189] absorbs at 1813 cm^{-1} and its 2 : 3-dichloro derivative [190] at 1822 cm^{-1}. Much interest has been aroused by *cyclo*propenone which has been studied by a number of workers. Two bands are shown by all derivatives. The 2 : 3-dichloro derivative [191] absorbs at 1886 and 1615 cm^{-1}, and the diphenyl derivative [192, 193] at 1850 and 1640 cm^{-1}. Both the C=C and the CO frequencies would be expected to be high in these compounds, so that they are difficult to identify in terms of individual group frequencies. Attempts have been made to do so using solvent shift techniques, and these suggest that the lower frequency band has more carbonyl character.

However there is little doubt from isotopic-substitution [194, 195] studies that the bands are strongly coupled and are more correctly described as symmetric and antisymmetric modes.

9.4(e). Large ring systems. The dependence of the carbonyl frequency upon the ring size has been studied by a number of workers, and recently these have been supplemented by intensity studies. Freiss and Frankenberg [147] showed that small changes took place in six-, seven- and eight-membered ring systems and that the frequencies did not parallel the carbonyl reactivities. Prelog [148] and Günthard [149, 173] have confirmed this over the series $(CH_2)_n C=O$, $n = 3$ to $n = 14$, and have discussed the variations in frequency, dipole moment and intensity with ring sizes. Frequency data are also given by Castinel [183] and by Renema [180]. It has already been noted that seven-membered ring ketones absorb at lower frequencies than their open chain counterparts. *Cyclo*heptanone absorbs at 1705 cm^{-1} and this low value is maintained in *cyclo*octanone, *cyclo*nonanone and *cyclo*decanone (1703 to 1704 cm^{-1}). Thereafter ν_{CO} rises towards the normal value, with *cyclo*dodecanone absorbing at 1712 cm^{-1}. All of these changes are attributable to small bond angle changes.

Although the frequencies vary in this way the absorption coefficient falls steadily up to about $n = 9$ before reaching a stable value, whilst the half-band width shows an alternative rise and fall as the ring size increases.

Parallel results, in so far as the frequency is concerned, have been obtained in very large ring systems derived from benzoyl *cyclo*-anones [107, 150].

9.4(f). *Trans*-annular effects in large ring systems. Abnormal carbonyl frequencies have been found in certain large ring systems which contain a nitrogen atom even though this is well separated from the carbonyl group by (CH_2) groups. This effect is operative only in rings of a certain size and configuration. It was first observed in some alkaloid derivatives which gave rather lower carbonyl frequencies than were expected [151] and has been fully confirmed by extensive studies by Leonard *et al.* [152–158]. In a system such as (I) below it is found that the C=O frequency (R=CH_3 or C_2H_5) is near 1665 cm^{-1} for $n = 3$ or $n = 2$ and 3, but near 1700 cm^{-1} for values of $n = 2$ or 4 or for ring sizes of 11, 13, 15, 17, 19 or 23. The origin of this effect becomes clear when the formula of the nine-membered ring system is written as in (II), and it is now reasonably certain that the close proximity of the nitrogen atom

with a lone pair of electrons is resulting in a direct effect upon the carbonyl group across intramolecular space. The proximity of the nitrogen and carbonyl groups in this system is supported by the production of a bridged-ring derivative with perchloric acid, and the fact that the carbonyl frequency rises with increased size of the R group which exerts a steric effect. These findings are of particular interest from the point of view of the origins of frequency shifts. Further examples of this effect have also been noted by Marion *et al.* [158, 159].

9.5. C=O Stretching vibrations in aromatic cyclic systems

9.5(a). Quinones. The infra-red spectra of quinones present a number of interesting features. *p*-Benzoquinone shows two carbonyl frequencies in solution (1669 cm^{-1} and 1656 cm^{-1}), although the lower-frequency band is appreciably weaker [91]. Many other quinones in which the quinonoid ring is fused to only one other ring behave similarly, but only one band is shown in compounds such as 9 : 10-anthraquinone, in which the quinonoid ring is central. Similarly, some substituted *p*-benzoquinones show two frequencies and others do not [160, 161]. This behaviour has led to much discussion, which is not yet finally resolved, as to whether the second band represents a C=C mode, is due to interaction or to a carbonyl absorption appearing independently for each of the two C=O groups. However, the main peak in most cases occurs in the expected position for an $\alpha\beta$, $\alpha'\beta'$-unsaturated ketone, although its position is modified by the nature and number of attached rings and by the electrical character of any substituents [174]. The carbonyl frequencies in polycyclic quinones can be related to the calculated electron densities at the carbonyl carbon atoms [91].

Flett [32] has examined a series of anthraquinones with various substituents on the ring. Anthraquinone itself absorbs at 1681 cm^{-1}, which is a little above the normal for an $\alpha\beta$-$\alpha'\beta'$-diaryl ketone. Substitution by various groups (apart from hydroxyl and amino-groups which could form chelates) resulted in an overall variation of 1692 cm^{-1} (1 : 8-dichloroanthraquinone) to 1675 cm^{-1} (1-methoxyanthraquinone). Similar small changes were found with anthrones and oxanthrones. Substitution other than of conjugated chains or rings therefore does not appear to affect the carbonyl frequency of multiple-ring quinones appreciably so long as no chelating groups are present, although the nature and number of such rings have an influence on the carbonyl frequency. The extent to which this occurs is not fully predictable. Thus, in solution, benzoquinone absorbs at 1664 cm^{-1}, anthraquinone at 1681 cm^{-1}, anthrone at 1653 cm^{-1} and oxanthrone at 1676 cm^{-1}. Josien and

Fuson [48] find the carbonyl absorption of pyrenequinone at 1639 cm^{-1}, and they suggest that this large shift may be related to the size of the highly conjugated molecule. Chrysenequinone is more normal in absorbing at 1658 cm^{-1}, and in general there is an increase in the carbonyl frequency as the number of rings is increased, provided that the additional rings do not make an angle with that containing the carbonyl group [73]. Polycyclic quinones have also been examined by Hadži and Sheppard [49]. In general, they find that the carbonyl frequencies are lower when two quinonoid carbonyls are in different rings than they are when they are in the same ring. Compounds of the first type [49] absorb in the range $1655-1635 \text{ cm}^{-1}$, and the second type [49, 65] in the range $1680-1660 \text{ cm}^{-1}$. Josien and Fuson [65, 73, 74, 90] have confirmed the latter range. Bergmann and Pinchas [75, 81] have also discussed the spectra of polycyclic quinones.

ortho-Quinones have received less attention, although a number of these have been studied by Josien and Fuson [90], and by Otting and Staiger [162]. The frequencies do not appear to differ very widely from the *para* compounds, although the latter workers report rather low values in some instances in the solid state. In diphenoquinones the increased resonance leads to lower carbonyl frequencies [90], and compounds of this type absorb near 1640 cm^{-1}.

9.5(b). Tropolones. In addition to aromatic six-membered rings, a good deal of work has been carried out on the carbonyl frequencies of seven-membered rings, which have three double bonds in a state of resonance [25, 34, 42, 50, 57, 76–78, 163]. A typical compound of this type is tropolone, which absorbs at 1615 cm^{-1}. This could be due in part to hydrogen bonding to the hydroxyl group which is also present in the molecule, but the degree of hydrogen bonding is less than in β-diketones, and the OH stretching vibration occurs at 3100 cm^{-1}. The lowering of the carbonyl frequency must therefore arise from some other cause, and Koch [50] attributes it to the ring size. Just as a five-membered ring results in a strengthening of the C=O bond and a weakening of the C–C bonds of the ring, he believes that a seven-membered aromatic type of ring will show the reverse effect with a weakening of the C=O bond and a strengthening of the ring. Tropolones can also form chelates with copper compounds, and in these the carbonyl frequency shows a further fall to 1590 cm^{-1} and 1595 cm^{-1}. Somewhat surprisingly the carbonyl frequency is reduced even further than this [114] by the presence of *ortho*-carboxylic acid groups, when the ring carbonyl absorbs at 1580–1570 cm^{-1}.

A number of analogous cases are known in the colchicine series, in which a similar type of structure is believed to exist. Scott and Tarbell [25] have found a strong 1620–1612 cm^{-1} band in six colchicine derivatives which they attribute to the carbonyl group, and this is confirmed to some extent by the fact that this band vanishes in tetrahydrocolchicine and is replaced by a 1710 cm^{-1} band which corresponds to the normal carbonyl frequency of a seven-membered saturated ring. A similar band near 1610 cm^{-1} has been found by Nicholls and Tarbell [42] and by Fabian, Delaroff, Poirier and Legrand [165] in a series of benzotropolones and in colchicines [79].

9.6. The intensities of carbonyl absorptions

The intensity of the carbonyl absorption renders it particularly suitable for quantitative work, so that there has been a good deal of work carried out on the molecular extinction coefficients of various carbonyl compounds. Very extensive measurements have been made on the absolute intensities of various types of carbonyl group [10, 21, 24, 55, 56, 58, 80, 82, 149, 199–201]. There are clear cut differences in the intensities of different classes of carbonyl compounds, so that for example aldehydes absorb less strongly than ketones which are in turn weaker than esters. Carbamates are especially strong absorbers. However there is sufficient overlap between the ranges to make this an unsatisfactory method of diagnosis. Conjugated aldehydes for example have much the same intensities as saturated ketones. Within any single class of compound the differences are negligible except where intensification occurs following conjugation. Aromatic ketones have been shown to give linear relationships between their intensities and the Hammett σ values but the slopes of the lines are very small.

9.7. Characteristic ketonic frequencies other than the C=O stretching absorption

The presence of the carbonyl group in a molecule often gives rise to the appearance of a medium intensity band in the single bond stretching region, in addition to the carbonyl absorption itself. Hadži and Sheppard [49] have observed a strong band between 1350 cm^{-1}

and 1200 cm^{-1}, for example, in polycyclic quinones which is absent from the corresponding hydrocarbons, and have suggested that it may arise from some motion of the carbonyl group coupled with the rest of the molecule. Certainly most ketones give a band in this region, and Colthup [59] has made the broad classification of aliphatic ketones 1325–1215 cm^{-1} and aromatic ketones 1225–1075 cm^{-1}. However, this is such a broad range, and so many other vibrations absorb at comparable frequencies that it is of no use for the recognition of ketones as such, although a study of this region for a limited group of very closely related compounds might give useful information in certain cases. These and other possible carbonyl absorptions at low frequencies have been discussed by Jones *et al.* [52, 53].

One other possibility for the identification of ketones is the deformation vibration of the C=O group. The group –CO–CH$_3$ gives rise to three bands associated with this mode [51], near 600, 500 and 400 cm^{-1}, but considerable alterations follow chain branching if this occurs near to the carbonyl group, so that the bands cannot be identified in methyl *iso*propyl ketone. This correlation is therefore mainly of academic interest.

9.8. The C=O stretching absorption in aldehydes

Less information is available for aldehydes than for ketones, but such evidence as there is indicates that the factors influencing frequency shifts are very much the same in both cases.

Normal unconjugated alkyl aldehydes show their carbonyl band at a slightly higher frequency than the corresponding ketones, and absorb in the approximate range 1740–1730 cm^{-1}. As with ketones, values for this frequency obtained from vapour-phase studies are notably higher, and Hartwell *et al.* [12], for example, quote 1752 cm^{-1} for acetaldehyde and 1757 cm^{-1} for propionaldehyde. Similarly there is little interaction between two adjacent aldehyde groups, and glyoxal absorbs at very much the same frequency as formaldehyde [28]. Renema [180] discusses the conformations of alkyl aldehydes in some detail. Some bands are asymmetric and this would suggest by analogy with ketones that C.H and H.H conformations occur. Renema suggests on the basis of graphical resolution of the bands that there is a separation of about 10 cm^{-1} between these two forms but there is some difficulty in the assignments. Propanal, and butanal, both absorb at 1730 cm^{-1} and both are known from N.M.R and infra-red studies to exist wholly in the C.H conformation. However, if this is the frequency of C.H forms it is difficult to account for the fact that acetaldehyde, which must have an H.H form absorbs at the same point. However the point is not of great importance.

Unsaturated aldehydes with the double bond in the $\alpha\beta$-position show a fall in the carbonyl frequency of the same order as occurs with ketones, and absorb in the range $1705-1685$ cm^{-1}. The effect of lengthening the conjugated chain has been fully studied by Blout *et al.* [16]. In a series of polyene aldehydes they find that the first member, crotonaldehyde, absorbs at 1685 cm^{-1} in solution, whilst the remainder of the series, which contain from two to seven double bonds in conjugation, absorb between 1677 cm^{-1} and 1664 cm^{-1}. They conclude, therefore, that the influence of the second $\gamma\delta$ conjugated bond is small and that the effect of any additional conjugation is insignificant. Similarly, a series of α-furyl aldehydes containing long polyene chains and a furyl residue did not show any marked frequency shift in the carbonyl absorption after the first member of the series Similar results have been obtained by Scrocco and Salvetti [54], who have discussed them from a theoretical point of view. Other data on conjugated aldehydes are given by Bowles *et al.* [202] and by Renema [180]. Values above 1700 cm^{-1} correspond to *s cis* conformations and values below this to *s trans*. 2-Furyl aldehyde for example exists in both forms and shows two bands at 1701 and 1682 cm^{-1}. Conjugation with an acetylenic bond produces a slightly greater shift. $CH_3C\equiv CCHO$ absorbs at 1673 cm^{-1}.

Aromatic rings in conjugation with the aldehyde groups have a less marked effect on the frequency, and aryl aldehydes absorb in the range $1710-1695$ cm^{-1} in solution. Benzaldehyde absorbs at 1704 cm^{-1} and α- and β-naphthaldehydes absorb [17] at 1700 cm^{-1} and 1702 cm^{-1}, respectively. Here also, the substitution of strongly electron attracting or donating groups in the ring has a marked effect on the carbonyl frequency. Many benzaldehydes show two bands in the carbonyl region, but solvent studies by Brookes *et al.* [203] have shown that this is due to Fermi resonance. The stronger, higher frequency band is due to the aldehyde group.

α-Halogen substituents have not been widely studied, but there is every reason to expect that field effects will operate, as in the corresponding ketones, to raise the carbonyl frequency. In chloral, for example, the carbonyl absorption is at 1762 cm^{-1} in solution.

Chelation effects in β-hydroxy-$\alpha\beta$-unsaturated aldehydes are also common. Hunsberger [17] has found such effects with 1-hydroxy-2-naphthaldehyde and with 2-hydroxyl-1-naphthaldehyde. The effect is smaller with 3-hydroxy-2-naphthaldehyde, which he attributes to some degree of bond fixation in naphthalene. The methyl ethers of this series show normal carbonyl frequencies and are not chelated. Parallel effects are found in aldehydes of the indane series [133], and in aldehydes related to camphor [26]. On the other hand, as in α-diketones, no interaction occurs between neighbouring carbonyl groups. Cosgrove *et al.* [166] have studied a number of compounds, such as phenylglyoxal and similar ketoaldehydes, in which both carbonyl frequencies are essentially normal.

The intensities of aldehydic carbonyl absorptions have not been widely studied, but Cross and Rolfe [56] have given a few values for the extinction coefficients of aldehydes which can be compared with those of the corresponding ketones. Propional and heptanal (1735 cm^{-1} and 1736 cm^{-1}) have a molecular extinction coefficient of 130 and 148 as compared with about 180 for dialkyl ketones. Crotonaldehyde has a value of 234 and benzaldehyde of 324. This latter value is comparable with the value of 310 given by acetophenone. Parallel data are given by Renema [180]. From the limited data available, therefore, it would appear that the intensities of the carbonyl bands of aldehydes will vary with structural features in much the same way as do those of ketones.

In alcoholic solution there is a distinct fall in the intensity of aldehyde carbonyl absorptions, due to the reaction $RCHO + ROH \longrightarrow R \cdot CH(OH)OR$.

Ashdown and Kletz [60] have reported a number of such cases, and have shown that a corresponding reduction in the intensities of the aldehydic C—H stretching and deformation modes also occurs, whilst a new band appears near $1020-1110 \text{ cm}^{-1}$ which may be associated with the C—O— linkage.

9.9. The C—H stretching absorption in aldehydes

The highly characteristic nature of C—H stretching vibrations has already been discussed in an earlier section, and it is to be expected that aldehydes will show a characteristic absorption in this region arising from the valence vibration of the hydrogen atom attached to the carbonyl group. This would be expected to be at a frequency different from that of a CH vibration attached to a methylene group. Colthup [59] quotes the range $2900-2700 \text{ cm}^{-1}$ for this vibration, but Pozefsky and Coggeshall [61] found two bands in this region from a number of aldehydes. These were near 2720 cm^{-1} and 2820 cm^{-1}. Later more detailed studies, both on individual aldehydes and on groups, have established [167, 175, 204, 205] that the doubling is due to a Fermi resonance between ν_{CH} and the first overtone of the CH deformation. The higher frequency band has the main CH stretching character. Nevertheless the reasons why this should be at a so much lower frequency than other sp^2 CH bonds are not yet fully understood.

With the substitution of strongly electronegative groups ν_{CH} is moved to a somewhat higher frequency. It then ceases to interact with the CH bending mode and only a single band is shown. Examples of this are FCHO (2980 cm^{-1}), CH$_3$OCHO (2935 cm^{-1}) and CCl$_3$CHO (2851 cm^{-1}).

9.10. Other characteristic aldehyde vibrations

The in plane hydrogen deformation of the CHO group is much more difficult to identify with certainty, but it probably falls near 1400 cm^{-1}. As it is not a strong band, it is difficult to identify amongst the many others present in the same region, and it is not therefore especially useful for analysis. The out-of-plane deformation is also difficult to locate. Colthup [59] gives the range 975–825 cm^{-1} for this vibration, which reflects the variability in position, whilst Ashdown and Kletz [60] quote values of 780 cm^{-1} for *n*-butaldehyde and 795 cm^{-1} and 905 cm^{-1} for *iso*butaldehyde.

As in the case of ketones the presence of the carbonyl grouping appears also to activate certain adjacent C–C vibrations, so that additional bands appear in the carbon–carbon stretching region. They are of only moderate intensity, and can never be used to supply more than rather weak confirmatory evidence for the presence of the aldehyde group. Colthup [59] quotes the ranges as follows: Aliphatic aldehydes, 1440–1325 cm^{-1}; aromatic aldehydes, 1415–1350 cm^{-1}, 1320–1260 cm^{-1} and 1230–1160 cm^{-1}. The limited numbers of aldehydes which we have ourselves examined all show bands in these regions, but we have noted marked shifts within the given ranges for aromatic compounds which differ only in the relative positions of another ring substituent. There are clearly a number of different factors involved in determining these frequencies, and until they are better understood it would be unwise to employ this region to any extent in the identification of aldehydes.

9.11. Bibliography

1. Weniger, *Phys. Rev.*, 1910, **31**, 388.
2. Barnes, Gore, Liddel and Van Zandt Williams, *Infra-red Spectroscopy* (Reinhold, 1944).
3. Thompson and Torkington, *J. Chem. Soc.*, 1945, 640.
4. Rasmussen, Tunnicliff and Brattain, *J. Amer. Chem. Soc.*, 1949, **71**, 1068.
5. Bonino and Scrocco, *Atti accad. Naz. Lincei*, 1949, **6**, 421.
6. Cromwell, Miller, Johnson, Frank and Wallace, *J. Amer. Chem. Soc.*, 1949, **71**, 3337.
7. Jones, Van Zandt Williams, Whalen and Dobriner, *ibid.*, 1948, **70**, 2024.
8. Jones, Humphries and Dobriner, *ibid.*, 1949, **71**, 241.
9. *Idem, ibid.*, 1950, **72**, 956.
10. Jones, Ramsay, Keir and Dobriner, *ibid.*, 1952, **74**, 80.
11. Jones and Dobriner, *Vitamins and Hormones*, Vol. 7 (N.Y. Academic Press, 1949), p. 294.
12. Hartwell, Richards and Thompson, *J. Chem. Soc.*, 1948, 1436.
13. Grove and Willis, *ibid.*, 1951, 877.
14. Kohlrausch and Pongratz, *Z. Physik. Chem.*, 1934, **B27**, 176.
15. Turner and Voitle, *J. Amer. Chem. Soc.*, 1951, **73**, 1403.
16. Blout, Fields and Karplus, *ibid.*, 1948, **70**, 194.
17. Hunsberger, *ibid.*, 1950, **72**, 5626.
18. Soloway and Friess, *ibid.*, 1951, **73**, 5000.
19. Gutsche, *ibid.*, 1951, **73**, 786.
20. Cheng, *Z. physikal Chem.*, 1934, **B24**, 293.
21. Hampton and Newell, *Analyt. Chem.*, 1949, **21**, 914.
22. Woodward, *J. Amer. Chem. Soc.*, 1941, **63**, 1123.
23. Djerassi and Scholz, *ibid.*, 1948, **70**, 1911.
24. Barrow, *J. Chem. Phys.*, 1953, **21**, 2008.
25. Scott and Tarbell, *J. Amer. Chem. Soc.*, 1950, **72**, 240.
26. Bonino and Mirone, *Gazz.*, 1954, **89**, 1058.
27. Kohlrausch and Pongratz, *Ber.*, 1934, **67**, 976.
28. Thompson, *Trans. Faraday Soc.*, 1940, **36**, 988.
29. Randall, Fowler, Fuson and Dangl, *Infra-red Determination of Organic Structures* (Van Nostrand, 1949).
30. Rasmussen and Brattain, *J. Amer. Chem. Soc.*, 1949, **71**, 1073.
31. Gordy, *J. Chem. Phys.*, 1940, **8**, 516.
32. Flett, *J. Chem. Soc.*, 1948, 1441.
33. Woodward, Stork, Wineman, Nelson and Bothner-By. To be published. Woodward and Kovach, *J. Amer. Chem. Soc.*, 1950, **72**, 1009.
34. Aulin-Erdtman and Theorell, *Acta Chem. Scand.*, 1950, **4**, 1490.
35. Cromwell, Barker, Wankel, Vanderhorst, Olson and Anglin, *J. Amer. Chem. Soc.*, 1951, **73**, 1044.
36. Lecomte, *Discuss. Faraday Soc.*, 1950, **9**, 125.
37. Morgan, U.S. Atomic Commission, 1949, AECD 12659, 16.
38. Lecomte, *Rev. Optique*, 1949, **28**, 353.
39. Duval, Lecomte and Douville, *Ann. Physique*, 1942, **17**, 5.
40. Munday, *Nature*, 1949, **163**, 443.
41. *The Chemistry of Penicillin* (Princeton University Press, 1949), p. 404.
42. Nicholls and Tarbell, *J. Amer. Chem. Soc.*, 1952, **74**, 4935.
43. Gutsche, *ibid.*, 1949, **71**, 3513.
44. Biguard, *Bull. Soc. Chem.*, 1940, (5), **7**, 894.
45. Godchot and Canquil, *Compt. rend. Acad. Sci. (Paris)*, 1939, **208**, 1065.

46. Whiffen and Thompson, *J. Chem. Soc.*, 1946, 1005.
47. Coulson and Moffitt, *Phil. Mag.*, 1949, **40**, 1.
48. Josien and Fuson, *J. Amer. Chem. Soc.*, 1951, **73**, 478.
49. Hadži and Sheppard, *ibid.*, 5460.
50. Koch, *J. Chem. Soc.*, 1951, 512.
51. Lecomte, Josien and Lascombe, *Bull. Soc. Chim. France*, 1956, 163.
52. Jones, Herling and Katzenellenbogen, *J. Amer. Chem. Soc.*, 1955, **77**, 651.
53. Jones, Nolin and Roberts, *ibid.*, 6331.
54. Scrocco and Salvetti, *Boll. Sci. Fac. Chim. Ind. Bologna*, 1954, **12**, 93.
55. Marion, Ramsay and Jones, *J. Amer. Chem. Soc.*, 1951, **73**, 305.
56. Cross and Rolfe, *Trans. Faraday Soc.*, 1951, **47**, 354.
57. Bartels-Keith and Johnson, *Chem. and Ind.*, 1950, 677.
58. Richards and Burton, *Trans. Faraday Soc.*, 1949, **45**, 874.
59. Colthup, *J. Opt. Soc. Amer.*, 1950, **40**, 379.
60. Ashdown and Kletz, *J. Chem. Soc.*, 1948, 1454.
61. Pozefsky and Coggeshall, *Analyt. Chem.*, 1951, **23**, 1611.
62. Jones, Humphries, Herling and Dobriner, *J. Amer. Chem. Soc.*, 1952, **74**, 2820.
63. Jones, Ramsay, Herling and Dobriner, *ibid.*, 2828.
64. Hunsberger, Ketcham and Gutowsky, *ibid.*, p. 4839.
65. Josien and Fuson, *Compt. rend. Acad. Sci. (Paris)*, 1952, **234**, 1680.
66. Duval, Freymann and Lecomte, *Bull. Soc. Chim. (France)*, 1952, 106.
67. Bellamy, Spicer and Strickland, *J. Chem. Soc.*, 1952, 4653.
68. Haszeldine, *Nature*, 1951, **168**, 1028.
69. Barnes and Pinkney, *J. Amer. Chem. Soc.*, 1953, **75**, 479.
70. Leonard and Robinson, *ibid.*, 2143.
71. Hergert and Kurth, *ibid.*, 1622.
72. Birch, *J. Chem. Soc.*, 1951, 3026.
73. Josien, Fuson, Lebas and Gregory, *J. Chem. Phys.*, 1953, **21**, 331.
74. Josien and Fuson, *Bull. Soc. Chim. (France)*, 1952, 389.
75. Bergmann and Pinchas, *Bull. Res. Council Israel*, 1952, **1**, 87.
76. Doering and Knox, *J. Amer. Chem. Soc.*, 1952, **74**, 5683.
77. *Idem, ibid.*, 1953, **75**, 297.
78. Doering and Hiskey, *ibid.*, 1952, **74**, 5688.
79. Nicholls and Tarbell, *ibid.*, 1953, **75**, 1104.
80. Francis, *J. Chem. Phys.*, 1951, **19**, 942.
81. Bergmann and Pinchas, *J. Chim. Phys.*, 1952, **49**, 537.
82. Lecomte, *Bull. Soc. Chim. France*, 1955, 1026.
83. Jones and Sandorfy, *Chemical Applications of Spectroscopy* (Interscience, New York), 1956.
84. Jones and Herling, *J. Org. Chem.*, 1954, **19**, 1252.
85. Lord and Miller, *Applied Spectroscopy*, 1956, **10**, 115.
86. Shigorin, *Doklady Akad. Nauk. S.S.R.*, 1954, **96**, 769.
87. Bellamy, *J. Chem. Soc.*, 1955, 4221.
88. Kagarise, *J. Amer. Chem. Soc.*, 1955, **77**, 1377.
89. Margoshes, Fillwalk, Fassel and Rundle, *J. Chem. Phys.*, 1954, **22**, 381.
90. Fuson, Josien and Shelton, *J. Amer. Chem. Soc.*, 1954, **76**, 2526.
91. Josien and Deschamps, *J. Chim. Phys.*, 1955, **52**, 213.
92. Scrocco and Liberti, *Ricerca. Sci.*, 1954, **24**, 1687.
93. Bellamy and Williams, *J. Chem. Soc.*, 1957, 4294.
94. Lascombe and Josien, *Bull. Soc. Chim. France*, 1955, 1227.
95. Hirota, Hagiwara and Satonaka, *Bull. Chem. Soc. Japan*, 1967, **40**, 2439.
96. Dubois, Massat and Guillaume, *J. Chim. Phys.*, 1968, **65**, 731.

97. Dubois and Massat, *Compt. Rend. Acad. Sci. Paris*, 1967, **265B**, 757.
98. Mecke and Noack, *Chem. Ber.*, 1960, **93**, 210.
99. Barlet, Montagne and Arnaud, *Spectrochim. Acta*, 1969, **25A**, 1081.
100. Josien, Lascombe, Lecomte and Mathieu, *Compt. Rend. Acad. Sci. (Paris)*, 1955, **240**, 1982.
101. Wheland, *Resonance in Organic Chemistry* (Wiley, New York, 1955), p. 52.
102. Mecke and Noack, *Angew. Chem.*, 1956, **18**, 150.
103. Heilmann, De Gaudemaris and Arnaud, *Compt. Rend. Acad. Sci. (Paris)*, 1955, **240**, 1995.
104. Franzen, *Chem. Ber.*, 1955, **88**, 717.
105. Theus, Surber, Colombi and Schinz, *Helv. Chim. Acta*, 1955, **38**, 239.
106. Gamboni, Theus and Schinz, *ibid.*, 255.
107. Schubert and Sweeney, *J. Amer. Chem. Soc.*, 1955, **77**, 4172.
108. Tanaka, Nagakura and Kobayashi, *J. Chem. Phys.*, 1956, **24**, 311.
109. Wiberley and Bunce, *Analyt. Chem.*, 1952, **24**, 623.
110. Fuson, Josien and Cary, *J. Amer. Chem. Soc.*, 1951, **73**, 4445.
111. Josien and Fuson, *Bull. Soc. Chim. (France)*, 1952, 389.
112. Cromwell and Hudson, *J. Amer. Chem. Soc.*, 1953, **75**, 872.
113. Dickson and Page, *J. Chem. Soc.*, 1955, 447.
114. Jones, *J. Amer. Chem. Soc.*, 1953, **75**, 4839.
115. Corey, *ibid.*, 1953, **75**, 2301, 3297; 1954, **76**, 175.
116. *Idem, Experimentia*, 1953, **9**, 329.
117. Corey and Burke, *J. Amer. Chem. Soc.*, 1955, **77**, 5418.
118. Corey, Topie and Wozniak, *ibid.*, 5415.
119. Sandris and Ourisson, *Bull. Soc. Chim. (France)*, 1956, 958.
120. Inayama, *Pharm. Bull. Japan*, 1956, **4**, 198.
121. Kummler and Huitric, *J. Amer. Chem. Soc.*, 1956, **78**, 3369.
122. Brutcher, Roberts, Barr and Pearson, *ibid.*, 1507.
123. Bellamy, Thomas and Williams, *J. Chem. Soc.*, 1956, 3704.
124. Mizushima, Shimanouchi, Miyazawa, Ichishima, Kuratani, Nakagawa and Shido, *J. Chem. Phys.*, 1953, **21**, 815.
125. Daasch and Kagarise, *J. Amer. Chem. Soc.*, 1955, **77**, 6156.
126. Fuson, House and Melby, *J. Amer. Chem. Soc.*, 1953, **75**, 5952.
127. Bartell, Guillory and Parkes, *J. Phys. Chem.*, 1965, **64**, 3043.
128. Alder, Schafer, Esser, Kriger and Reubke, *Ann.*, 1955, **593**, 23.
129. Bellamy and Beecher, *J. Chem. Soc.*, 1954, 4487.
130. Shigorin, *Zhur. Fiz. Khim.*, 1954, **28**, 584.
131. Park, Brown and Lacher, *J. Amer. Chem. Soc.*, 1953, **75**, 4753.
132. Eistert and Reiss, *Chem. Ber.*, 1954, **87**, 92.
133. Hunsberger, Lednicer, Gutowsky, Bunker and Tunsoig, *J. Amer. Chem. Soc.*, 1955, **77**, 2466.
134. Hadži and Sheppard, *Trans. Faraday Soc.*, 1954, **50**, 911.
135. Shaw and Simpson, *J. Chem. Soc.*, 1955, 655.
136. Bellamy and Branch, *ibid.*, 1954, 4491.
137. Bender and Figueras, *J. Amer. Chem. Soc.*, 1953, **75**, 6304.
138. Sweeney, Curran and Quagliano, *ibid.*, 1955, **77**, 5508.
139. Susz and Cooke, *Helv. Chim. Acta*, 1954, **37**, 1273.
140. Cooke, Susz and Herschmann, *ibid.*, 1280.
141. Jewell and Spur, *Ohio State Symposium* 1956, cf. *Spectrochim. Acta*, 1956, **8**, 305.
142. Bellamy and Williams, *J. Chem. Soc.*, 1957, 861.
143. Duncanson, *ibid.*, 1953, 1207.

144. Tarbell and Hoffman, *J. Amer. Chem. Soc.*, 1954, **76**, 2451.
145. Pierre and Arnaud, *Bull. Soc. Chim. France*, 1966, 1690.
146. Jones, Sandorfy and Trucker, *J. Phys. Radium*, 1954, **15**, 320.
147. Freiss and Frankenberg, *J. Amer. Chem. Soc.*, 1952, **74**, 2679.
148. Prelog, *J. Chem. Soc.*, 1950, 420.
149. Bürer and Günthard, *Helv. Chim. Acta*, 1956, **39**, 356.
150. Schubert, Sweeney and Latourette, *J. Amer. Chem. Soc.*, 1954, **76**, 5462.
151. Anet, Bailey and Robinson, *Chem. and Ind.*, 1953, 944.
152. Leonard, Fox, Oki and Chiavarelli, *J. Amer. Chem. Soc.*, 1954, **76**, 630.
153. Leonard and Oki, *ibid.*, 3463.
154. Leonard, Fox and Oki, *ibid.*, 5708.
155. Leonard, Oki and Chiavarelli, *ibid.*, 1955, **77**, 6234.
156. Leonard, Oki, Brader and Boaz, *ibid.*, 6237.
157. Leonard and Oki, *ibid.*, 6241, 6245.
158. Mottus, Schwarz and Marion, *Can. J. Chem.*, 1953, **31**, 1144.
159. Anet and Marion, *ibid.*, 1954, **32**, 452.
160. Barchewitz, Tatibouet and Souchay, *Compt. Rend. Acad. Sci. (Paris)*, 1953, **236**, 1652.
161. Yates, Ardas and Fieser, *J. Amer. Chem. Soc.*, 1956, **78**, 650.
162. Otting and Staiger, *Chem. Ber.*, 1955, **88**, 828.
163. Kurantani, Tsuboi and Shimanouchi, *Bull. Chem. Soc. Japan*, 1952, **25**, 250.
164. Tarbell, Smith and Boekelheide, *J. Amer. Chem. Soc.*, 1954, **76**, 2470.
165. Fabian, Delaroff, Poirier and Legrand, *Bull. Soc. Chim. France*, 1955, 1455.
166. Cosgrove, Daniels, Whitehead and Goulden, *J. Chem. Soc.*, 1952, 4821.
167. Pinchas, *Analyt. Chem.*, 1955, **27**, 2.
168. Heilmann, Gaudemaris and Arnaud, *Bull. Soc. Chim. France*, 1957, 112, 119.
169. Jones, Forbes and Mueller, *Can. J. Chem.*, 1957, **35**, 504.
170. Mohrbacher and Cromwell, *J. Amer. Chem. Soc.*, 1957, **79**, 401.
171. Allen, Davis, Stewart and Van Allen, *J. Org. Chem.*, 1955, **20**, 306.
172. Allen and Van Allen, *ibid.*, 323.
173. Bürer and Günthard, *Chimica*, 1957, **11**, 96.
174. Josien and Deschamps, *J. Chim. Phys.*, 1957, **54**, 885.
175. Pinchas, *Analyt. Chem.*, 1957, **29**, 334.
176. Whiffen, *private communication*.
177. Susz and Wuhmann, *Helv. Chim. Acta.*, 1957, **40**, 722, 971.
178. Rao, *Chemical Applications of Infrared Spectroscopy* (Academic Press, New York, 1963).
179. Bellamy, *Advances in Infrared Group Frequencies* (Methuen, London, 1968).
180. Renema, Thesis, University of Utrecht, 1972.
181. Pierre, Barlet and Arnaud, *Spectrochim. Acta*, 1967, **23A**, 2297.
182. Lappert, *J. Chem. Soc.*, 1962, 542.
183. Castinel, Chiurdoglu, Josien, Lascombe and Vanlanduyt, *Bull. Soc. Chim. France*, 1958, 807.
184. Dollish, Fateley and Bentley, *Characteristic Raman Frequencies in Organic Compounds* (Wiley, New York, 1974).
185. Cataliotti and Jones, *Spectrochim. Acta*, 1971, **27A**, 2011.
186. Ginzburg and Khlevnyuk, *Optics and Spectroscopy*, 1971, **31**, 169.
187. Davis and Kim, *Theor. Chim. Acta*, 1972, **25**, 89.
188. Miller, Kiviat and Matsubara, *Spectrochim. Acta*, 1968, **24A**, 1523.

189. Turro and Hammond, *J. Amer. Chem. Soc.*, 1966, **88**, 3672.
190. Parsons and Green, *J. Amer. Chem. Soc.*, 1967, **89**, 1030.
191. Mitchell and Merritt, *Spectrochim. Acta*, 1971, **27A**, 1643.
192. Volpin, Koreshkov and Kursanov, *Izvest. Akad. Nauk. SSSR Otel Khim. Nauk.*, 1959, 560.
193. Krebs, *Angew. Chem. Int. Ed. Eng.*, 1965, **4**, 10.
194. Krebs and Schrader, *Zeit. Naturforsch.*, 1966, **213**, 184.
195. Krebs and Schrader, *Ann. Chemie*, 1967, **709**, 46.
196. Brown, *Spectrochim. Acta*, 1962, **18**, 1065.
197. Bagli, *J. Amer. Chem. Soc.*, 1962, **84**, 177.
198. Becker, Ziffer and Charney, *Spectrochim. Acta*, 1963, **19**, 1871.
199. Thompson, Needham and Jameson, *Spectrochim. Acta*, 1957, **9**, 208.
200. Thompson and Jameson, *Spectrochim. Acta*, 1958, **13**, 236.
201. Flett, *Spectrochim. Acta*, 1962, **18**, 1537.
202. Bowles, George and Maddams, *J. Chem. Soc.* (B), 1969, 810.
203. Brookes and Morman, *J. Chem. Soc.*, 1961, 3372.
204. Lucczeau and Sandorfy, *Can. J. Chem.*, 1970, **48**, 3694.
205. Rock and Hammaker, *Spectrochim. Acta*, 1971, **23A**, 1899.
206. Gaset, Lafaille and Lattes, *Bull. Soc. Chim. France*, 1968, 4108.

10

Carboxylic Acids

10.1. Introduction and table

Carboxylic acids exist normally in dimeric form with very strong hydrogen bridges between the carbonyl and hydroxyl groups of the two molecules. Even in the vapour state, and in dilute solution in certain solvents, this association persists to some extent, so that the carbonyl frequencies as normally measured are considerably modified. For this reason the infra-red spectra of these materials are usually measured in the solid or liquid state. To some extent this masks the differences produced by changes in the structure immediately around the carbonyl group, so that the differentiation of $\alpha\beta$-unsaturated or α-halogen acids is not always as clear cut as is the case with ketones. Nevertheless it is generally true that the same factors which influence the carbonyl frequencies of ketones apply to carboxylic acids, and that the shifts are in the same directions. The carbonyl frequencies of carboxylic acids appear in much the same spectral region as aldehydes and ketones, but the acids can be identified by the study of other regions of the spectrum.

The abnormally strong hydrogen bonding in these acids is an advantage in one way, as the O–H stretching vibrations are so distorted from the normal as to be characteristic, so that observations in this region give a valuable indication of the presence of carboxylic acids.

There are also a number of other regions of the spectra from which some data may be obtained, although with less certainty than from those mentioned above. There are the regions near 1400 cm^{-1}, 1250 cm^{-1} and 920 cm^{-1}. The first two arise from strongly

183

coupled vibrations involving the in plane OH deformation, and the C–O stretch respectively. The third is due to the out of plane OH deformation of the dimer form.

The study of all of these regions will generally provide a reasonably reliable identification of carboxylic acids, especially if account is taken of the intensity of the carbonyl absorption in relation to that of known acids. In addition, a limited amount of data on the immediate environment of the COOH group may sometimes be obtained from a study of the C=O frequency.

Confirmation of the identification of a carboxylic acid can be readily obtained by the examination of a salt, or of a solution of the acid in water in which ionisation has occurred. Ionisation of the acid results in equilibration of the two oxygen atoms attached to the carbon with the disappearance of the carbonyl absorption, and the appearance of two new bands near 1550 cm^{-1} and 1400 cm^{-1} arising from the symmetrical and anti-symmetrical vibrations of the COO– grouping. The addition of mineral acids to ionised aqueous solutions reduces the ionisation, and the carbonyl absorption reappears.

The identification of carboxylic acids by infra-red methods is therefore usually possible with reasonable certainty and, as will be seen, it is sometimes possible to obtain some additional data on the environment of the carboxyl group.

Table 10

Carboxylic acid.	OH stretching vibrations (free)	$3560-3500 \text{ cm}^{-1}$ (m.)
	OH stretching vibrations (bonded)	$2700-2500 \text{ cm}^{-1}$ (w.)
C=O vibrations:		
	Saturated aliphatic acids	$1725-1700 \text{ cm}^{-1}$ (s.)
	α-halogen-substituted aliphatic acids	$1740-1705 \text{ cm}^{-1}$ (s.)
	$\alpha\beta$-unsaturated acids	$1705-1690 \text{ cm}^{-1}$ (s.)
	Aryl acids	$1700-1680 \text{ cm}^{-1}$ (s.)
	Acids showing internal hydrogen bonding	$1670-1650 \text{ cm}^{-1}$ (s.)
C–O-stretching vibrations or OH deformation vibrations		$\begin{cases} 1440-1395 \text{ cm}^{-1} \text{ (w.)} \\ 1320-1211 \text{ cm}^{-1} \text{ (s.)} \end{cases}$
OH. Deformation (out of plane)		$950-900 \text{ cm}^{-1}$ (v.)
$-C\diagup^{O}_{\diagdown O^-}$		$\begin{cases} 1610-1550 \text{ cm}^{-1} \text{ (s.)} \\ 1420-1300 \text{ cm}^{-1} \text{ (s.)} \end{cases}$
Band progression in solid fatty acids		$1350-1180 \text{ cm}^{-1}$ (w.)

These correlations are listed in Table 10 and discussed in detail below.

The special case of amino-acids, in which zwitterion structures arise, will be considered separately in the section dealing with carbon nitrogen compounds.

All correlations relate to samples examined in the solid or liquid state.

10.2. The OH stretching vibration of the carboxyl group

The OH stretching frequencies of acids have been extensively studied, as they are anomalous in showing no absorption in the normal free OH or bonded OH regions when examined in the solid state.

A number of aliphatic acids have been studied as vapours at various temperatures [1, 2, 52, 53] and as solutions in carbon tetrachloride and other solvents [3–8, 54–56]. The results of this work show that whereas the OH stretching absorption of the monomers lies near 3550 cm^{-1}, the dimeric form gives rise to a broad absorption region with many sub-maxima, between 3000 cm^{-1} and 2500 cm^{-1}. Analogous changes occur in the carbonyl region. The dimeric form is not readily broken, and even in relatively dilute solutions in carbon tetrachloride, and in the vapour state at low temperatures, bands due to both the monomer and dimer are shown.

The monomer bands under these conditions have been studied in detail by Goulden [54], who finds an overall range from 3504 cm^{-1} in trifluoroacetic acid, the strongest studied, to 3545 cm^{-1} in p-N-dimethylaminobenzoic acid, which was the weakest. This dependance of the frequency upon the acid strength is a reflection of the changing inductive effects of the substituents of the carboxyl group which affects both frequencies and $_pK_a$ values. Simple linear relationships therefore connect OH frequencies and $_pK_a$ values for the separate series of aliphatic, aromatic and $\beta\gamma$-unsaturated acids. The monomeric OH frequencies of substantial numbers of benzoic acid derivatives have been given by Brookes [30] and by Josien [37]. Insofar as the *meta* and *para* derivatives are concerned, there appears to be an inverse relationship between ν_{OH} and ν_{CO}, so that the latter rises as the former falls. The relationship with acidity constants has been studied by Guilleme *et al.* [45] and by Katon [57]. Guilleme has shown that there is a good linear relationship between the $_pK_a$ values and the equilibrium constants for dimerisation. This shows less scatter than the corresponding plot of $_pK_a$ versus ν_{OH}. The frequencies show a major discrepancy in the series propionic acid (3538 cm^{-1}), acrylic acid (3550 cm^{-1}) and 3-butynolic acid (3528 cm^{-1}), in which there is a progressive fall in the $_pK_a$ values. The equilibrium constants on the other hand fall regularly in line with the acidities. Other examples in which the detailed validity of the frequency/acidity relation are questioned are given by Katon *et al.* [57]. They find that alkyl acids with a single halogen atom at the α- or β-carbon atom absorb between 3534 and 3530 cm^{-1} without regard either to the nature of the halogen or to the $_pK_a$ changes. They quote examples with major $_pK_a$ differences with virtually identical frequencies. Relationships of this kind have therefore no more than a broad general validity and cannot be applied in detail.

In considering monomeric OH frequencies one must always consider the possibility of internal hydrogen bonding leading to lower frequencies. α-Hydroxy acids for example give four OH stretching bands in dilute solution, corresponding to free alcohol, free acid, alcohol bonded to acid, and acid bonded to alcohol. The frequency of this last is 3440 cm^{-1}.

In the liquid or solid states, however, no monomer bands occur, and Flett [9] and Shreve *et al.* [10] have confirmed that only dimers (or higher polymers) are found. This is true also of solutions in polar solvents such as dioxane or acetone unless the concentration is very low [95]. The infra-red spectra of acids are therefore best determined on materials in this state, when complications due to the presence of monomer do not arise; the greater proportion of the data available on carboxylic acid frequencies therefore relates to spectra obtained on solids or liquids.

Under these conditions the examination of a very considerable number of acids has [9, 10, 55, 56, 58, 89, 91] shown that the OH absorptions occur as broad bands with a series of minor peaks over the range 3000–2500 cm^{-1}. In most cases the main peak is near 3000 cm^{-1} with a main satellite band near 2650 cm^{-1}. Although this is usually overlaid to some extent by the CH absorptions, the pattern is nevertheless highly characteristic. Dimerisation explains the considerable OH shift, but does not account for either the appearance of the satellite bands or for the overall breadth of the absorption. These topics have been discussed by many workers. One suggestion has been that the fundamental OH stretching mode is accompanied by a series of supplementary peaks arising from its interaction with low-frequency vibrations of the dimer involving stretching of the hydrogen bond [3, 59–63]. A closely related suggestion has been elaborated in quantum mechanical terms by Bateuv [61] and Stepanov [62, 63]. An alternative due to Fuson and Josien [58] visualises the broad peaks as arising from a series of hydrogen bonds of different discreet lengths. The whole problem has been fully discussed recently by Bratoz, Hadži and Sheppard [55, 56], who have pointed out that there are difficulties associated with either of these explanations, particularly in respect of the lack of temperature dependence of the OH bands and in the results of deuteration studies. They go on to point out that all the observable peaks in the range can be satisfactorily identified as combination bands of lower-frequency vibrations involving the COOH group, the intensity being enhanced by Fermi resonance with the OH fundamental. For example, in the H acids the coupled C—O and OH frequencies near 1420 cm^{-1} and 1300 cm^{-1} account for the main 2700 cm^{-1} satellite. This satellite shifts to 2100 cm^{-1} in the D acids, where it corresponds to a summation of the B_u and A_g bands, which are coincident at 1050 cm^{-1}. Satellite bands which are found at higher frequencies than the fundamental can also be satisfactorily explained as summation bands of one or other of these modes with the carbonyl fundamental. A parallel Fermi-resonance interpretation has been used by Odinokov et al. [94] to explain the minor peaks which appear on the main OH stretching bands of acids associated with strong bases. The problem of the great breadth of the main OH absorption is, however, not yet fully resolved.

Pauling [11] has accounted for the increased strength of the hydrogen bonds in fatty acid dimers on the basis of the large

contribution of an ionic resonance structure $\mathrm{RC} \underset{\displaystyle \mathrm{OH}_+ \cdots \mathrm{O}_-}{\overset{\displaystyle \mathrm{O}^- \cdots \mathrm{HO}^+}{\big\langle}} \overset{\displaystyle }{\big\rangle} \mathrm{CR},$

and this enables him to explain the greater shift in the OH frequency of these dimers over that in hydrogen-bonded alcohols in which the OH linkages are not loosened by the influence of the charges.

Since few other compounds absorb in the $2700-2500 \text{ cm}^{-1}$ range, this is a particularly valuable region for the identification of carboxylic acids, although the bands are often weak, so that their detection in acids of high molecular weight may be difficult. The separation of these bands from the CH stretching vibrations is sufficiently great with a rock-salt prism to prevent any possibility of confusion. In general, therefore, absorption between 2700 cm^{-1} and 2500 cm^{-1} can be taken as strong evidence for the presence of a dimeric carboxylic acid, but this identification should never be made without taking account also of other regions of the spectrum, and especially of the carbonyl region. Absorption between 2700 cm^{-1} and 2500 cm^{-1} is indicative only of a strongly hydrogen-bonded OH group, and although bonds of this strength are unusual in compounds other than carboxylic acids, they do occur in a limited number of cases. Also it is possible in exceptional cases for the OH bands of intramolecularly bonded acids to occur at still lower frequencies. In potassium hydrogen maleate [64], for example, the hydrogen atom is probably symmetrically placed between the two oxygen atoms. This compound has no strong bands which show any alteration on deuteration at frequencies above 1600 cm^{-1}.

Examples of other materials showing hydrogen bonds of comparable strength are found in the enolic β-diketones and similar materials [12] in which a strong chelate ring occurs. Similarly neither o-hydroxydiphenylsulphoxide nor α-nitroso-β-naphthol show any normal OH vibration, and this has been attributed to chelation [13]. We have also found that sulphonic acids and the organo-phosphoric acids show OH absorption in this region [14]. The carbonyl absorption of β-diketones is, however, shifted to considerably lower frequencies than ever occur with carboxylic acids, so that they are not a real complication, provided account is taken of the C=O vibrations as well as of this O—H vibration. Also chelated hydroxyl groups of this type do not normally show the band structure which is so typical of carboxylic acids.

These bands are unlikely to be confused with SH or SiH absorptions, which also appear in this region [38], as the latter are characteristically sharp and well defined, in contrast to the broad, rather diffuse bands given by acids.

As a final check deuterium exchange can be employed when the OH band shows the expected degree of shift [39, 55, 56]. The OH stretching band of CD$_3$COOH is, for example, at 3125 cm^{-1} and the OD band of CD$_3$COOD at 2299 cm^{-1}.

These characteristic OH bands can also be used to give information on the crystal structure of solid acids. Kuratani [15] has studied the infra-red dichroism of cinnamic, adipic and other acids crystallising as needles, over the wave-lengths 3570—2850 cm^{-1}, and found that the absorption is stronger when the electric vector of the polarised radiation is parallel to the long axis of the needles, from which it follows that the O—H . . . O bonds are parallel to this axis.

10.3. The C=O stretching vibration

10.3(a). The influence of physical state. The physical state in which carboxylic acids are examined has a direct effect on the carbonyl frequency. As has been mentioned, a number of workers have examined some of the simpler acids in the vapour phase [1, 2], when, depending on the temperature, varying proportions of the monomeric and dimeric forms occur. Hartwell, Richards and Thompson [16] have examined a number of acids in this way and compared their results with those obtained from the liquids. In a typical case they found acetic acid at 20°C as a vapour absorbed at 1735 cm^{-1} and 1785 cm^{-1}, whilst at 60°C it absorbed at 1735 cm^{-1} and 1790 cm^{-1}, but then showed a considerably greater intensity in the higher frequency band, indicating a greater proportion of monomer. In the liquid state it absorbed at 1717 cm^{-1}.

In dilute solutions in non-polar solvents, such as carbon tetrachloride, two carbonyl frequencies corresponding to the monomer and dimer are found. The frequency difference between the two bands is always close to 45 cm^{-1}. Studies on the intensity changes in these bands on dilution can be carried out in the same way as on the OH frequencies to determine equilibrium coefficients [65]. The monomer frequency measured in this way is usually a little lower than that obtained from the vapour. In trifluoroacetic acid, for example, the monomer in the vapour absorbs at 1820 cm^{-1} and falls to 1810 cm^{-1} in carbon tetrachloride [66]. With unflorinated alkyl acids the monomer frequency in carbon tetrachloride has an average value [17] of 1760 cm^{-1}.

In certain solvents, such as dioxane, only a single carbonyl frequency is shown. This was originally attributed to the existence of the acids as free monomers in this state [9]. However, the frequencies found in this solvent and in ether are consistently some 20 cm^{-1} below the monomer values in carbon tetrachloride [67], and it therefore seems likely that strong solvent interaction effects are taking place. This has been confirmed by Fraenkel *et al.* [68], who demonstrated a gradation of the frequency with the polarity of the solvent. At higher concentrations the dimer form takes over with a corresponding fall in the carbonyl frequency [95]. In compounds such as pyridine, 1 : 1 and 1 : 2 complexes are formed leading to the occurrence of the monomeric carbonyl frequency accompanied by the carboxylate ion absorptions [68].

However, the main bulk of data on the carbonyl frequencies of these acids has been obtained from the examination of the solid or liquid phases or of solutions of sufficient concentration for hydrogen bonding to persist. It is therefore the frequencies for the dimeric materials which are of value for correlation purposes. In the spectra of such materials, changes of crystal form which influence the degree of hydrogen bonding can result in marked differences in the low-frequency region, but the carbonyl frequencies are not appreciably altered. A typical case is that of the cinnamic acids. Lecomte and Guy [18] quote values of 1702 cm^{-1}, 1707 cm^{-1} and 1709 cm^{-1} for the C=O frequency of three different forms of *cis*-cinnamic acid, and $1699 \text{ cm}^{-\cdot}$ for each of two forms of *trans*-cinnamic acid.

However, changes of as much of 30 cm^{-1} towards higher frequencies occur on melting due to the disruption of the strong hydrogen bonds of the crystal and their replacement by looser forms of association [67, 89], and marked changes also occur in some instances of acids examined in pressed discs due to interaction with the alkali halide [78].

Barnes *et al.* [19] have given values for the C=O frequencies of a number of acids, and Lecomte [20] has discussed the relationship between the carbonyl frequency and its surrounding structure, on the basis of the examination of a limited number of materials. This latter aspect has been further studied by Gillette [21], and especially by Flett [9, 22], whilst data on individual groups of carboxylic acids have been given by a number of other workers [23–25, 96–99]. The general findings are discussed below.

10.3(b). Saturated aliphatic acids. Saturated monobasic aliphatic acids which do not carry electron-attracting substituents absorb between 1725 cm^{-1} and 1705 cm^{-1} when examined in the solid or liquid state. This is in line with Flett's findings on some sixty carboxylic acids [9], and agrees with the findings of the other workers mentioned and of ourselves. Sinclair, McKay and Jones [42] find the carbonyl absorption of saturated fatty acids ($C_{14}-C_{21}$) in solution to be 1708 cm^{-1}, and the same value is obtained for unsaturated fatty acids other than $\alpha\beta$-unsaturated acids [43]. Freeman [44] quotes $1712 \pm 6 \text{ cm}^{-1}$ for this frequency in twenty-seven branched chain acids he has examined in solution. In the solid state these frequencies are a few cm^{-1} lower. Confirmatory results have been obtained by subsequent workers [56, 67, 89, 90, 97]. Thioacetic acid [27] absorbs a little below this range at 1696 cm^{-1}.

10.3(c). The influence of α-halogen groups. As in the case of ketones, the substitution of α-halogen or cyano-groupings results in the carbonyl absorption shifting towards higher frequencies. α-Chloro-propionic acid, for example, absorbs at 1730 cm^{-1}, as against 1710 cm^{-1} for β-chloropropionic acid [9]. In 2-bromostearic acid the shift is smaller and the band is at 1716 cm^{-1}, as against 1708 cm^{-1} for stearic acid [43]. The monomeric forms show a similar although smaller shift in solution. Similarly, in the vapour phase chloroacetic acid absorbs at 1794 cm^{-1}, whilst acetic acid absorbs at 1785 cm^{-1}. Gillette [21] observed a similar correlation for the methyl and halogen acetic acids, and found a progressive frequency rise in the carbonyl absorption through the series mono ⟶ trichloroacetic acid. Work on other types of carbonyl frequencies has indicated, however, that the shift which follows the introduction of the second and third α-halogen atoms is appreciably smaller than with the first [70], and a similar effect is to be expected in acids. Rotational isomerism is also to be expected in mono- and di-α-halogen acids, in line with the similar effects found in ketones and in esters [71], leading to multiple carbonyl frequencies. In this connection it is interesting to note that the liquid state spectrum of dichloroacetic acid illustrated by Bratoz *et al.* [56] shows two carbonyl bands despite the absence of any monomer as indicated by free OH bands. However, isomerism of this type would be more difficult to detect in acids, due to the complications of monomer/dimer equilibria in solution and to the fact that the acids themselves are highly polar, so the concentrations of less polar configurations may be small in the liquid state. In general, the shift found for α-chlorine substitution is of the order of 20 cm^{-1}, so that we would suggest the range 1740–1720 cm^{-1} for liquid and solid acids of this type as a general approximation. Acids with α-fluorine substitution are, however, likely to absorb at even higher frequencies [46, 47, 66, 72]. However as noted above, the shifts from α-bromo substitution are negligible and the carbonyl band appears near 1715 cm^{-1}. This has been confirmed by Katon *et al.* [100] who point out that the great bulk of the data on α-halogen substitution relates to chlorine compounds, and that in fact there is no significant shift with either bromine or iodine. They conclude that no generally useful correlation is possible for α-halogenated acids and quote an overall range of 1740 to 1705 cm^{-1}.

10.3(d). The influence of conjugation. In the series of αβ-unsaturated acids examined by Flett [9] the C=O frequency lies in the general range 1710–1700 cm^{-1}, except for cinnamic acid, where the aryl group provides γδ-conjugation and the frequency falls below

1700 cm^{-1}. Otherwise the effect is a slight shifting of the frequency towards the lower end of the normal frequency range. With aryl substitution, on the other hand, the C=O frequency lies between 1700 cm^{-1} and 1680 cm^{-1} and only rises above this value in substituted benzoic acids which have heavily electron-attracting nitro-groups.

However, as many of the alkyl materials were liquids, this apparent difference could be due, in part at least, to alterations in the phase rather than to some fundamental difference. Many cases are known of $\alpha\beta$-unsaturated acids which absorb in the range 1680--1700 cm^{-1}. Freeman [48] has quoted a range of 1700--1692 cm^{-1} for one such series, and corresponding values have been obtained by other workers [73, 74]. It is therefore clear that both aryl and olefinic conjugation cause a frequency fall, but it is doubtful if the two can be distinguished in this way.

Extension of the conjugation beyond the $\alpha\beta$-position does not appreciably affect the carbonyl frequency in olefinic compounds. With acetylenic conjugation a slightly greater shift, often to below 1680 cm^{-1}, is observed, but in this case extension of the conjugation to three or more acetylenic bonds causes a small frequency rise above the minimum value. This has been discussed in relation to the acidities of these compounds by Allan et al. [74].

Benzoic acids have been studied extensively by Josien et al. [96], Eglington et al. [98], and Flett [99], and relationships sought between changes in ν_{CO} and in the $_pK_a$ value. It would seem that apart from ortho substituted benzoic acids where steric factors may come into play, there is a good general relation of the form $\nu_{CO} = 1785 - 10.5\,_pK_a$. This equation is derived from data on monomeric acid frequencies, but in view of the almost constant difference of 45 cm^{-1} between these and their dimers, the same equation can be used to derive the latter.

10.3(e). The influence of chelation. Some acids are capable of forming internal hydrogen bonds, and the carbonyl frequency is reduced accordingly. Of nine acids recorded by Flett [9] as having carbonyl frequencies below 1680 cm^{-1}, seven are of this type. The simplest case is that of fumaric acid, which absorbs at 1680 cm^{-1}, in contrast to the normal value of 1705 cm^{-1} of maleic acid [9]. ortho-Hydroxybenzoic acids [8] can chelate in a similar way to the ortho-hydroxyacetophenones and ortho-hydroxybenzaldehydes, with a correspondingly large shift of the carbonyl frequency.

Salicylic acid absorbs at 1655 cm^{-1}, and this is comparable with the shifts experienced with the β-hydroxy-$\alpha\beta$-unsaturated ketones. β-Amino-$\alpha\beta$-unsaturated acids also show this effect, as is to be expected. 3-Amino-2-naphthoic acid absorbs at 1665 cm^{-1}. Flett [9], and Musso [75] quote a number of other similar cases. We have also observed a case of this type with N-phenylanthranilic acid, in which the carbonyl absorption occurs at 1660 cm^{-1}.

In the consideration of carbonyl frequencies of hydroxy-, amino- or ketonic acids, allowances must also be made for the possibility that ring closure has occurred with the formation of a lactone, lactam or lactol. The characteristic carbonyl frequency of the carboxylic acid then disappears. Penicillic acid [28] has been shown to be such a case, and it exists as a free acid only under certain limited physical conditions. Similarly, acetophenone *ortho*-carboxylic acid absorbs at 1732 cm^{-1}. This corresponds to a five-membered ring lactol frequency rather than that of an aryl acid. *ortho*-Formylbenzoic acid (1738 cm^{-1}) also exists as the lactol rather than as the free acid [17], and numerous other cases are known. In some cases, such as benzil-*ortho*-carboxylic acid, both forms are known. The keto-form has absorptions at 1698 cm^{-1} and 1683 cm^{-1}, corresponding to the aryl ketone and to the acid, whilst the lactol form absorbs at 1692 cm^{-1} and 1745 cm^{-1}, corresponding to the aryl ketone and the five-membered ring lactol carbonyl absorption.

10.3(f). Dicarboxylic acids.

Flett [9] and Schonmann [26] have studied a number of dicarboxylic acids. Oxalic acid absorbs very strongly between 1710 cm^{-1} and 1690 cm^{-1}, but in malonic acid two frequencies are shown at 1740 cm^{-1} and 1710 cm^{-1}. This interaction effect is reduced in succinic acid, which shows only a weak band at 1780 cm^{-1}, and its main carbonyl absorption at 1700 cm^{-1}. Higher members, such as adipic acid, show only a single band near 1700 cm^{-1} and this has been confirmed by Corish and Davison [67] for other longer-chain fatty acids. Phthalic acid is also normal in showing a single CO band at 1695 cm^{-1}, whilst terephthalic acid absorbs at 1690 cm^{-1}. Dicarboxylic acids can therefore be regarded as being essentially normal in their carbonyl frequencies, provided the first members of the series are ignored.

Schotte and Rosenberg [76, 77] have also studied dicarboxylic acids, particularly in relation to the stereoisomerism of $\alpha\alpha'$-di-substituted compounds. Minor differences occur between the carbonyl frequencies of optically active and racemic forms — probably due to differences in crystal structure — and these enable the two to be differentiated.

10.3(g). The intensity. Carbonyl absorptions from carboxylic acids are generally more intense than those of ketones. Cross and Rolfe [29] quote values for the molecular extinction coefficients of a number of acids, and show that they are about twice the values for the corresponding ketones. Flett [9, 99] has also given a good deal of data on intensities of the dimer bands. It is clear that there are distinct differences between various classes, but the assessment of absolute intensities is especially difficult in view of the problems of monomer/dimer equilibrium and of the effects of changes of phase or of solvent. It is therefore very doubtful whether intensity measurements have much to offer for diagnostic purposes.

10.4. Other characteristic carboxylic acid vibrations

10.4(a). Near 1400 cm^{-1}. A band near 1400 cm^{-1} has been observed by Flett [9] in forty-five out of sixty carboxylic acid spectra. It is therefore less generally useful than the two correlations already discussed. The band is of variable intensity and this may account for its absence in some cases. The band occurs in the overall range 1440 to 1375 cm^{-1} but is usually found near 1430 cm^{-1}. Hadži and Sheppard [49] have also found a band within the same frequency range in fifteen carboxylic acids. These include diacetylene carboxylic acid and trichloracetic acid, and as neither the C≡C nor the CCl$_3$ structures could possibly have any fundamental near 1400 cm^{-1}, this clearly identifies this absorption with the carboxyl group. In an attempt to establish the type of vibration involved, they have compared their spectra with those of all the corresponding deutero derivatives, and been able to show that this absorption, together with another near 1300 cm^{-1}, arises from a C–O vibration coupled with an OH in-plane deformation vibration to such an extent that neither can be specifically assigned in the original acids. Because of the similarity of the 1300 cm^{-1} band with a strong C–O stretching band at a similar frequency in esters, it is usual to call this the C–O stretch and assign the 1300 cm^{-1} band to the in plane OH deformation. In the deuterated materials the interaction is much reduced and the individual absorptions can be more clearly seen. In dicarboxylic acids the corresponding band at 1435 cm^{-1} has been identified specifically with a vibration of the dimerised carboxyl group as it disappears in the molten acids [67, 89], in which the hydrogen bonds are more randomly distributed. Possibly, the absence of this band reported in some acids is due to this cause. Francis [50] and Sinclair *et al.* [42, 43] have recently pointed out that saturated fatty acids and esters which have a methylene group adjacent to the carbonyl group, all absorb at 1410 cm^{-1}, whereas no comparable band is found in $\alpha\beta$-unsaturated acids. They associated it therefore with a CH_2 deformation which has been modified by the adjacent CO group. This absorption is of course quite distinct from the carboxyl group absorption discussed above, and it occurs along with it in acids containing the group $-CH_2COOH$ [49, 67]. This absorption is more intense than a normal CH_2 group, and in short-chain fatty acids it is useful in the recognition of α-chain branching. For acids of chain length less than C_{14}, the 1410 cm^{-1} band is stronger than the 1475 cm^{-1} methylene deformation absorption unless there is α-branching, when the CH_2COOH group is lost.

10.4(b). Near 1250 cm^{-1}. A band near 1250 cm^{-1} appears in all the acids examined by Flett [9], and Shreve *et al.* [10] have noted the presence of a doublet between 1280 cm^{-1} and 1250 cm^{-1} in a large number of long-chain fatty acids which they believe may be characteristic of such compounds. These results indicate that the position of this band varies widely from compound to compound, and that it can occur anywhere within the range 1320–1210 cm^{-1}. It has usually been identified as being the strongest band in the spectrum between 1600 cm^{-1} and 700 cm^{-1}. Thus Flett [9] lists eighteen acids as absorbing above 1280 cm^{-1}, of which fourteen are aromatic and two more are unsaturated, whilst of ten absorbing below 1240 cm^{-1} six are substituted acetic acids and others, such as α-chloropropionic acid, contain electronegative substituents.

On the other hand, other workers studying saturated fatty acids and those in which any double bonds are well removed from the carboxyl group [42, 43, 44] all find this absorption very close to 1290 cm^{-1}, and Hadži and Sheppard [49] quote the relatively narrow range of 1300 ± 15 cm^{-1} for this absorption in all the acids which they have examined. Freeman [44] has pointed out that in the fatty acids series this band is invariably accompanied by a second absorption at lower frequency. These two bands are remarkably consistent in their frequencies (1285 ± 5 cm^{-1} and 1235 ± 5 cm^{-1})

and in their relative intensity patterns, so long as there is no chain branching nearer to the carboxyl than the δ carbon atom. With a nearer approach of the point of branching to the carboxyl group a more pronounced disturbance of these frequencies and of their intensity patterns is observed, so that in some cases the intensity of the second band is greater than that of the first. Guertin et al. [79] and Corish and Chapman [89] have extended this work to cover lower members of the fatty acid series with essentially similar results. This observation would appear to offer a reasonable explanation of the considerable disparity between the wide frequency ranges quoted by one group of workers and the narrow ranges given by others. It seems likely that all acids will show an absorption in the region of 1300 cm^{-1}, but that in many cases it will be accompanied by a second band at lower frequencies which may be of greater intensity. The identification of the carboxyl groups with the strongest band in this region will therefore lead to a much wider frequency range than the direct search for bands close to 1300 cm^{-1}. The higher frequency absorption is almost certainly derived from the coupled C—O and OH in-plane deformation modes [49], but the origin of the second is less certain. Shreve et al. [10] associate their band near 1200 cm^{-1} with the C—O valence and point out that esters and similar materials also absorb near this frequency [31]. Kuratani [51] also assigns a band in trichloracetic acid to this cause on the grounds that the absorption persists in the acetate ion, but in view of the findings of Hadži and Sheppard on interaction effects these suggestions must be treated with caution until further information is available.

O'Connor et al. [24] also report a band in the range 1170–1150 cm^{-1} in the spectra of a large number of even-numbered carbon atom fatty acids which is not present in the corresponding ester spectra. The lower frequency suggests, however, that this is probably due to some other cause and that it is not directly related to the bands under discussion.

10.4(c). The OH deformation vibration of the carboxyl group. Davies and Sutherland [3] have suggested that in the lower fatty acids the out-of-plane deformation mode of the hydroxyl of the carboxyl group may arise near 935 cm^{-1}. Shreve et al. [10] have found strong absorption between 939 cm^{-1} and 926 cm^{-1} in the spectra of a series of long-chain fatty acids which vanishes on esterification, and which they suggest may be tentatively assigned to this cause.

Flett [9, 99] has found a band in the 930 cm^{-1} range in very many acids. O'Connor *et al.* [24] and Sinclair *et al.* [42] confirm this and show that it vanishes on ester formation. Hadži and Sheppard find a band at 935 ± 15 cm^{-1} in all the fifteen acids they have examined. They have assigned it to the OH out-of-plane deformation mode and have shown that it exhibits the expected shift to 675 ± 15 cm^{-1} on deuteration.

This band is broad and reasonably intense. The breadth suggests an origin in a hydrogen bonded system, and it is in fact characteristic of the dimer form and is not found in the spectra of the monomers [1, 48]. The general range is that suggested by Hadži and Sheppard [49] but divergencies must be expected in some cases due to local interactions in the solid.

10.4(d). Band progression in fatty acids. Long-chain fatty acids examined in the solid or crystal state exhibit a regular series of evenly spaced absorptions in the region 1350–1180 cm^{-1}. Jones *et al.* [41] have pointed out that the number and appearance of these bands can afford information as to the length of the carbon chain involved. Thus lauric acid (C_{12}) shows only three regularly spaced bands in this region, the lowest being at 1195 cm^{-1}. As the carbon chain lengthens the number of such bands increased and heneicosoic acid (C_{21}) shows nine such bands, the lowest being at 1184 cm^{-1}. Between C_{16} and C_{21} the number of bands increased by one for each unit increase in the length of the carbon chain. Similar bands are shown by fatty acid esters and also by paraffins, although in the latter case the intensities are very much reduced. They are therefore to be associated with rocking or twisting motions of the CH_2 groups in the chains. These motions are primarily associated with the *trans* arrangement of the methylene groups such as occurs in *n*-paraffins. In consequence, anything which interferes with this arrangement modifies the spectrum in this region. Corish and Davison [67] have suggested that the virtual disappearance of these bands in molten acids is due to a continuous and random distribution of the $(CH_2)_n$ chains. The changes in structure in this region following changes of state have been fully confirmed by other workers [80, 81, 89–92]. Similarly, chain branching or the presence of double bonds tends to modify these patterns. In tetrabromostearic acid the characteristic pattern is lost, but the bands are clear and sharp in 2-bromostearic acid. A study of this region may therefore give data of some value in specific cases. It may provide, for example, a simple means of differentiating between saturated fatty acids of different chain lengths, as the spectra of these compounds as a class are otherwise largely indistinguishable. Meiklejohn *et al.* [91] have shown that there are also substantial differences between the spectra of fatty acid salts in this region which can equally well be used for identification. They have also made suggestions on the use of these absorptions for chain-length determinations.

10.5. The ionised carboxyl group; salts

Lecomte and his co-workers have made a very extensive study of the infra-red spectra of the salts of organic acids and have examined almost a thousand such materials [32—35]. When ionisation occurs, giving the COO^- group, resonance is possible between the two C–O bonds. In consequence the characteristic carbonyl absorption vanishes and is replaced by two bands between $1610 \ cm^{-1}$ and $1550 \ cm^{-1}$ and between $1400 \ cm^{-1}$ and $1300 \ cm^{-1}$, which correspond to the anti-symmetrical and symmetrical vibrations of the COO^- structure.

Of these bands, the former is very much more characteristic, as it is generally more constant in frequency whilst many other skeletal vibrations occur in the wide range $1400-1300 \ cm^{-1}$. Lecomte's general finding has been confirmed by later workers, and, as will be seen later, a somewhat similar state of affairs has been found to occur with amino-acids in which the zwitterion form permits resonance in the same way. Raman data [40] on numbers of carboxylic acids also indicate that salt formation results in the disappearance of the C=O absorption and its replacement by a band near $1430 \ cm^{-1}$. The fact that in this case the group is identified by the symmetric vibration is in accordance with theory which requires this mode to be strong in the Raman and weak in the infra-red. The reverse is true of the asymmetric frequency, and Ehrlich [80] estimates the intensity ratio for polymeric acids in the infra-red as about 7.6 : 1.

In the solid state the two frequencies show minor variations with the nature of the metallic ion and also with the nature of the group to which the ionised carboxyl group is attached. Kagarise [83] has shown that for mono- and di-valent elements there is a linear relationship between the electro-negativity of the element and the asymmetric stretching frequency of the salt, and Stimpson [84] has also noted variations of this type. Changes in the nature of the substituent group also have marked effects, and there are, for example, well-defined differences between the characteristic frequencies of formate, acetate and oxalate ions [85, 99]. Similarly, in substituted benzoic acids the carboxylate frequencies vary with the nature of the aromatic substituents [84]. In general ν_{as} occurs in the region of 1550 cm^{-1} and is of more diagnostic value than ν_s. However it is significantly more sensitive than ν_{CO} to substituent effects [101], and this results in a wide overall range. A halogen atom on the α-carbon atom, can shift ν_{as} by as much as 100 cm^{-1}. Hammett σ type relationships are reasonably well obeyed by ν_{as} in substituted benzoates [102]. This is not true if one uses the mean values of ν_{as} and ν_s suggesting that as in the case of the corresponding modes in the nitro group, the symmetric mode is involved to a much greater extent in coupling. In sodium trifluoroacetate the influence of the CF_3 group is particularly marked, and the COO$^-$ frequencies occur [86] at 1680 cm^{-1} and 1457 cm^{-1}. The former is close to the normal range for un-ionised acids, but must be compared with the value of 1825 cm^{-1} shown by the unassociated free acid.

Hydrogen bonding of the carboxylate group can lead to a small reduction in the asymmetric frequency. In the salt $CF_3COONa.2CF_3COOH$, for example [86], the bonded carboxylate ion absorbs at 1625 cm^{-1} (*as*) as compared with 1667 cm^{-1} for trifluoroacetic acid in pyridine solution [67]. In disodium ethylenediamine tetracetic acid the asymmetric carboxylate frequency of 1637 cm^{-1} is higher than in the tetrasodium salt (1597 cm^{-1}). This has also been ascribed to hydrogen bonding [85], but other factors may be involved in this special case.

In general, the spectra of fatty acid salts in the solid state show more distinctive differences than do the original acids, and Childers and Struthers [88] have made use of this in the analysis of acid mixtures of various types.

The carbonyl frequencies of ionised acids have been studied in solution by Gore, Barnes and Petersen [36], and by Dunn and McDonald [102] who have used solutions in water and in heavy water, in silver chloride cells to scan the whole spectral range. In a typical case sodium acetate was found to show the COO$^-$ absorptions near 1560 cm^{-1} and 1410 cm^{-1}. Using heavy water which avoids the obscuring of the carbonyl region by the water bands, it is possible to observe the reappearance of the unionised acid, and consequently of the carbonyl group, on adding DCl, when a strong band appears at 1730 cm^{-1}. This change can be reversed by the addition of NaOD. This characteristic shifting of the carbonyl frequency on passing from the un-ionised to the ionised acid is highly characteristic of carboxylic acids, and provides a neat and simple way in which the presence of such materials can be confirmed.

10.6. Bibliography

1. Bonner and Hofstadter, *J. Chem. Phys.*, 1938, **6**, 531.
2. Herman and Hofstadter, *ibid.*, p. 534.
3. Davies and Sutherland, *ibid.*, p. 755.
4. Davies, *J. Chem. Phys.*, 1940, **8**, 577.
5. Buswell, Rodebush and Roy, *J. Amer. Chem. Soc.*, 1938, **60**, 2239.
6. Klotz and Gruen, *J. Phys. and Colloid Chem.*, 1948, **52**, 961.
7. McCutcheon, Crawford and Welsh, *Oil and Soap*, 1941, **18**, 9.
8. Martin, *Nature*, 1950, **166**, 474.
9. Flett, *J. Chem. Soc.*, 1951, 962.
10. Shreve, Heether, Knight and Swern, *Analyt. Chem.*, 1950, **22**, 1498.
11. Pauling, *The Nature of the Chemical Bond* (O.U.P., 1940), p. 306.
12. Rasmussen, Tunnicliff and Brattain, *J. Amer. Chem. Soc.*, 1949, **71**, 1068.
13. Amstutz, Hunsberger and Chessick, *ibid.*, 1951, **73**, 1220.
14. Bellamy and Beecher, *J. Chem. Soc.*, 1952, 1701.
15. Kuratani, *J. Chem. Soc. Japan*, 1950, **71**, 401.
16. Hartwell, Richards and Thompson, *J. Chem. Soc.*, 1948, 1436.
17. Grove and Willis, *ibid.*, 1951, 877.
18. Lecomte and Guy, *Compt. rend. Acad. Sci. Paris*, 1948, **227**, 54.
19. Barnes, Gore, Liddel and Williams, *Infra-red Spectroscopy* (Reinhold, 1944).
20. Lecomte, *Traite de Chimie Organique*, Ed. Grignard et Baud (Masson et Cie., Paris, 1936), vol. 2, p. 143.
21. Gillette, *J. Amer. Chem. Soc.*, 1936, **58,**, 1143.
22. Flett, *Trans. Faraday Soc.*, 1948, **44**, 767.
23. Bartlett and Rylander, *J. Amer. Chem. Soc.*, 1951, **73**, 4275.
24. O'Connor, Field and Singleton, *J. Amer. Oil Chem. Soc.*, 1951, **28**, 154.
25. Rao and Daubert, *J. Amer. Chem. Soc.*, 1948, **70**, 1102.
26. Schonmann, *Helv. Phys. Acta*, 1943, **16**, 343.
27. Sheppard, *Trans. Faraday Soc.*, 1949, **45**, 693.
28. Munday, *Nature*, 1949, **163**, 443.
29. Cross and Rolfe, *Trans. Faraday Soc.*, 1951, **47**, 354.
30. Brookes, Eglington and Morman, *J. Chem. Soc.*, 1961, 106.
31. Thompson and Torkington, *J. Chem. Soc.*, 1945, 640.
32. Lecomte, *Rev. Optique*, 1949, **28**, 353.
33. Duval, Lecomte and Douvillé, *Ann. Physique*, 1942, **17**, 5.
34. Douvillé, Duval and Lecomte, *Bull. Soc. chim.*, 1942, **9**, 548.
35. Duval, Gerding and Lecomte, *Rev. trav. chim.*, 1950, **69**, 391.
36. Gore, Barnes and Petersen, *Analyt. Chem.*, 1949, **21**, 382.
37. Josien, Peltier and Pichevin, *Compt. Rend. Acad. Sci. Paris*, 1960, **250**, 1643.
38. Colthup, *J. Opt. Soc. Amer.*, 1950, **40**, 397.
39. Herman and Hofstadter, *J. Chem. Phys.*, 1939, **7**, 460.
40. Edsall, *ibid.*, 1937, **5**, 508.

41. Jones, McKay and Sinclair, *J. Amer. Chem. Soc.*, 1952, **74**, 2575.
42. Sinclair, McKay and Jones, *ibid.*, p. 2570.
43. Sinclair, McKay, Myers and Jones, *ibid.*, p. 2578.
44. Freeman, *ibid.*, p. 2523.
45. Guilleme, Chabanel and Wojtkowiak, *Spectrochim. Acta*, 1971, **27A**, 2355.
46. Haszeldine, *Nature*, 1951, **168**, 1028.
47. Husted and Ahlbrecht, *J. Amer. Chem. Soc.*, 1953, **75**, 1607.
48. Freeman, *ibid.*, 1859.
49. Hadži and Sheppard, *Proc. Roy. Soc.*, 1953, **A216**, 247.
50. Francis, *J. Chem. Phys.*, 1951, **19**, 942.
51. Kuratani, *J. Chem. Soc. Japan*, 1952, **73**, 758.
52. Fuson, Josien, Jones and Lawson, *J. Chem. Phys.*, 1952, **20**, 1627.
53. Josien and Fuson, *Compt. Rend. Acad. Sci. (Paris)*, 1952, **235**, 1025.
54. Goulden, *Spectrochim. Acta*, 1954, **6**, 129.
55. Bratoz, Hadži and Sheppard, *Bull. Sci. Acad. RPF. Yougoslavie*, 1953, **1**, 71.
56. *Idem, Spectrochim. Acta*, 1956, **8**, 249.
57. Katon and Sinho, *App. Spectroscopy*, 1971, **25**, 497.
58. Fuson and Josien, *J. Opt. Soc. Amer.*, 1953, **43**, 1102.
59. Davies and Evans, *J. Chem. Phys.*, 1952, **20**, 342.
60. Chulanovski and Simova, *Doklady, Akad. Nauk. S.S.S.R.*, 1949, **68**, 1033.
61. Bateuv, *Izvest, Akad. Nauk. S.S.S.R., Otdel Khim. Nauk.*, 1950, 402; *ibid.*, *Ser. Fiz.*, 1950, **14**, 429.
62. Stepanov, *J. Phys. Chem. (U.S.S.R.)*, 1945, **19**, 507; 1946, **20**, 907.
63. *Idem, Nature*, 1956, **157**, 808.
64. Cardwell, Dunitz and Orgel, *J. Chem. Soc.*, 1953, 3740.
65. Barrow and Yerger, *J. Amer. Chem. Soc.*, 1954, 5428.
66. Josien, Fuson, Lawson and Jones, *Compt. Rend. Acad. Sci. (Paris)*, 1952, **234**, 1163.
67. Corish and Davison, *J. Chem. Soc.*, 1955, 6005.
68. Fraenkel, Belford and Yankwich, *J. Amer. Chem. Soc.*, 1954, **76**, 15.
69. Barrow, *ibid.*, 1956, **78**, 5802.
70. Bellamy, Thomas and Williams, *J. Chem. Soc.*, 1955, 3704.
71. Josien and Callas, *Compt. Rend. Acad. Sci. (Paris)*, 1955, **240**, 1641.
72. Haszeldine, *J. Chem. Soc.*, 1954, 4026.
73. Harrand and Tuernal-Vatran, *Ann. Phys.*, 1955, **10**, 5.
74. Allan, Meakins and Whiting, *J. Chem. Soc.*, 1955, 1874.
75. Musso, *Chem. Ber.*, 1955, **88**, 1915.
76. Schotte and Rosenberg, *Arkiv, Kemi.*, 1956, **8**, 551.
77. Rosenberg and Schotte, *Acta. Chem. Scand.*, 1954, **8**, 867.
78. Farmer, *Chem. and Ind.*, 1955, 586.
79. Guertin, Wiberley, Bauer and Goldenson, *Analyt. Chem.*, 1956, **28**, 1553.
80. Neuilly, *Compt. Rend. Acad. Sci. (Paris)*, 1954, **238**, 65.
81. Rigaux, *ibid.*, 63, 783.
82. Ehrlich, *J. Amer. Chem. Soc.*, 1954, **76**, 5263.
83. Kagarise, *J. Phys. Chem.*, 1955, **59**, 271.
84. Stimpson, *J. Chem. Phys.*, 1954, **22**, 1942.
85. Ito and Bernstein, *Can. J. Chem.*, 1956, **34**, 170.
86. Klemperer and Pimentel, *J. Chem. Phys.*, 1954, **22**, 1399.
87. Chapman, *J. Chem. Soc.*, 1955, 1766.
88. Childers and Struthers, *Analyt. Chem.*, 1955, **27**, 737.
89. Corish and Chapman, *J. Chem. Soc.*, 1957, 1746.

90. Wensel, Schiedt and Breusch, *Z. Naturforsch*, 1957, **12B**, 71.
91. Meiklejohn, Meyer, Aronovic, Schuette and Meloch, *Analyt. Chem.*, 1957, **29**, 329.
92. Von Sydow, *Acta Chem. Scand.*, 1955, **9**, 1119.
93. George, Green and Pailthorpe, *J. Mol. Structure*, 1971, **10**, 297.
94. Odinokov and Iogansen, *Spectrochim. Acta*, 1972, 28A, 2343.
95. Odinokov, Maximov and Dzizenko, *Spectrochim. Acta*, 1969, 25A, 131.
96. Peltier, Pichevin, Dizabo and Josien, *Compt. Rend. Acad. Sci. Paris*, 1959, **248**, 1148.
97. Josien, Lascombe and Vignalou, *Compt. Rend. Acad. Sci. Paris*, 1960, **250**, 4146.
98. Eglington, Morman and Brookes, *J. Chem. Soc.*, 1961, 106.
99. Flett, *Spectrochim. Acta*, 1962, **18**, 1537.
100. Katon, Carll and Bentley, *App. Spectroscopy*, 1971, 25, 229.
101. Spinner, *J. Chem. Soc.*, 1964, 4217.
102. Dunn and McDonald, *Can. J. Chem.*, 1969, 47, 4577.

11

Esters and Lactones

11.1. Introduction and table

Esters have two characteristic absorptions arising from the C=O and C–O– groups. The carbonyl frequency is notably raised above that of normal ketones by the influence of the adjoining oxygen atom, so that differentiation of the two is usually possible. There is, however, a certain amount of overlap between, for example, unsaturated esters in which the CO frequency is lowered, and ketones such as α-halogen-substituted materials in which the CO frequency is raised. It is therefore necessary to take account of the intensity of the CO band, and also of the single bond C–O band which, in esters, is very strong and which can usually be differentiated from the weaker C–C bands of ketones which appear in the same spectral region. A similar enhancement of the carbonyl intensity under the influence of the adjacent oxygen atom also occurs, of course, in acids; but in practice the existence of these in polymeric form in the state in which they are usually examined results in a compensating shift to a lower frequency. There is still some degree of overlapping between the carbonyl frequency ranges of esters and acids, and the C–O stretching bonds also absorb in similar ranges, but, as has been shown (Chapter 10), the identification of acids from the OH region or by salt formation will usually resolve any difficulties, whilst again there are marked intensity differences in the carbonyl absorptions of the two classes.

The ester carbonyl absorption follows closely in behaviour the ketonic carbonyl band insofar as frequency shifts arising from environmental changes are concerned, and, as will be seen, the shifts

203

are usually of the same order of magnitude in both cases. This applies also to ring systems in which the changes in the CO frequencies of cyclic ketones with alteration of the ring size are closely paralleled by the comparable behaviour of lactones. As the various structural factors causing frequency shifts have been fully discussed under aldehydes and ketones in Chapter 9, they will be considered in rather less detail here.

As in the case of ketones, no discussion of carbonyl frequencies of esters with nitrogen attached to the carbonyl group is included in this section, as such materials will be dealt with in Chapter 12.

The single bond C–O absorption falls in the same general region as other C–O stretching vibrations of unsaturated ethers and alcohols, and is not particularly significant as such, especially as the frequency shows less stability than is the case with the carbonyl absorption. However, sufficient work has been done in some cases to enable the smaller and more specific frequency ranges of individual classes of

Table 11*

C=O *Stretching Vibrations*

Normal saturated esters	$1750-1730$ cm^{-1}
$\alpha\beta$-unsaturated and aryl esters	$1730-1717$ cm^{-1}
Vinyl ester type compounds	$1775-1755$ cm^{-1}
Esters with α-electronegative substituents	$1770-1745$ cm^{-1}
α-keto esters	$1755-1740$ cm^{-1}
β-keto esters (enolic)	Near 1650 cm^{-1}
Salicylates and anthranilates	$1690-1670$ cm^{-1}
γ-keto esters and higher	$1750-1735$ cm^{-1}
δ-lactones	$1750-1735$ cm^{-1}
δ-lactones, saturated	$1780-1760$ cm^{-1}
δ-lactones, $\alpha\beta$-unsaturated	$1760-1740$ cm^{-1}
δ-lactones, $\beta\delta$-unsaturated	Near 1800 cm^{-1}
β-lactones	Near 1820 cm^{-1}
Thio-esters, all types	Near 1695 cm^{-1}

C–O– *Stretching Vibrations*

Formates	$1200-1180$ cm^{-1}
Acetates	$1250-1230$ cm^{-1}
Phenolic acetates	Near 1205 cm^{-1}
Propionates and higher	$1200-1150$ cm^{-1}
Acrylates, fumarates, maleates	$1300-1200$ and $1180-1130$ cm^{-1}
Benzoates and phthalates	$1310-1250$ and $1150-1100$ cm^{-1}

*All bands shown in this table are strong.

esters, such as acetates, butyrates, benzoates, etc., to be charac-terised, whilst certain esters show two distinct absorption bands in this region, which increases considerably the possibilities of their identification.

In the case of the acetates, a very extensive study of the structure of the C—O band of a large number of sterol acetates has enabled Jones *et al.* [1] to obtain evidence as to the stereochemical structure of C_3 acetoxy-sterols in relation to the hydrogen atom at C_5.

The correlations dealt with are shown in Table 11.

11.2. C=O Stretching vibrations in esters

11.2(a). Saturated esters. The carbonyl frequencies of some thirty-six simple esters, including formates, acetates, butyrates, etc., were examined by Thompson and Torkington in 1945 [2]. In the case of formates, which, as the first members of a series, might be expected to be anomalous, the carbonyl frequency was in the range 1724–1722 cm^{-1}, but with all the others it fell very close to 1740 cm^{-1}. This result has been fully confirmed by later workers [12, 29, 69, 70]. Apart from the formates almost all alkyl esters absorb between 1748 and 1730 cm^{-1} in carbon tetrachloride solution. There is a small progressive fall in the frequency as the length of either the acid or alcohol residue is lengthened or branched. Thus, methyl acetate absorbs at 1748 cm^{-1}, ethyl propionate at 1739 cm^{-1}, and s-butyl *iso*butyrate at 1730 cm^{-1}.

The bands of acetates are always narrower than those of other esters [12] which suggests that rotational isomerism involving two forms is taking place as in the ketones.

The carbonyl frequencies of esters of inorganic alcohols can vary within very wide limits. Freeman [71] has given values ranging from 1798 cm^{-1} for CH_3COONO_2 to 1580 cm^{-1} for $CH_3COOHgC_2H_5$.

Hydrogen bonding has only a small effect unless resonance stabilisation occurs. Searles *et al.* [41] have shown the carbonyl shift in methanol solution is about 8 cm^{-1}. With lactones the proton-accepting power can be a little greater, leading to shifts of up to 15 cm^{-1}, depending upon the ring size. Henbest and Lovell [68] have also noted a few unusual instances in which hydrogen bonding takes place preferentially on the alcohol oxygen atom, and the carbonyl frequency then rises about 10 cm^{-1} above the usual value.

These general findings have been confirmed by many other workers. Hampton and Newell [4] have described the carbonyl frequencies of nineteen esters of various kinds. They note that the frequency is more subject to variation with increasing molecular complexity, but this is primarily due to known causes, such as α-halogenation or $\alpha\beta$-unsaturation and their value of 1738 cm^{-1}, for example, for ethyl palmitate, is substantially the same as for simpler saturated materials. These workers have also given values for the molecular extinction coefficients of esters. The values show considerable variations from class to class, but within a given group they are reasonably constant and consistent. The values quoted for simple saturated esters of the type under discussion, for example, range from 569 to 610 units. Anderson and Seyfried [5] have also given data on the positions and intensities of ester carbonyl absorptions which are generally in line with these.

With more complex molecules the correlation still holds good. Rasmussen and Brattain [6] find the carbonyl frequencies of methyl pivalate, ethyl diphenyl acetate and methyl sarcosinate to be all close to 1739 cm^{-1}, indicating that there is little or no inductive effect on the C=O arising from α-methyl, aryl or amino-substitution. With ethyl cyanoacetate, however, the frequency rises to 1751 cm^{-1}, reflecting the influence of the C$\equiv$N group. Shreve *et al.* [7] and Sinclair *et al.* [33, 34] have examined the methyl esters of long-chain fatty acids, and find the carbonyl frequency falls in the range 1748–1739 cm^{-1}, whilst a number of triglycerides absorb [7] between 1751 and 1748 cm^{-1}. O'Connor *et al.* [8] report similar findings in the methyl and ethyl esters of fatty acids.

The most extensive study of ester carbonyl absorptions is that made by Jones and co-workers [9–11, 28, 42, 43] who have investigated the carbonyl frequencies of over a hundred sterol acetates, propionates, etc. With certain exceptions which will be dealt with later, all the saturated materials absorbed in the range 1742–1735 cm^{-1}, regardless of where the ester group was situated in the steroid nucleus, although there were indications that similarly substituted materials each had an even narrower range of characteristic frequencies.

Intensity studies on esters have been reported by many workers [13, 35–44, 45, 72, 73]. In general ester carbonyl bands are about twice as intense as the bands of comparable ketones. This applies equally to conjugated esters, which like the ketones show an intensification compared with the saturated examples. This assists the differentiation of esters and of their types, but it would be unwise to place too much reliance on it.

11.2(b). The influence of conjugation. The expected lowering of the carbonyl frequency by $\alpha\beta$-unsaturation has been confirmed by many workers [2–4, 6, 10, 46, 47]. However, the weakening of the carbonyl bond by $\alpha\beta$-unsaturation is, in general, less than in the case of ketones, and the extent of the shift is of the same magnitude for aryl substitution as for a simple double bond. Methyl methacrylate absorbs at 1718 cm^{-1} and propyl methacrylate [13] at 1721 cm^{-1}, whilst a series of eleven steroid benzoates [10] absorbed in the range 1724–1719 cm^{-1}. Direct confirmation that the reduction in the frequency is due to the influence of the double bond is given by the fact that these steroid benzoates absorbed at 1739–1735 cm^{-1} after hydrogenation to the hexahydrobenzoate [10]. The frequency fall is therefore of the order of 20 cm^{-1}. Further conjugation in the $\gamma\delta$-position has little, if any, appreciable effect, and ethyl cinnamate [4], for example, absorbs at 1717 cm^{-1}, against 1724 cm^{-1} for methylbenzoate [6]. Methyl esters of naphthalene and phenanthrene carboxylic acids also absorb [30] at 1724 cm^{-1}. These values relate to studies on liquids or solutions, and the vapour frequencies are likely to be a little higher [3].

These frequencies are, of course, subject also to other changes arising from additional structural features. For example, the *ortho-*, *meta-* and *para*-nitrobenzoates [6] absorb at 1733 cm^{-1}, reflecting a marked lessening of the influence of the aryl group. Compounds such as ethyl $\beta\beta$-diethoxyacrylate [6] (1736 cm^{-1}), diallyl maleate [4] (1738 cm^{-1}) and diethyl-1-2-dicyanoethane dicarboxylate [46] (1755 cm^{-1}) also show abnormally high carbonyl frequencies. These are probably due to field effects similar to those operating in *a*-halogenated compounds. In simple conjugated esters such as acrylates or crotonates the frequency reduction from conjugation is about 15 cm^{-1}. With alkyl benzoates there is a regular fall in frequency as the size or complexity of the alkyl group is increased [76, 77]. This is precisely parallel to the effect already noted in saturated esters, and is of comparable magnitude.

Conjugation with acetylenic links produces an enhanced shift, and compounds of this type absorb [47] in the range 1720–1708 cm^{-1}.

11.2(c). Vinyl esters. One case of frequency alteration in esters which is not paralleled in ketones is that arising from the structure $CO-O-C=C$. Compounds of this type show a marked enhancement of the carbonyl frequency, regardless of whether the double bond is normal or part of an aromatic ring. Barnes *et al.* [14] noted this effect in vinyl acetate, which absorbs at 1776 cm^{-1}. Hartwell *et al.* [3] found a similar effect with phenyl acetate and Jones *et al.* have reported it in steroid phenolic esters [28]. Other instances of this effect are given by Grove and Willis [15] and by Walsh [16], and more recently by Freedman [77], McManis [79] and others. The rise in the carbonyl frequency of vinyl esters is usually about 20 cm^{-1}. With phenyl esters the difference is slightly higher (25 cm^{-1}) but the difference is not sufficient to have any diagnostic value. Bellamy [48] has suggested that as vinyl acetate has no

resonance energy the normal mesomerism to forms such as $-\overset{\displaystyle O^-}{\underset{+}{\underset{|}{C}}}=OR$

is suppressed in such cases, so that the frequency of the carbonyl absorption is determined solely by the high electronegativity of the oxygen atom. Freedman [77] has proposed that the effect arises primarily from the increased electronegativity of the ester oxygen atom, and has produced good evidence in support of this from studies on phenyl acetates. In $\alpha\beta$-unsaturated vinyl or phenyl esters the two opposing effects cancel out, and these absorb at the same frequencies as saturated esters.

11.2(d). The influence of α-halogen groups. The influence of α-halogen substitution in esters is closely parallel to the cases of ketones already discussed. There is a very substantial amount of data available [36–38, 49–53, 12, 80], which was indeed used for the early studies which established the occurence of rotational isomerism and of field effects. Conformations which have the halogen atom at the α-carbon atom eclipsed with the carbonyl oxygen show a frequency rise. When the halogen is in a *gauche* conformation there is little or no effect. In consequence, esters with one or two halogen atoms at the α-carbon atom always show two carbonyl bands in the vapour or solution states but only one in the crystal. Variations in the relative intensities of the two bands occur depending on the polarity of the solvent which helps to determine the relative proportions of the two forms. Tri-α-halogenated esters show only one band as only one form is allowed.

Fluorine substitution shows the largest frequency rise [25, 49, 51] as is to be expected. This is usually about 45 cm^{-1} for the *cis* form as compared with the *gauche*. With α-chlorine substitution the difference is about 30 cm^{-1}, for bromine about 20 cm^{-1}, and for iodine about 10 cm^{-1}. The nature of this effect as a dipolar interaction is well established, and as it depends upon the proximity in space of the halogen and oxygen atoms, a similar effect is to be expected in *ortho* halogenated benzoates. This is indeed the case, and many such examples have been described. Even with *ortho* iodo-benzoates there is a difference of 9 cm^{-1} between the *cis* and *trans* forms [75]. A parallel effect occurs with other polar groups at the α-carbon atom. An α-cyano group results in twin carbonyl bands, one of which is about 25 cm^{-1} higher than the other. An α-oxygen atom behaves similarly, and 1 : 1-diacetoxy propane [54] absorbs at 1761 cm^{-1}.

In a few special cases frequency shifts of ester carbonyl group have been observed in structures with the grouping $-CO-O-CH_2-CF_3$, despite the apparent separation of the halogen atoms from the carbonyl oxygen [50]. No similar effect is found with β-flourination on the other side of the carbonyl, and compounds with the group $CF_3CH_2CO-O-R$ show a normal ester frequency. The rise in the frequency of the first group is not well understood, but may be connected with the *cis* arrangement of esters which could allow the fluorine atoms to approach the oxygen of the carbonyl group.

11.3. The C=O stretching vibration in diesters and keto esters

11.3(a). α-Diesters and α-keto esters. It is not easy to assess the effects of interactions in α-diketones or keto esters. These occur in *cis* and *trans* forms, but in some cases such as dibutyl oxalate [4] (1746 cm^{-1}) and methyl pyruvate [17] (1747 cm^{-1}) the ester band is essentially normal. In others, such as dimethyl oxalate two bands appear at 1780 and 1752 cm^{-1}. These cannot be simple symmetric and antisymmetric modes, as the former would be forbidden by symmetry in the *trans* isomer. It is usually assumed that the higher frequency band is in some way associated with the *cis* form in which some dipolar interaction between the oxygen atoms might be expected.

11.3(b). β-keto esters and related compounds, The phenomenon of conjugate chelation has already been described in some detail in the case of diketones (Chapter 9), and essentially similar effects are observed with esters of comparable structure. Rasmussen and Brattain [6] have studied a number of esters in which this effect is found. Ethyl $\alpha\alpha$-dimethylaceto-acetate cannot enolise and in solution is normal in showing absorptions at 1718 cm^{-1} and 1742 cm^{-1} due to the ketone and ester carbonyl groups. Both ethyl-α-methyl-acetoacetate and ethylacetoacetate, however, show an additional band at 1650 cm^{-1}, which is ascribed to the ester carbonyl group after chelation to the enolic hydroxyl group. The methyl ester also shows absorption at 1632 cm^{-1}, which has been assigned [30] to the C=C linkage. On the other hand, there does not seem to be much possibility of enolisation in diethyldiacetylsuccinate, as this shows only normal ketone and ester bands. The β-diester, diethylmalonate, is essentially normal in its carbonyl frequency [2], although in carbon tetrachloride solution some splitting into a doublet is observed [46].

In the aromatic series n-butylsalicylate shows its ester absorption [6] at 1675 cm^{-1} and methylsalicylate [4] at 1684 cm^{-1}; this is again ascribed to conjugate chelation. The shift is less than occurs with enolised chelate β-diketones or alkyl-β-keto esters, indicating that the chelation is not so strong. This probably reflects the weaker double-bond character of the ring linkages. The influence of chelation is clearly shown by the examination of acetylated n-butylsalicylate, in which the carbonyl frequency returns to 1723 cm^{-1}, which is normal for an aryl ester. The acetate band is also normal and appears at 1770 cm^{-1}, reflecting its vinyl ester structure.

With compounds such as methyl-10-hydroxy-9-phenanthrene carboxylate [30], the increased degree of fixation of the ring double bonds results in stronger chelation, so that the frequency shift is increased to about 75 cm^{-1}. As in the case of the corresponding ketones, the frequency shifts of chelated esters have been used to measure double-bond character in the indane series [55].

Chelation also occurs in cyclic β-keto esters, such as derivatives of enolisable keto esters of *cyclo*hexanone and *cyclo*pentanone. Leonard *et al.* [31] and Bellamy and Beecher [56] have examined a number of such cases. Non-enolisable keto esters of *cyclo*hexanone show absorptions near 1735 cm^{-1} and 1718 cm^{-1} corresponding to the ester and ketonic carbonyl groups. However, compounds such as ethyl-*cyclo*hexanone-2-carboxylate show these bands and two others at 1656 cm^{-1} and 1618 cm^{-1} which must arise from the chelate structure of the enol form. Leonard *et al.* [31] associate the first of these new bands with the chelated carbonyl absorption and the second with the double-bond absorption. A similar effect is observed in *cyclo*pentanone derivatives, but the intensities of the bands are then much reduced.

Chelation is also to be expected in β-amino-αβ-unsaturated esters [6] by analogy with the ketones. This has been found experimentally in the case of methyl-*N*-methylanthranilate, which absorbs at 1685 cm^{-1}, indicating a chelation of about the same strength as occurs with the salicylate. Removal of the chelating hydrogen atom results in reversion of the ester carbonyl frequency to 1730 cm^{-1}.

β-Diesters such as diethyl malonate do not enolise. This ester shows two carbonyl bands at 1745 and 1732 cm^{-1} which are probably due to the presence of two conformers. The higher frequency band is probably due to a form in which some dipolar interaction occurs, as the diethyl derivative in which steric hindrance is to be expected, shows only a single band at 1732 cm^{-1}.

11.3(c), γ- and δ-diesters and keto esters. γ- and δ-Diesters do not interact to any appreciable extent. Diethylsuccinate [17] absorbs at 1733 cm^{-1}, whilst phthalates usually absorb near 1730 cm^{-1}. This latter value is slightly high for aryl esters, and this may be due to the fact that the electronic influence of the ring is shared between the two carbonyl groups. Ester groups even further apart show no evidence of interaction. Diethyladipate is completely normal, and absorbs at 1739 cm^{-1}.

In considering γ- and δ-keto esters and similar materials the possibility of lactol formation should not be overlooked, as the ester frequencies will then appear as those of the corresponding five- or six-membered ring lactones. The characteristic frequencies of these will be discussed below. The use of the number and position of carbonyl frequencies in the identification of lactol structures has been demonstrated in a number of cases by Grove and Willis [15].

11.4 C=O Stretching vibrations in lactones, lactols, etc.

11.4(a). Six-membered rings (δ-lactones). Based on data by Rasmussen and Brattain [6, 18], Jones [10. 42], and Grove and Willis [15], which has been further substantiated by later work by Hall and Zbinden [81], the carbonyl band in unfused six-membered ring lactones is generally reported as falling in the 1750 to 1735 cm^{-1} range [6, 10, 18, 42, 81], close to the frequencies of normal esters. However, there does seem to be a tendency for the bands to appear towards the upper end of this range. A double bond at the $\alpha\beta$-position lowers the frequency by about 20 cm^{-1}, but at the $\gamma\delta$-position it raises the frequency as in vinyl esters. This correlation is quoted by Rasmussen and Brattain [18] on the basis of unpublished work, but they have also [6] given a value for δ-valerolactone (1738 cm^{-1}). Jones et al. [10, 42] have confirmed this correlation in a number of steriod δ-lactones, and it has also been found to hold good for a number of six-membered ring lactols by Grove and Willis [15]. Unsaturated rings of this type have not been extensively studied, but it is to be expected, by comparison with the behaviour of γ-lactones, that the carbonyl frequency will be subject to the same degrees of shift from $\alpha\beta$-conjugation, α-electronegative substituents or vinyl ester type unsaturation as with open-chain esters.

This correlation does not apply to lactones with two carbonyl groups in the same six-membered ring. Wasserman and Zimmerman [23] have examined mesolactide and benzilide in which the carbonyl frequencies occur between 1767 cm^{-1} and 1757 cm^{-1} and Randall et al. [17] quote a value of 1721 cm^{-1} for dehydracetic acid. There are compounds in which interaction effects might well be expected and are too complex for any interpretation of these shifts to be attempted.

Another factor which might be expected to lead to frequency shifts is the introduction of ring strain by the fusion of the lactone to other cyclic systems. In certain spiro-type δ-lactones of the steroid series this is certainly the case, as absorption occurs in the 1793—1786 cm^{-1} range [42]. This is a higher frequency even than the corresponding spiro γ-lactones (1781—1777 cm^{-1}), which are essentially normal. Wilder and Winston [57] have also reported carbonyl frequencies near 1764 cm^{-1} in tricyclic lactones which are believed on other grounds to be six-ring systems. However, the extent of shift will clearly depend upon the degree of additional strain, if any, which is introduced.

11.4(b). Five-membered rings. Saturated lactones have been extensively studied [6, 10, 15, 18—22, 27, 12, 80]. They absorb near 1780 cm^{-1} showing the upward frequency shift to be expected from the ring strain. $\alpha\beta$-Unsaturated γ-lactones would be expected to absorb about 20 cm^{-1} lower than the saturated materials at 1750 cm^{-1}, and this has been confirmed by Grove and Willis [15], who have examined a number of aromatic lactols such as acetophenone-*ortho*-carboxylic acid, the lactol of benzil-*ortho*-carboxylic acid, phthalide and a number of similar products. A doubling of the carbonyl band sometimes occurs in lactones. This cannot be due to rotational isomerism and must arise from Fermi-resonance interactions [84]. With unsaturation $\beta\gamma$ to the carbonyl group of five-membered ring lactones, the vinyl ester type of structure is produced, and the frequency consequently rises to near 1800 cm^{-1}. A typical case is that of $\beta\gamma$-angelica lactone, which absorbs at 1799 cm^{-1}, and 3-methylenephthalide and phthalidylidene acetic acid [15] absorb at 1780 cm^{-1} and 1800 cm^{-1}, respectively. Woodward and Kovach [22] regard 1792 cm^{-1} as characteristic of enol lactones which have this structure. In instances in which one of the double bonds is exocyclic it is possible for both types of conjugation to occur together. In proto-anemonin and its homo derivative an $\alpha\beta$-conjugated γ-lactone system occurs which has also a vinyl ether structure due to the presence of an exocyclic double bond on the γ carbon atom. These compounds [58] absorb in the liquid state at 1776 cm^{-1}. In the converse case [59], in which the vinyl ether system is in the ring and the $\alpha\beta$ conjugation is exocyclic, the corresponding frequency is 1750 cm^{-1}.

Chelate systems can also occur in certain cases. Rasmussen and Brattain [6] have observed this in α-acetyl-γ-butyrolactone, which has the β-keto ester structure. The non-enolised form gives rise to bands at 1773 cm^{-1} (five-membered ring lactone) and 1718 cm^{-1} (ketone), but there is also a band at 1656 cm^{-1}, which the authors attribute to a chelated carbonyl group similar to those observed in open-chain products. Halogen substitution in the α-position also raises the carbonyl frequencies. In the extreme case, perflouro-butyrolactone absorbs at 1873 cm^{-1} due to the influence of the fluorine atoms [29], 2-bromobutyrolactone [12] absorbs at 1797 cm^{-1}. Anomalous frequencies also occur when electronegative substituents are present on the carbon atom of γ-lactones. Brugel *et al.* [60] have cited a number of such cases, of which γ-acetoxy-γ-valerolactone which absorbs at 1797 cm^{-1} is a typical example

Other factors which might be expected to influence these carbonyl frequencies are ring strain arising from fused-ring systems and hydrogen bonding effects. However, in contrast to the behaviour of δ-lactones, tri- and tetra-cyclic γ-lactones absorb normally [61] near 1770 cm^{-1}, and this is true also of spiro-type steroid γ-lactones [42]. Hydrogen bonding, on the other hand, appears to have rather more effect upon both γ- and δ-lactones than it has upon the corresponding esters, and shifts of up to 15 cm^{-1} can occur from this cause [41].

11.4(c). Four-membered rings. These have not been studied to any extent, and are very uncommon. The Raman spectrum of β-butyrolactone has been obtained by Taufen and Murray [32] and shows a band at 1818 cm^{-1}. Remena [12] finds twin carbonyl bands in the infra-red spectrum of this compound, at 1796 and 1782 cm^{-1}. The spectrum of β-propionolactone in the infra-red has been studied by Searles *et al.* [41], and this compound absorbs at 1841 cm^{-1} in carbon tetrachloride solution. These values confirm the expectation that the increased ring strain will result in a further shift towards higher frequencies.

11.5. Thiol esters

Thiol esters have been studied by Rasmussen [18], and more recently by Nyquist and Potts [82] and Baker and Harris [83]. Dialkyl thiolesters show their carbonyl bands in the narrow range 1698 to 1690 cm^{-1}. The reasons for this substantial fall as compared with normal esters are not fully understood and have been the subject of some controversy [80]. However, it is probably a direct consequence of the low electronegativity of sulphur. αβ-Conjugation in aryl alkyl thiolesters lowers the frequency to 1665 to 1670 cm^{-1} which is a rather larger shift than is found in esters. On the other hand the rise in frequency when an aromatic ring is attached to the sulphur atom is less than in the phenyl esters, and such compounds absorb near 1710 cm^{-1}.

α-Halogen substitution with one or two halogen atoms results in two carbonyl bands as in normal esters. However in this case there is no frequency rise. The higher frequency band occurs in the normal range for thiol esters and it is the lower frequency band which is displaced. The doubling must originate from rotational isomerism, but the difference in behaviour from esters is not understood. A second curious effect in thiolesters is the lowering of the carbonyl frequency which follows sulphur substitution at the α-carbon atom [85]. The band then falls close to 1675 cm^{-1}. This has been explained as due to a transfer of charge from the CO group into a vacant orbital on the α-carbon sulphur atom.

11.6. C—O Stretching vibrations

11.6(a). General. The C—O stretching mode gives rise to strong absorption bands in the region 1300—1000 cm^{-1} in acids, alcohols, ethers and esters. As a skeletal mode it is much less stable in position than the corresponding carbonyl vibration, and it is very sensitive to changes in the mass and nature of the attached groups. Despite its very considerable intensity, the band is often difficult to recognise, as it occurs in a region of the spectrum where many strong bands commonly occur. Moreover, it is not to be expected that a band will arise from a pure C—O stretching motion, and the movements of other atoms are undoubtedly involved. Fowler and Smith [62] have studied a series of esters and concluded that no bands in this region can be unequivocally assigned. Nevertheless, useful data can be obtained from this region by a study of the most intense bands, which are the most likely to be primarily associated with the oxygen function.

Formates, acetates, butyrates and other simple esters have been studied in some detail by Thompson and Torkington [2], who were able to show that the strongest band, which commonly occurs near 1200 cm^{-1}, was relatively constant in frequency in each group, whilst in each case there was one other strong band in the 1200—1000 cm^{-1} range which was more subject to frequency shift following changes in the nature of the alcohol residue. The first of these they assigned tentatively to the C—O band contiguous to the carbonyl group, and its intensity has been studied in some detail by Francis [45] and by Russell and Thompson [63]. The second, more variable band was assigned to the C—O link of the alcohol residue. Later workers have found that $\alpha\beta$-unsaturated esters show two strong bands in the 1300—1100 cm^{-1} region, whilst others have been able to employ the C—O frequencies for the study of steric effects in steroid molecules. In view of the variations from class to class, these correlations will be considered individually.

11.6(b). Formates. Methyl formate has its strongest band in the 1250—1050 cm^{-1} region at 1214 cm^{-1}, falling to 1195 cm^{-1} in ethyl formate and to 1185 cm^{-1} in higher homologues [2, 63]. Disregarding the first member of the series, this band is likely to occur in the range 1200—1180 cm^{-1}, but the frequency will almost certainly be influenced by conjugation in phenolic formates and similar materials. The extent to which this will occur is not known, so that this correlation and, indeed, all those relating to the C—O stretching mode, must be applied with caution. Jones *et al.* [1] have confirmed the presence of the 1180 cm^{-1} band in two sterol formates. Formates also show a second absorption of much lower intensity [63] in the range 1161—1151 cm^{-1}.

11.6(c). Acetates. Acetates have been much more fully studied than other esters. Thompson and Torkington [2] found a characteristic band near 1245 cm^{-1} in the lower eight acetates up to butyl acetate, and the band did not change in position in the branched-chain isomers. Jones *et al.* [1, 65] have recently examined the position of this band in over one hundred 3-acetosteroids with most interesting results. Firstly, the complete group of spectra confirmed that a very strong band was present in this region in all the acetates studied. In many cases it was a single band, when it occurred in the range 1247—1236 cm^{-1}; in other cases splitting into three components with bands between 1250 cm^{-1} and 1200 cm^{-1} was found. Correlation of these findings with the known structures of the sterols revealed that the single-band structure was shown by all those compounds in which the C_3 acetate group and the C_5 hydrogen atom were in the *trans*-configuration and also in those compounds from which the C_5 hydrogen atom was absent and a Δ^5-double bond was present. Similarly the triplet structure was found in all cases in which these two groups were in the *cis*-configuration. Each of the *trans*- and *cis*-forms have of course two stereoisomers, but these are readily differentiated by precipitation of the alcohol with digitonin, which determines the configuration at C_3. This method therefore provides a means of establishing the relative configurations of the C_3-hydroxy- and the C_5-hydrogen atom of any given steroid. The complication of the single-band structure being shown also by the Δ^5-sterols is not a real difficulty, as the presence of the double bond can be readily identified spectroscopically. The method is not, however, applicable to compounds with elements or groups other than hydrogen at C_5. Comparable results have been obtained by other workers both in sterols [39, 64] and in decanol acetates [40].

The authors suggest that the origin of this differentiation may lie in the possibility that the *cis*-compounds showing multiple bands may be able to exist as a mixture of labile isomers resulting from hindered rotation about the C—O bond of the acetate group. However, this is rendered somewhat improbable by the finding of Allsop *et al.* [66] that no similar phenomenon occurs in triterpenoid-

3-acetates in which both equatorial and axial substitution leads to only a single band at 1243 cm^{-1}. This must be attributed to the presence in these cases of the *gem* dimethyl group at the C_4 carbon, but they point out that this in itself would not prevent rotational isomerism taking place. Doubling of the 1250 cm^{-1} band also occurs in simple acetates in solution [63].

Jones *et al.* [1] have also examined three steroid acetates in which ring I is fully aromatic. As is to be expected, this results in a shift of the C—O stretching frequency, and the strongest band in this region is now found near 1205 cm^{-1}. The position of the C—O absorption in these, and in saturated steroid acetates, is inversely tied to that of the carbonyl absorption, and a simple linear relation connects the two [65].

The second absorption arising from acetates occurs in the 1060—1000 cm^{-1} range, and is much more difficult to identify owing to its weaker intensity. However, it can be identified with reasonable certainty in steroid and triterpenoid acetates, and like the 1250 cm^{-1} absorption, it is then valuable in assigning the steric configuration. Thus Rosenkrantz and Skrogstrom [64] and Jones and Herling [65] have shown that the precise position of this band within this range can be used to differentiate axial and equatorial configurations, and in general the band appears to follow closely the behaviour of the related OH band found in the hydroxy steroid compounds in the same region. A similar differentiation can also be achieved in the triterpene series [66], although it is interesting to note that in this case the direction of shift on passing from an axial to an equatorial configuration appears to be the opposite to that which occurs in the sterols.

11.6(d). Propionates, butyrates and higher homologues. Thompson and Torkington [2] quote the following mean values for the C—O stretching modes of the series of simpler higher esters which they have examined:

Propionates   1190 cm^{-1},   normal   butyrates   1190 cm^{-1}, *iso*butyrates 1200 cm^{-1}, and *iso*valerates 1195 cm^{-1}.

The frequencies found within each group were very consistent, and comparable values have been quoted by Kendall *et al.* [67] for higher members of the series. They list the following as characteristic: adipates 1175 cm^{-1}, ricinoleates 1174 cm^{-1}, 2-ethyl-hexoates 1168 cm^{-1}, laurates 1163 cm^{-1}, oleates, 1172 cm^{-1}, sebacates 1172 cm^{-1}, stearates 1174 cm^{-1}, citrates 1183 cm^{-1} and benzoates 1280 and 1120 cm^{-1}. Apart from the last, the distinction between the others is probably too fine to be of much value, but it strongly supports the correlation of this band with an ester frequency. Shreve *et al.* [7] have commented on the position of the C—O— vibration in the methyl esters of long-chain fatty acids, in which they find a common pattern of three bands near 1250 cm^{-1}, 1205 cm^{-1} and 1175 cm^{-1} which they associate with the C—O linkage, and this has been confirmed by Sinclair *et al.* [33]. The band near 1175 cm^{-1} is, however, the strongest in each case, so that the assignment is generally in agreement with the earlier correlations.

A somewhat similar pattern is found [7] in triglycerides, but in this case the absorption pattern is of a strong band near 1163 cm^{-1} flanked by weaker bands near 1250 cm^{-1} and 1110 cm^{-1}. It is therefore reasonable to expect a strong band to be present between 1250 cm^{-1} and 1160 cm^{-1} in esters higher than propionates provided, of course, there are no other interfering structures present such as $\alpha\beta$-unsaturation either of the C—O— link or the C=O group.

11.6(e). $\alpha\beta$-unsaturated esters. Only a limited number of spectra of esters of this type are available in the literature, and no systematic study of this group has been published. Colthup [26] indicates that acrylates, fumarates, maleates, benzoates and phthalates all give two strong bands in the approximate regions 1310—1250 cm^{-1} and 1200—1100 cm^{-1}. However, we can confirm that both benzoates and phthalates show two strong and usually recognisable bands between 1300 cm^{-1} and 1250 cm^{-1} and 1150 cm^{-1} and 1100 cm^{-1} which are often helpful in identifying these groups. Jones *et al.* [1] report finding a strong band at 1270 cm^{-1} in five steroid benzoates which also supports this correlation.

11.7. Bibliography

1. Jones, Humphries, Herling and Dobriner, *J. Amer. Chem. Soc.*, 1951, 73, 3215.
2. Thompson and Torkington, *J. Chem. Soc.*, 1945, 640.
3. Hartwell, Richards and Thompson, *ibid.*, 1948, 1436.
4. Hampton and Newell, *Analyt. Chem.*, 1949, 21, 914.
5. Anderson and Seyfried, *ibid.*, 1948, 20, 998.
6. Rasmussen and Brattain, *J. Amer. Chem. Soc.*, 1949, 71, 1073.
7. Shreve, Heether, Knight and Swern, *Analyt. Chem.*, 1950, 22, 1498.
8. O'Connor, Field and Singleton, *J. Amer. Oil Chem. Soc.*, 1951, 28, 154.
9. Jones, Humphries and Dobriner, *J. Amer. Chem. Soc.*, 1949, 71, 241.
10. *Idem, ibid.*, 1950, 72, 956.

11. Jones and Dobriner, *Vitamins and Hormones* (N.Y. Academic Press,) 1949, 294.
12. Renema, Thesis, University of Utrecht, 1972.
13. Cross and Rolfe, *Trans. Faraday Soc.*, 1951, **47**, 354.
14. Barnes, Gore, Liddel and Williams, *Infra-red Spectroscopy* (Reinhold, 1944).
15. Grove and Willis, *J. Chem. Soc.*, 1951, 877.
16. Walsh, *Trans. Faraday Soc.*, 1947, **43**, 75.
17. Randall, Fowler, Fuson and Dangl, *Infra-red Determination of Organic Structures* (Van Nostrand, 1949).
18. Rasmussen and Brattain, *The Chemistry of Penicillin* (Princeton Univ. Press, 1949), p. 404.
19. Whiffen and Thompson, *J. Chem. Soc.*, 1946, 1005.
20. Richards and Thompson, *C.P.S. Reports*, 442, 511.
21. *Idem, The Chemistry of Penicillin* (Princeton Univ. Press, 1949), p. 386.
22. Woodward and Kovach, *J. Amer. Chem. Soc.*, 1950, **72**, 1019.
23. Wasserman and Zimmerman, *ibid.*, p. 5787.
24. Hurd and Kreuz, *ibid.*, p. 5543.
25. Hauptschein, Stokes and Nodiff, *ibid.*, 1952, **74**, 4005.
26. Colthup, *J, Opt. Soc. Amer.*, 1950, **40**, 397.
27. Marion, Ramsay and Jones, *J. Amer. Chem. Soc.*, 1951, **73**, 305.
28. Jones, Humphries, Herling and Dobriner, *ibid.*, 1952, **74**, 2820.
29. Josien, Lascombe and Vignalou, *Compt. Rend. Acad. Sci. Paris*, 1960, **250**, 4146.
30. Hunsberger, Ketcham and Gutowsky, *J. Amer. Chem. Soc.*, 1952, **79**, p. 4839.
31. Leonard, Gutowsky, Middleton and Peterson, *ibid.*, p. 4070.
32. Taufen and Murray, *ibid.*, 1945, **67**, 754.
33. Sinclair, McKay and Jones, *ibid.*, 1952, **74**, 2570.
34. Sinclair, McKay, Myers and Jones, *ibid.*, 2578.
35. Stahl and Pessen, *ibid.*, 5487.
36. Haszeldine, *Nature*, 1951, **168**, 1028.
37. Husted and Ahlbrecht, *J. Amer. Chem. Soc.*, 1953, **75**, 1607.
38. Rappaport, Hauptschein, O'Brien and Filler, *ibid.*, 2695.
39. Fürst, Kuhn, Scotoni and Günthard, *Helv. Chim. Acta*, 1952, **35**, 951.
40. Dauben, Haerger and Freeman, *J. Amer. Chem. Soc.*, 1953, **74**, 5206.
41. Searles, Tamres and Barrow, *J. Amer. Chem. Soc.*, 1953, **75**, 71.
42. Jones and Herling, *J. Org. Chem.*, 1954, **19**, 1252.
43. Jones and Sandorfy, *Chemical Applications of Spectroscopy* (Interscience, New York), 1956.
44. Barrow, *J. Chem. Phys.*, 1953, **21**, 2008.
45. Francis, *ibid.*, 1951, **19**, 942.
46. Felton and Orr, *J. Chem. Soc.*, 1955, 2170.
47. Allan, Meakins and Whiting, *ibid.*, 1874.
48. Bellamy, *ibid.*, 4221.
49. Bender, *J. Amer. Chem. Soc.*, 1954, **75**, 5986.
50. Filler, *ibid.*, 1376.
51. McBee and Christman, *ibid.*, 1955, **77**, 755.
52. Josien and Calas, *Compt. Rend. Acad. Sci. (Paris)*, 1955, **240**, 1641.
53. Bellamy and Williams, *J. Chem. Soc.*, 1957, 4294.
54. *Idem, ibid.*, 1957, 861.
55. Hunsberger, Lednicer, Gutowsky, Bunker and Tunsoig, *J. Amer. Chem. Soc.*, 1955, **77**, 2466.
56. Bellamy and Beecher, *J. Chem. Soc.*, 1954, 4487.

57. Wilder and Winston, *J. Amer. Chem. Soc.*, 1955, **76**, 5598.
58. Grundmann and Kober, *ibid.*, 2332.
59. Brügel, Dury, Stengel and Suter, *Angew. Chem.*, 1956, **68**, 440.
60. Brügel, Stengal, Reicheneder and Suter, *ibid.*, 441.
61. Berson, *J. Amer. Chem. Soc.*, 1954, **76**, 4975.
62. Fowler and Smith, *J. Opt. Soc. Amer.*, 1953, **43**, 1054.
63. Russell and Thompson, *J. Chem. Soc.*, 1955, 479.
64. Rosenkrantz and Skrogstrom, *J. Amer. Chem. Soc.*, 1955, **77**, 2237.
65. Jones and Herling, *J. Amer. Chem. Soc.*, 1956, **78**, 1152.
66. Allsop, Cole, White and Willis, *J. Chem. Soc.*, 1956, 4868.
67. Kendall, Hampton, Hausdorff and Pristera, *App. Spectroscopy*, 1953, 179.
68. Henbest and Lovell, *J. Chem. Soc.*, 1957, 1965.
69. Laato and Isotalo, *Acta Chem. Scand.*, 1967, **21**, 2119.
70. Katritzky, Lagowski and Beard, *Spectrochim. Acta*, 1960, **16**, 964.
71. Freeman, *J. Amer. Chem. Soc.*, 1958, **80**, 5954.
72. Thompson and Jameson, *Spectrochim. Acta*, 1958, **13**, 236.
73. Brown, *Spectrochim. Acta*, 1961, **18**, 1615.
74. Bowles, George and Cunliffe Jones, *J. Chem. Soc.*, 1970, 1070.
75. Brookes, Eglington and Morman, *J. Chem. Soc.*, 1961, 661.
76. Eglington, Brookes and Morman, *J. Chem. Soc.*, 1961, 106.
77. Freedman, *J. Amer. Chem. Soc.*, 1960, **82**, 2454.
78. McManis, *App. Spectroscopy*, 1970, **24**, 495.
79. Feairhalter and Katon, *J. Mol. Structure*, 1968, **1**, 238.
80. Bellamy, *Advances in Infrared Group Frequencies* (Methuen, London, 1968).
81. Hall and Zbinden, *J. Amer. Chem. Soc.*, 1958, **80**, 6428.
82. Nyquist and Potts, *Spectrochim. Acta*, 1959, **15**, 514.
83. Baker and Harris, *J. Amer. Chem. Soc.*, 1960, **82**, 1923.
84. Jones, Angell, Ito and Smith, *Can. J. Chem.*, 1959, **37**, 2007.
85. Wladislaw, Viertler and Demant, *J. Chem. Soc.* (B), 1971, 565.

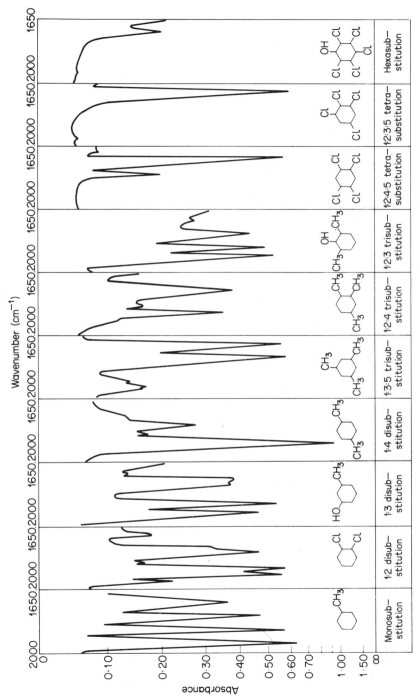

Figure 9. Aromatic substitution patterns 2000—1600 cm^{-1}

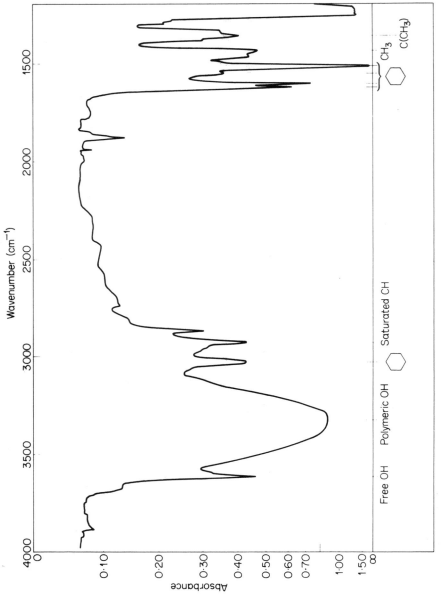

Figure 10. *p*-Cresol thin film

222

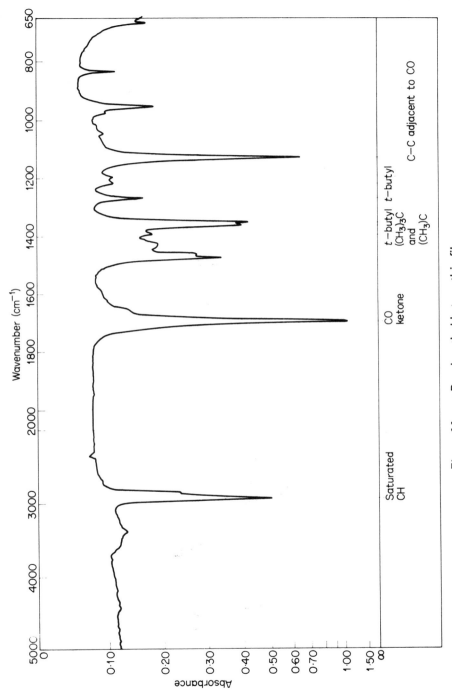

Figure 11. *tert*-Butyl methyl ketone thin film

223

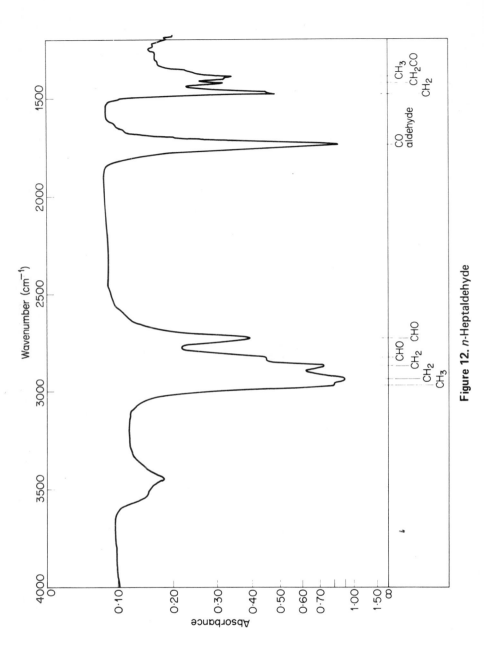

Figure 12. *n*-Heptaldehyde

224

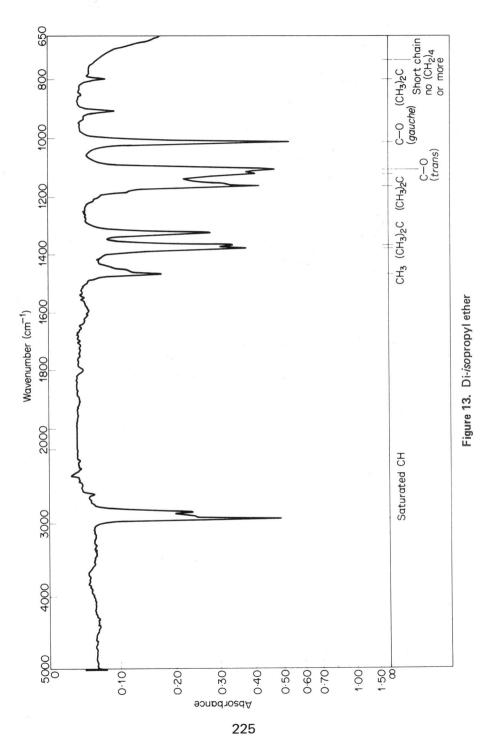

Figure 13. Di-*isopropyl* ether

225

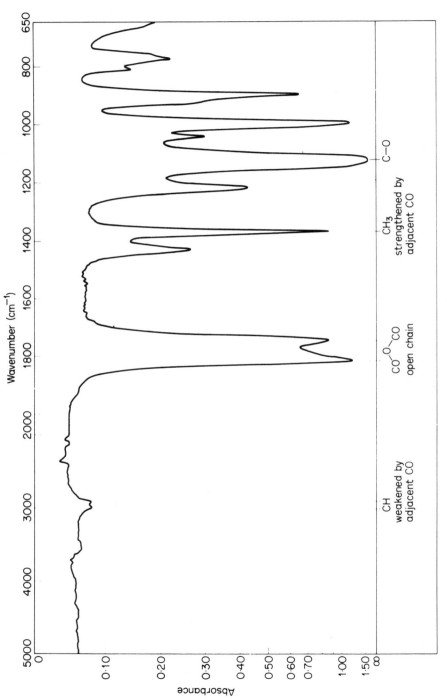

Figure 14. Acetic anhydride thin film

226

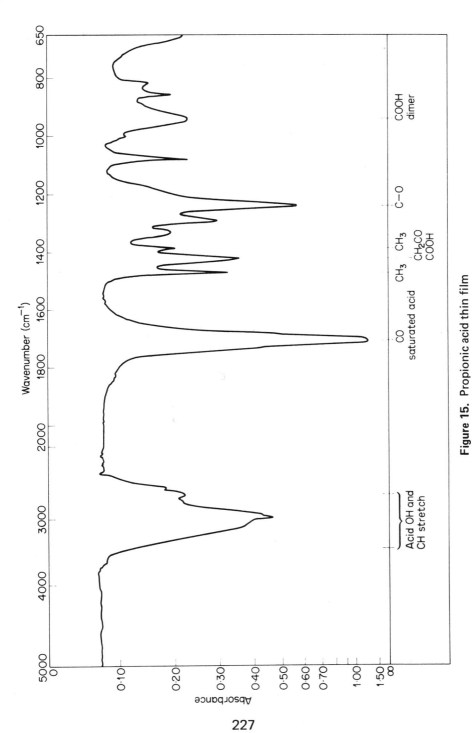

Figure 15. Propionic acid thin film

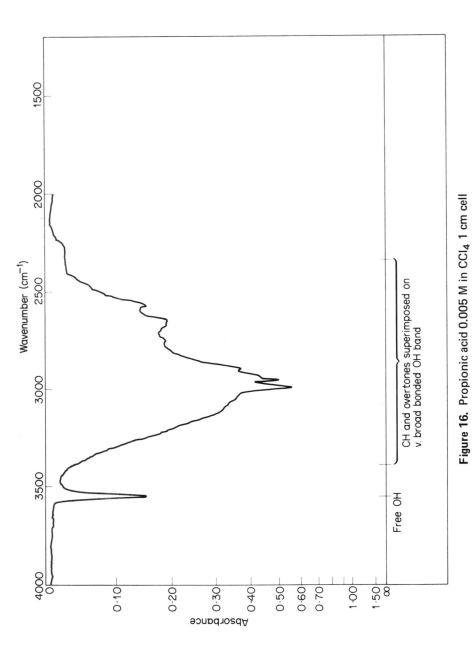

Figure 16. Propionic acid 0.005 M in CCl$_4$ 1 cm cell

228

Part Three | Vibrations involving mainly C—N and N—H linkages

12

Amides, Proteins and Polypeptides

12.1. Introduction and table

Although there is a good deal of resonance towards a dipolar form, amides do not enolise so that they show a carbonyl absorption, the frequency of which will be influenced to some extent by the dipolar structure with which it can resonate. It will also be affected to a lesser extent by the nature of the R group adjacent to the carbonyl. This absorption is termed the amide I band and is common to all types of amide, including cyclic forms. Its frequency is, however, markedly affected by hydrogen-bonding effects, so that considerable shifts can occur on passing from the solid to solution. This applies with even greater force to the other characteristic amide bands, so that in correlation work on amides and related materials particular attention must be paid to the physical state in which the material is examined.

In addition to the carbonyl absorption, primary and secondary amides show characteristic bands originating in NH modes which, taken together with the first, are usually sufficient to characterise an amide grouping with reasonable certainty. Primary amides show two NH stretching modes corresponding to the asymmetric and symmetric motions of the hydrogen atoms, whilst secondary amides usually show only a single absorption, the position of which depends on whether the compound exists in the *cis-* or *trans-*form. Most secondary amides show frequencies corresponding to both forms in solution, although that from the *trans-*isomer predominates in nearly all cases [12, 18, 26, 74, 97, 99]. Again, hydrogen bonding effects give rise to considerable shifts in the solid state, and absorptions from both free and bonded NH vibrations can often be observed together in concentrated solutions. The position of the bonded NH absorptions varies also with the nature of the hydrogen bonds concerned, and there are, for example, differences between the NH frequencies of secondary amides bonded in the *cis-* and *trans*-forms. With many secondary amides a second NH absorption is found at lower frequencies, whilst with more complex materials such as polypeptides and proteins, multiple absorptions occur in this region. Di-ketopiperazine, for example, in the solid state shows five bands, all of which are believed to be associated with NH stretching modes. The interpretation of these complex absorptions will be considered separately in the section dealing with polypeptides and proteins. It should be noted, however, that marked differences exist between the behaviour of OH and NH groups with changes of state. For example, N-ethyl acetamide and similar materials [75, 76] show a progressive change in the NH frequency with increasing concentration, in contrast to the more discreet shifts of the OH absorptions. This has led various workers to suggest that differences in mechanism may be involved. It has been suggested, for example, that proton transfer may take place in amides [77], although it is difficult to see how this could fail to lead to the existence of keto-enol forms. An alternative suggestion, due to Cannon [76, 78], is that dipolar interactions between the $-OCN^+$ groups play a more considerable part than direct hydrogen bonds in some secondary amides. It seems more probable that the drift in frequency which occurs with changes in concentration is due to the multiplicity of polymeric forms which are possible with *trans* amides. Nyquist [99] regards the band at 3366 cm^{-1} in N-methylacetamide in dilute solution as originating in the dimer. The frequency falls as the concentration is increased and reaches a final value of 3305 cm^{-1} in solutions where the degree of polymerisation is about seven units. In systems in which steric hindrance prevents such extensive polymerisation the frequency fall with increasing concentration is much reduced.

The NH$_2$ deformation mode in primary amides is to be expected in the 1600 cm^{-1} region, and the absorption which is found here in all such compounds (amide II band) is generally ascribed to this cause. However, secondary amides also exhibit a strong characteristic band at rather lower frequencies, the origin of which has been the subject of much controversy. It has been variously assigned as an NH deformation mode, a C–N stretching mode and as a mixed vibration involving both types of motion. The band is accompanied in

secondary amides by a weaker absorption (Amide III), near
1300 cm^{-1}. This is found in both cyclic and open-chain compounds.
In primary amides it falls near 1400 cm^{-1} and is assignable to the
C–N stretching mode.

Table 12

NH *Stretching Modes*

Primary amides (free NH)	Near 3520 cm^{-1} and 3410 cm^{-1} (m.)
„ (bonded NH)	Near 3350 cm^{-1} and 3180 cm^{-1} (m.)
Secondary amides (free NH) (*trans-*)	3480–3440 cm^{-1} (m.)
„ „ (*cis-*)	3435–3395 cm^{-1} (m.)
„ (bonded NH) (*trans-*)	3320–3270 cm^{-1} (m.)
„ „ (*cis-*)	3180–3140 cm^{-1} (m.)
„ „ (*cis-* and *trans-*)	3100–3070 cm^{-1} (w.)

CO *Absorption (Amide I)*

Primary amides	Solid, near 1650 cm^{-1}.	Dilute solutions, near 1690 cm^{-1} (s.)
Secondary amides	Solid, 1680–1630 cm^{-1}.	Dilute solution, 1700–1680 cm^{-1} (s.)
Tertiary amides	Solid and dilute solutions, 1670–1630 cm^{-1} (s.)	
Cyclic amides	(*a*) Large rings, near 1680 cm^{-1} (solution) (s.)	
	(*b*) γ-lactams (1) Unfused, near 1700 cm^{-1} (s.)	
	„ (2) Fused, 1750–1700 cm^{-1}	
	(*c*) β-lactams (1) Unfused, 1760–1730 cm^{-1} (solution) (s.)	
	„ (2) Fused (to thiazolidine rings), 1780–1770 cm^{-1} (solution) (s.)	

NH$_2$ *Deformation (Amide II)*

Primary amides only	1650–1620 cm^{-1} (solid) (s.), 1620–1590 cm^{-1} (solution) (s.)

Amide II

Secondary, non-cyclic amides only	1570–1515 cm^{-1} (solid) (s.), 1550–1510 cm^{-1} (solution) (s.)

Amide III

Secondary amides only	Near 1270 cm^{-1} (m.)

NH *Deformation (Amide V)*

Bonded secondary amides only	Near 720 cm^{-1} (m.) broad.

Other Correlations (Amide IV and VI Absorptions)

Secondary amides only	Near 620 cm^{-1} and 600 cm^{-1}
Primary amides	1420–1400 cm^{-1} (m.)

Changes in the C=O frequencies of amides corresponding to ring strain, occur in lactams with small rings, and the frequency is also altered by fusion of the lactam ring with another cyclic structure in which the nitrogen atom loses its attached hydrogen atom. There are also differences in the hydrogen-bonded NH stretching modes of lactams and small rings, due to the fact that in these the amide group must exist in the *cis*-form.

In addition, the substitution of other functional groups in the immediate vicinity of the amide link often gives rise to marked spectral changes. The spectra of compounds containing the CO—NH—CO structure, for example, are sufficiently different from those of normal amides to enable them to be differentiated. Urethanes and anilides correspond more closely with normal amides, but in the substituted ureas the interpretation of the 1600 cm^{-1} region becomes extremely difficult due to the complexity of the NH absorptions. Insofar as it can be recognised, however, the amide carbonyl absorption appears at a normal frequency. Amido-acids have been considered elsewhere (Chapter 13) on account of the affinities they show to amino-acids.

Other characteristic absorptions of secondary amides have been described at lower frequencies, but are of less diagnostic value. These include the out-of-plane NH deformation [79] frequency which occurs near 700 cm^{-1} and is exceptionally broad in the spectra of solids and concentrated solutions. This has been termed the Amide V absorption by Mizushima *et al.* [80]. Assignments of Amide IV and Amide VI bands at still lower frequencies have also been made. These are essentially skeletal vibrations, and they have been discussed further by Miyazawa [81].

These correlations are listed in Table 12 on p. 233.

12.2. NH Stretching vibrations

12.2(a). Primary amides. In dilute solutions simple primary amides show two free NH stretching absorptions near 3520 cm^{-1} and 3410 cm^{-1}, and so are clearly similar to normal amines. Acetamide [3] in dilute chloroform solution, for example, absorbs at 3538 cm^{-1} and 3420 cm^{-1}.

A number of workers [4–9, 82] have studied these absorptions in both the solid state and in solution. The most detailed study is by Cleverley [83], who has listed both the frequencies and intensities of the free and bonded NH bands of a wide range of amides in chloroform solution. The length and nature of the alkyl side-chain have little effect on these frequencies, which occur in essentially the same positions throughout. Most simple alkyl amides absorb near 3528 and 3413 cm^{-1} in dilute solution, with a band separation close to 115 cm^{-1}. For such monomeric cases there is a direct frequency relationship between ν_{as} and ν_s. Puranik and Ramiah [101] give the equation $\nu_s = 1.1214\nu_{as} - 542.5$ and Venkataramiah et al. [71] the variant $\nu_s = [(1.12 \mp 0.06)(\nu_{as} - 3650) + 3415]$. Both equations fail if one of the hydrogen atoms is bonded and the other is not, or if they are unequally bonded. Oxamide for example does not follow these equations.

In solutions of moderate concentration both free and bonded absorptions appear and a complex pattern results. N-octanamide, for example, in chloroform has its free NH_2 absorptions at 3530 and 3415 cm^{-1}, but shows additional bands at 3498, 3345, 3300 and 3182 cm^{-1}. This suggests that different types of association are occurring simultaneously. In the solid state the pattern simplifies again, and two broader NH peaks are found near 3350 and 3180 cm^{-1}.

The behaviour of thioamides is remarkably similar to that of the amides themselves. The bands appear at slightly lower frequencies [1, 84] with mean values of 3503 and 3384 cm^{-1} in dilute solution. They follow the equation of Venkataramiah [71] for normal amides and the band separation is likewise close to 115 cm^{-1} suggesting that the hybridisation of the nitrogen atom and the consequential bond angles are closely similar in the two cases.

12.2(b). Secondary amines. In very dilute solutions simple secondary amides studied under low resolution show only one band in the 3460–3420 cm^{-1} range [4, 5, 6, 7]. Under conditions of high resolution this band can be split into two components in many cases. Amides such as γ-butyrolactam and δ-valerolactam, which can exist only in the cis-form, show only one band in the range 3440–3420 cm^{-1}, whilst others such as benzanilide, which has a trans-configuration, show a single band at slightly higher frequencies (3460–3440 cm^{-1}). The twin bands shown by most secondary amides have therefore been assigned to the cis- and trans-rotational isomers [74], and it becomes possible to estimate the relative proportions of the two by a comparison of their relative intensities. In normal secondary amides such as N-methylacetamide the trans-form predominates to the extent of 95%, but in sterically hindered compounds, such as *N-tert.*-butylphenylacetamide, the *cis*-form is present to the extent of approximately 70%. The frequency range of the *cis* and *trans* forms of secondary amides has been discussed by Hallam *et al.* [12, 18, 26] and by Rao *et al.* [155]. *Trans* amides absorb in the 3480 to 3440 cm^{-1} range and *cis* amides between 3435 and 3395 cm^{-1}. As most open chain secondary amides exist almost wholly in the *trans* form, the lower frequency band is often not seen. These are of course monomeric frequencies. In the solid state the absorption frequency is determined largely by the type of hydrogen bonding involved, and in some cases, especially with dipeptides, polypeptides and proteins, two or, in some cases, multiple absorptions are found. With open-chain secondary amides the main NH absorption occurs near 3270 cm^{-1} in the solid state [4, 6, 7, 88]. In solution both free and bonded NH absorptions are shown, and the frequency of the latter depends very much upon the nature of the solvent and upon the concentration employed [75, 76, 82]. In solids the band is near 3280 cm^{-1}, whilst in solution in carbon tetrachloride it ranges in the case of *N*-ethylacet-amide from 3372 to 3309 cm^{-1}, depending on the concentration [75, 76]. This is generally accepted as evidence that variations in the length of the polymer chain are taking place [99].

In addition to the main 3280 cm^{-1} absorption, a number of secondary amides absorb more weakly at 3080 cm^{-1}. This band is also associated with NH modes and, like the 3280 cm^{-1} band, it vanishes in dilute solution and is replaced by a single absorption at 3420 cm^{-1}. The 3080 cm^{-1} band in *trans* amides has been identified as originating in an overtone of the amide II band (see below). However the band persists in *cis* amides and lactams when it must be ascribed to a combination tone. A similar explanation is given for the appearance of this band in solid thioamides [85].

Cyclic lactams with ring sizes of seven or less behave [12, 154] like normal *cis* amides in dilute solutions [6, 10, 16], and absorb near 3420 cm^{-1}. However, in the condensed phase the band near 3280 cm^{-1} is not shown, but is replaced by another at 3175 cm^{-1}. In solution this band occurs at 3220 cm^{-1}, and the frequency is independent of concentration [75], except of course at very high dilutions when the monomer is released. This suggests that only a dimeric species is present normally. These materials also show the 3080 cm^{-1} band, which is usually more intense than in open-chain amides. The differences between the NH frequencies of these two types of amide in the condensed phase have led Darmon and Sutherland [6, 10] to suggest that the 3290 cm^{-1} absorption is due to amides which are hydrogen bonded in the *trans*-form I

I.

whereas the 3175 cm^{-1} absorption arises from the *cis* arrangement such as in II or III.

II.   III.

The fact that the 3175 cm^{-1} band is observed only in cyclic lactams in which the *cis*-configuration must exist, provides strong support for this view, particularly as there is now a good deal of evidence available from both infra-red and dipole moment studies which confirms that open-chain amides exist predominantly in the *trans*-configuration [3, 73, 74, 86, 87]. With lactams with larger rings (nine or more members) the puckering of the ring allows the preferred *trans* form and these then behave exactly like the open chain amides. Eight-membered rings represent a transitional state and these can exist in either the *cis* or *trans* form depending on the conditions [12, 102].

Departures from these generalised frequencies occur in special cases. Thus the attachment of a halogen atom to the nitrogen of a secondary amide displaces the NH stretching frequency by about 50 cm^{-1} [70]. This is not surprising in view of the changes in electronegativity. Thioamides absorb at slightly lower frequencies than normal secondary amides and are more susceptible to steric effects which can influence the *cis/trans* ratios. *Trans* thioamides [1, 2, 84] absorb between 3390 and 3430 cm^{-1} in dilute solution and the *cis* forms between 3400 and 3365 cm^{-1}.

12.3. The amide I band (the carbonyl absorption)

12.3(a). General. All amides show a strong absorption band near 1640 cm^{-1} when examined in the solid state. The fact that the absorption is at an appreciably lower frequency than the carbonyl absorption of normal ketones must be due to the resonance effect with the ionic form. This is enhanced by the strong association effects in the solid state, and the corresponding vapours absorb at considerably higher frequencies. Formamide, for example [89], absorbs at 1740 cm^{-1} as a vapour, falling to 1709 cm^{-1} in dilute solution in chloroform. Similarly, N-methyl [90] and N-ethylacetamide [75] absorb in the $1720-1715$ cm^{-1} range as vapours, fall to near 1700 cm^{-1} in dilute solutions, and to 1650 cm^{-1} in the liquids. These values suggest that the contribution of the ionic form is relatively small in the vapour state. In NN-disubstituted amides in which hydrogen bonding is impossible it is nevertheless likely that strong dipolar association effects play some part in the low frequencies found in the solid state [76]. The carbonyl band in this case is not a pure C=O stretching mode but is appreciably coupled with the C—N stretch and to some extent with the NH bend. This is well shown by N^{15} substitution and by deuteration [97, 144]. Normal coordinate studies indicate that the carbonyl frequency has more carbonyl stretching character in the secondary amides than in the primary cases. Suzuki [145] attributes only 64% carbonyl character to this band in formamide and 59% in acetamide, the remainder being largely C—N stretch. However in N-methylacetamide the carbonyl contribution has risen to 80% the remainder being contributions from both the C—N stretch and the NH bend [106].

The precise location of the amide I band is determined by the presence or absence of any substituents on the nitrogen atom and by their electronegativity, by the inclusion of the amide group in a strained ring, by the presence of α-halogen atoms in some cases, and particularly by the physical state in which the samples are examined. All these cases are considered individually below. The special cases such as —CO—NH—CO— and O—CO—NH— structures are not included in the immediate discussion, but are considered collectively later.

12.3(b). Primary amides. R—CO—NH$_2$. Simple primary amides examined in the solid state absorb near to 1650 cm^{-1}. No precise range can be given, as the number of samples examined is not large, and interest has focused mainly on the secondary amides. However, a sufficient number have been examined by various workers [4, 5, 9, 25, 29, 147] to substantiate the correlation. Acetamide appears to be an exception in absorbing at 1694 cm^{-1} in the solid state [5], despite the fact that X-ray evidence indicates association in this form. This amide I absorption is subject to considerable alteration on change of state in which hydrogen bonding is broken [4, 5, 13– 15, 25], and is also liable to variations in solution depending on the polarity of the solvent employed [4]. Thus five unsubstituted amides examined by Richards and Thompson [4] in dilute solution in dioxane all absorbed close to 1690 cm^{-1}, in contrast to the 1650 cm^{-1} value found for the corresponding solids. In concentrated solutions the frequency appeared at some intermediate value depending on the dilution. Thus hexoamide absorbs at 1655 cm^{-1} in the solid state, at 1668 cm^{-1} in concentrated solutions and at 1680 cm^{-1} in dilute solution (chloroform). The corresponding values of 1692 cm^{-1} and 1672 cm^{-1} given for this absorption band in dioxan and in methanol [4] indicate the degree of frequency shift likely to be associated with alterations in the type of solvent employed. In chloroform solution all the amides $CH_3(CH_2)_n CONH_2$ [91] from $n = 1$ to $n = 10$ absorb consistently at 1679 cm^{-1}. In carbon tetrachloride solution these values rise to 1700 to 1690 cm^{-1} [146, 148]. The influence of α-halogen substitution has not been very widely studied but Haszeldine [65] reports upward frequency shifts in fully fluorinated amides. Brown [146] gives data on such compounds quoting both frequencies and intensities. It would seem that a single halogen at the α-carbon atom has little effect on the carbonyl frequency, presumably because it is in a rotational configuration in which it is twisted away from the carbonyl oxygen. However, two bands appear in α-dichloroacetamide, the higher of which shows a frequency rise of 21 cm^{-1}. A further rise of 16 cm^{-1} occurs when a third chlorine is introduced.

12.3(c). Secondary amides (*N*-mono-substituted amides) (open-chain).

Until recently, the great majority of studies on secondary amides have been concerned only with their examination in the solid state. Simple *N*-mono-substituted amides [4, 5, 9, 146, 148], in this state, show the amide I band very close to 1640 cm^{-1}. Richards and Thompson [4] have pointed out, however, that electrophilic substituents on the nitrogen atom can give rise to an increase in this frequency up to 1680 cm^{-1}. This is confirmed by Lacher *et al.* [66] and by Gierer [92], and by Le Klein *et al.* [70] who have studied *N*-chlorinated compounds which show elevated frequencies. In dilute solution in inert solvents the carbonyl frequency rises to about 1685 cm^{-1} for simple mono-substituted amides and up to 1700 cm^{-1} for anilides and similar products [4, 5, 9]. Nyquist [149] finds values between 1690 and 1683 cm^{-1} for all the alkyl acetamides from methyl to tertiary butyl, and Beer *et al.* [97] find comparable values. Formamides have [148] slightly higher values near 1700 cm^{-1}. All these values relate to the *trans* configurations which are preferred for the acetamides. Conjugation lowers the carbonyl frequencies as is to be expected, and benzamides in dilute solution absorb near 1660 cm^{-1}. However it requires a high degree of dilution to achieve monomeric frequencies in the amides and the influence of effects such as conjugation are more usually masked by the effects of association. The influence of α-halogen substitution have been studied by several workers, and their results have been discussed by Bellamy [93, 100]. Unlike α-halogen ketones, the introduction of a single α-halogen atom does not alter the carbonyl frequency significantly [94], owing to the fact that the halogen atom takes up a *trans*-configuration in relation to the carbonyl oxygen. This is of course unexpected as the skew form would be the normal case. However, Nyquist [99, 149] has shown from a study of the CO and NH frequencies that an intramolecular hydrogen bond NH . . . Cl locks the chlorine atom in the *trans* position with respect to the carbonyl oxygen. When further halogens are substituted at the same carbon atom they are unable to take up a fully eclipsed position to the carbonyl oxygen. The second and third halogen atoms therefore each produce a small frequency rise. In the solid state it may be that intermolecular bonds may be preferred and the situation would then be very different. This has not been fully explored.

The rise in the carbonyl frequency which occurs when the oxygen atom is near in space to a halogen, is found also in N halogenated amides [66, 70, 92]. When the hydrogen atom of the NHCl group is *trans* to the carbonyl the halogen is brought even nearer in space to the oxygen than it is in α-halogenated ketones. However the impact on the carbonyl frequency is smaller as can be seen from a comparison of the *cis* and *trans* forms. Although the *trans* form is found exclusively in this series in polar solvents, both forms coexist in carbon tetrachloride. The carbonyl frequencies of *N*-chloroacetamide and N-chloropropionamide are 1728 (*trans*) and 1715 cm^{-1} (*cis*). The corresponding N-bromo compounds absorb at 1717 and 1700 cm^{-1}. With di- and tri-substitution at the α carbon, a nearer approach of the halogen to the oxygen becomes possible and elevated frequencies are found [88, 91, 95]. Alkyl trifluoromethyl amides of the type CF_3CONHR, for example, absorb in the range 1698–1718 cm^{-1} in the solid state.

12.3(d). Tertiary amides. NN'-Di-substituted amides are incapable of forming hydrogen bonds, and the carbonyl absorption band is consequently not much influenced by changes in state or in the complexity of the molecule. The band usually falls near 1650 cm^{-1} unless a phenyl group is substituted on to the nitrogen atom, when it is raised to 1690 cm^{-1}. As with acetanilides, this is due to the competitive effect of the ring for the lone pair electrons of the nitrogen atom. In consequence, the contribution of the ionic form of the amide is reduced and the carbonyl frequency is raised. A similar effect may account for the high frequencies shown by N-nitroso-amides, which absorb near 1740 cm^{-1} in solution [96], although alternative explanations are possible in this case. The frequency in the solid state, or in non-polar solvents usually falls near 1660 cm^{-1} with a few examples absorbing as low as 1630 cm^{-1}. This lower value than primary or secondary amides is consistent with the greater basicity and the greater availability of the lone pair for resonance. Tertiary formamides, like their secondary counterparts, absorb at appreciably higher frequencies, usually close to 1680 cm^{-1}. They also show an interesting anomaly in that phenyl substitution on the nitrogen atom does not raise ν_{CO} as it does in other cases. N-Phenyl-N-methylformamide absorbs at 1686 cm^{-1} within 1 cm^{-1} of N-dimethylformamide.

The values quoted above for solid or solution frequencies relate of course to situations in which there are no hydrogen bonds. The carbonyl group of tertiary amides is highly polar and sizeable shifts to lower frequencies can occur in bonding solvents.

α-Halogen substitution in tertiary amides has been studied by Letaw and Gropp [88] and by Speziale and Freeman [152]. A single α-halogen atom has no significant effect on the frequency, but when a second halogen is added two bands appear, one of which is at a higher frequency. With the addition of a third halogen atom the carbonyl band is again single but is at the higher frequency. In these cases no internal hydrogen bonding is possible, but there is dipole moment evidence to suggest that the preferred conformation of the mono-halogen compounds is one in which the halogen is nearer to the nitrogen than to the oxygen. When a second halogen is added, two conformations are allowed and two bands appear.

12.3(e). Lactams. The carbonyl absorption of lactams is normal in unstrained rings of six or more carbon atoms, but it shifts towards higher frequencies in smaller rings due to ring strain. This resembles the state of affairs in the lactones, except that in this case an additional variable is introduced by the possibility of their being parts of a fused system of rings with a tertiary nitrogen atom.

Six or more membered Rings. Six-membered rings absorb close to 1672 cm^{-1} [12, 150]. This is lower than the mean value of open chain secondary amides (1685 cm^{-1}) because the lactams have a *cis* arrangement in which ν_{CO} is slightly lower. Fusion into bicyclic ring systems does not affect the frequency as it does in the case of five-membered rings, because the additional strain can be more easily accommodated.

Seven-membered ring lactams also absorb near 1672 cm^{-1}, as do eight-membered rings in the solid state. However in solution eight-membered rings revert to the *trans* form and the frequency rises to 1680 cm^{-1}. A similar frequency is given by all larger ring sizes, corresponding to a stable unstrained *trans* form [12, 15, 104, 105, 114]. When seven-membered ring lactams are fused to other rings there is a frequency fall of about 30 cm^{-1}. This is an interesting contrast to the behaviour of four-membered rings where ring fusion results in a frequency rise.

Five-membered Rings. γ-Butyrolactam [5] absorbs in the liquid state and in solution [75, 85] at 1700 cm^{-1}, and at 1754 cm^{-1} in the vapour phase [85], whilst a series of fused γ-lactams examined in connection with penicillin studies, indicated that the carbonyl group has a frequency of about 1750–1700 cm^{-1} with the greater proportion lying towards the lower end of this range [9, 64]. Hall and Zbinden [150] and Hallam [12] have also studied five-membered ring lactams. The average frequency rise compared with the acyclic analogue is about 31 cm^{-1} for single rings, and 37 cm^{-1} for bicyclic systems. With N-alkyl lactams the effects of ring strain are larger and the rise is about 45 cm^{-1} for unfused systems.

Four-membered Rings. These show the expected high frequency shift of about 83 cm^{-1} [150]. Such systems, and the effects of ring fusion, have been much studied in relation to penicillin [19–24].

These show that a substantial further rise occurs on fusion with the thiazolidine ring so that the frequency approaches 1780 cm^{-1}.

12.3(f). The intensity of the amide I band. The great majority of studies on amides have been carried out on samples in the solid state, so that not much is known about factors which may influence the intensity of the carbonyl absorption. Studies in solution are made more difficult by the very considerable changes in intensity which result from solvent changes. Data on amide intensities has been given by Brown [146], Jones and Cleverley [91] and Thompson and Jameson [153]. The results do not allow any diagnostic conclusions to be drawn other than the generalisation that amide carbonyls are two to three times as intense as are the corresponding saturated ketones.

12.4. The amide II band

Primary and secondary amides show a second strong band in the 1600–1500 cm^{-1} region which is absent from the spectra of tertiary amides and also from those of small ring cyclic lactams under normal conditions. The origins of this band are different in the two cases. Nevertheless, the band is referred to generally as the amide II band.

12.4(a). Primary amides. $RCONH_2$. All primary amides show a second band of weaker intensity very close to the main carbonyl absorption. Usually its intensity is about half to one-third that of the CO absorption, and it lies on the low-frequency side in the range 1650–1620 cm^{-1} (solids) [4, 5, 9]. In some instances the two bands fall so close together in the spectrum of materials examined as solids that only a single band appears. However, owing to the fact that the two bands show frequency shifts in opposite directions following changes of state, the presence of both can be established in this way. Phenacetamide, for example seems to have a single broad band at 1645 cm^{-1} in the solid state which is resolved into two bands at 1678 cm^{-1} and at 1580 cm^{-1} in dilute solution [5]. There has been much controversy over this band, but there is now little doubt that it is principally associated with the NH_2 scissoring mode. Suzuki [145] calculates that in N-methylacetamide this band has 85% NH_2 bending character. This accounts for the sensitivity of the band to deuteration, and for the downwards frequency shift on dilution when hydrogen bonds are broken. This is in the opposite direction to the shifts of the NH stretching bands and is consistent with the bonds becoming easier to bend as they become harder to stretch, due to the increase in the s character of the nitrogen orbitals.

The results of the various workers in this field [4, 5, 9] indicate that the band occurs in the $1650-1620$ cm^{-1} region in the solid state, falling to $1620-1590$ cm^{-1} in dilute solution. In chloroform solution all primary amides $CH_3(CH_2)_n CONH_2$ show this band at $1590-1588$ cm^{-1}, with an ϵ_{max} of $180-210$ for all values of n from 1 to 10 [91]. The nature of the R group in the $RCONH_2$ structure does not materially affect this frequency.

12.4(b). Secondary amides. RCONHR. The amide II absorption band shown by all secondary amides in the region of 1550 cm^{-1} (solids) has been the centre of much controversy. Early workers attempted to assign this band to either the NH deformation mode [9, 10, 35, 58], or to the C–N stretch [5, 24, 34, 67] and good supporting evidence was adduced in favour of each proposition. It was also suggested that the band was in some way connected with the carbonyl group [27–31]. It has since been realised that none of these explanations can account for the whole of the data, and following Fraser and Price [59, 108], Miyazawa *et al.* [80], Becher *et al.* [111, 112], and Gierer [92] various interpretations involving coupled modes have been suggested. It was the detailed study of the *cis* forms of corresponding alkyl amides, in which this band does not appear which finally opened the way to an understanding of the origins of this band [106, 154].

In the *cis* form of *N*-methylacetamide a band appears at 1450 cm^{-1} which originates in the NH bending mode (78% from normal coordinate studies), and another at 1350 cm^{-1} which can be identified with the C–N stretch (71%). In the *trans* form it would appear that the NH bend is moved so that it couples more strongly with the C–N stretching mode. This results in two bands, at about 1550 cm^{-1} (amide II) and about 1270 cm^{-1} (amide III). The higher frequency band now has only about 60% NH bending character [145]. This explains the sensitivity of both bands to deuterium substitution.

12.4(c). The frequency range of the amide II band. The position of the amide II absorption in *N*-mono-substituted amides has been studied by many workers. Richards and Thompson [4] have given the spectra of a substantial number, whilst Randall *et al.* [5] give details of the band position in 72 mono-substituted amides. The results of several other groups of workers who find comparable results have also been summarised [9]. The frequency range found by these workers is $1570-1515$ cm^{-1} for materials examined in the solid state, a considerable proportion of which absorb near the mean value of 1540 cm^{-1}. Richards and Thompson [9] have shown that electrophilic substituents on the nitrogen atom give a small frequency shift towards longer wave-lengths in some cases, but in N-chloromethyl acetamide the amide II band appears to be absent [92, 103]. Flett [147] has tabulated data on secondary amides in KBr discs, and Beer *et al.* [97] and Nyquist *et al.* [99] have studied solutions. However gradations in the band positions occur with changes in concentration. This is due to changes in the degree of polymerisation, and makes it very difficult to quote realistic frequency ranges for other than solids or for very dilute solutions.

In solution the amide II band shifts towards lower frequencies within the range $1550-1510$ cm^{-1}; thus methylphenaceturate absorbs at 1555 cm^{-1} in the solid state, at 1535 cm^{-1} when molten, and at 1517 cm^{-1} in dilute solution [5]. This shift to lower frequencies on the breaking of hydrogen bonds is in the opposite direction to that which occurs with absorptions arising from stretching modes, such as the OH and NH absorptions near 3300 cm^{-1}, but is in the direction to be expected from changes in deformation frequencies. As in the case of the amide I band, the extent of the shift on passing from the solid state to dilute solution is influenced by the complexity of the molecule itself. In the complex materials in which considerations of geometry may inhibit the full formation of hydrogen bonds in the crystal lattice, the shifts observed on solutions are small. The 1515 cm^{-1} band of solid benzyl-penicillin, for example, does not alter appreciably in position on solution. Typical examples of frequency shifts on change of state are given by N-methylacetamide (1565 cm^{-1} liquid, 1534 cm^{-1} solution, 1490 cm^{-1} vapour [87, 90−92]) and by α-chloro-N-methylacetamide [94] (1565 cm^{-1} solid, 1550 cm^{-1} liquid, 1515 cm^{-1} vapour). However, in general, few data are available on the characteristic frequencies of this absorption in dilute solutions or in the vapour phase.

As indicated earlier, this absorption is absent from cyclic lactams with ring sizes smaller than eight [12, 114] but occurs in the related thiolactams. In the thioamides a very similar band appears near

1550 cm^{-1} but in this case it appears to be an almost pure NH deformation as the corresponding C—N stretch at 1290 cm^{-1} is unaffected by deuteration [85, 110].

12.5. Other correlations

12.5(a). Primary amides. Randall *et al.* [5] noted the presence of a band in the range 1418—1399 cm^{-1} in all unsubstituted amides they have examined, which is absent from the spectra of *N*-substituted amides. They assigned this to the C—N stretching absorption. This assignment has been confirmed by the detailed vibrational analysis of a number of primary amides, although there is of course a little coupling with the NH$_2$ bending mode, and the frequency range is a good deal wider than was originally supposed (1430—1310 cm^{-1}). The band is relatively weak in the infra-red and is of little use for diagnostic purposes. It is of medium intensity in the Raman spectrum where it can occasionally be more useful.

12.5(b). The amide III band. This absorption band occurs in secondary amides [80, 90, 108] in the region 1305—1200 cm^{-1}, and is usually notably weaker than either of the amide I or II bands. It is almost certainly due to a mixed vibration involving CN and N—H modes, and this aspect has already been discussed in relation to the amide II band above. The amide III absorption in normal amides is sensitive to deuteration changes, indicating mixing, but in thioamides it is essentially unchanged, and is then wholly attributable to a C—N mode [85]. Changes of state result in marked shifts in normal amides, but, as is to be expected, these are notably reduced in the thio-derivatives. The band is not easy to identify in the infra-red. It is of medium intensity but falls in a region where many other bands occur. The situation is much better in the Raman spectrum where the band is of considerable intensity and easily identified. In the Raman spectra of simple liquid amides the band occurs between 1295 and 1250 cm^{-1}, falling to 1250 to 1200 cm^{-1} in dilute solutions.

12.5(c). Skeletal vibrations. Low-frequency absorptions, termed by Mizushima *et al.* [80] the amide IV and amide VI bands, have been identified in secondary amides near 620 cm^{-1} and near 600 cm^{-1} respectively [80, 81]. They have their origins in skeletal modes, and are of limited use for characterisation purposes. The amide IV band is often of strong to medium intensity in the Raman spectrum where it is more readily identified. It originates in the N—C=O deformation.

12.5(d). The out-of-plane NH deformation mode. A characteristic feature of proteins and of secondary amides is an ill-defined absorption with a maximum near 700 cm^{-1}. On deuteration the intensity of this band falls and a new band appears near 530 cm^{-1}. This change by a factor of 1.36 is almost exactly what would be expected for a pure NH deformation mode, and thus characterises this absorption [79]. On dilution the band weakens, and in secondary amides eventually vanishes altogether at high dilution. It is therefore clearly connected with the associated form of these compounds, and precisely similar bands, behaving in the same way, are observed in thioamides [85]. The free γ NH frequency corresponding to this mode is not well defined. In N-methylacetamide it is assigned at 648 cm^{-1} in dilute solution [80], and in general it falls below the rock-salt region.

12.6. Amides containing specialised structures

12.6(a). Compounds containing the group CO—NH—CO. Open-chain systems containing the CO—NH—CO group are uncommon but a small number have been studied by Uno and Machida [156]. Two carbonyl bands appear, but unlike the corresponding anhydrides the band separation is small and they are not always resolved. They fall between 1740—1720 and 1720—1700 cm^{-1}. Both carbonyl groups can be expected to be *trans* to the NH bond and they should therefore be strongly coupled if the system is planar. The facts that the band separation is so small, and that the frequencies are high for amides, suggests that there is little resonance and that coupling is limited because the carbonyl groups do not lie in the same plane.

With cyclic systems such as succinimide and related compounds coplanarity is to be expected and now much greater separations are found. Succinimide shows bands at [157] 1781 and 1715 cm^{-1}, and other compounds behave similarly [156, 157]. The frequency ranges are quite wide 1790—1735 and 1745—1680 cm^{-1}. As with the anhydrides it is the antisymmetric mode which has the lower frequency and has the greater intensity in the infra-red. Sizeable frequency shifts in this band can occur if the nitrogen atom is also substituted [158]. It is at 1738 cm^{-1} in phthalimide, at 1720 cm^{-1}

in the N-methyl derivative, and at 1700 cm^{-1} in the N-silyl derivative.

Trimeric isocyanates also contain the CO—N—CO system. Alkyl derivatives show their main band around 1700—1680 cm^{-1} with a weaker shoulder at 1715—1710 cm^{-1}. Aromatics have a higher frequency with the main band near 1715 cm^{-1} and the weaker shoulder near 1780 cm^{-1}. This is parallel to the raised carbonyl frequencies of anilides as compared with alkyl amides. Isocyanate dimers are of some theoretical interest in that the carbonyl groups are in direct opposition so the symmetric mode is forbidden in the infra-red. Only a single carbonyl band is therefore seen, and this is near 1780 cm^{-1}.

With cyclic compounds containing three carbonyl groups, those immediately adjacent to the NH group behave as above, whilst the third shows a normal CO absorption. The spectra of six trialkyl pyrolidine triones examined by Skinner and Perkins [32] show absorptions at 1786 cm^{-1}, 1724 cm^{-1} and 1667 cm^{-1}. The absorption at 1724 cm^{-1} is attributed to the CO group in the 3 position.

12.6(b). Compounds containing the group R—NH—CO—OR. In compounds of this type the carbonyl absorption responsible for the amide I band is likely to be lowered in frequency by the NH group as in normal amides, but the C—O—R residue will exert an influence in the opposite direction, as it does in the case of esters the carbonyl absorption of which is higher than that of ketones. This expectation is realised in practice. Primary urethanes absorb in the 1730—1720 cm^{-1} range in very dilute solution in chloroform [159] and at rather higher frequencies in carbon tetrachloride. There is a progressive fall in the frequency as one passes from primary to secondary and then to tertiary systems. Secondary urethanes have been more widely studied [148, 149, 159—161]. Pinchas [159] gives the range as 1722—1705 cm^{-1} in chloroform, but Katritzky *et al.* [160] give 1739 to 1705 cm^{-1} for aryl ethyl urethanes in the same solvent. Nyquist [149] using carbon tetrachloride finds a higher range and quotes 1738—1730 cm^{-1}. Some frequency rise in aryl as compared with alkyl urethanes is to be expected by analogy with the anilides. Katritzky [160] has shown that the frequency rises with the electron-withdrawing powers of the ring substituents, but the overall shifts are small. Sato [161] in fact finds little difference between vinyl and ethyl groups on the nitrogen atom.

Renema [148]   quotes   1732 cm^{-1} for   the   compound
$NH_2 NHCOOC_2 H_5$ and 1793 cm^{-1} for $NO_2 NHCOOC_2 H_5$.

Tertiary urethanes absorb close to 1685 cm^{-1} in chloroform [159], rising to 1710 cm^{-1} in carbon tetrachloride [148].
Urethane itself absorbs at 1618 cm^{-1}, corresponding to the amide II
absorption of an unsubstituted amide, whilst the *NN*-di-substituted
products show no band in this region. The simple *N*-mono-
substituted urethanes [5] and carbamates show their NH stretching
absorption for the solid products in the 3300–3250 cm^{-1} region,
which is normal for open-chain amides.

This same residue occurs in cyclic systems, such as the oxazolones,
which have been studied by Gompper and Herlinger [115] and
compared with the corresponding thioketo compounds. The carbonyl frequency then falls near 1750 cm^{-1}, and the amide III
absorption is raised to 1380 cm^{-1}. Cyclic urethanes have also been
discussed by Hall and Zbinden [150].

12.6(c). Anilides Ph—NHCOR. The substitution of an aromatic
residue on the amide nitrogen does not have any very significant
influence on the spectrum, apart from raising the frequency of the
amide I band due to the carbonyl group by a small amount. Richards
and Thompson [4] found the CO band of a series of anilides to be at
1700 cm^{-1} in dilute solution, and this value was further increased to
1715 cm^{-1} in compounds in which the substituents on the ring
increase their electron attracting properties. The amide II band
appeared to be much less subject to alteration in this way, and only
small shifts to lower frequencies could be detected. Gierer [92]
records essentially similar results for a series of acetanilides with
different ring substituents. In the solid state the arylamides show the
CO absorption at rather higher frequencies than alkylamides. The
spectra of a considerable range of anilides of fatty acids have been
recorded by Dijkstra and De Jonge [116]. Much confirmatory data
on anilides has been given by Thompson [153], Flett [147],
Katritzky [162] and Freedman [163]. The general concensus of
their findings is that the carbonyl band appears in the 1710–
1695 cm^{-1} range in secondary anilides. When the NH group is
replaced by the NCH_3 group there is a frequency fall of about
25–40 cm^{-1}.

The NH stretching absorptions are normal for open-chain amides
and occur near 3270 cm^{-1} (solid), but the identification of any

possible additional absorption in the 3070 cm^{-1} region is prevented in this case by the CH aromatic absorptions which fall in the same region.

12.6(d). Substituted urea derivatives —NHCONH— etc. The spectra of urea derivatives have been examined by Thompson, Nicholson and Short [25], by Randall *et al.* [5], by Boivin and Boivin [117, 118], by Becker [111, 112], and more recently by Jose [164], Mido [165] and Kutepov [166]. Monoalkyl ureas show three bands in the 1650—1500 cm^{-1} region, which correspond to the carbonyl (amide I), NH, and NH$_2$ bendings (amide II). The carbonyl frequency is markedly lower than in normal amides and now falls well below the frequency of the amide II band from the NH$_2$ group. ν_{CO} is in the 1605 cm^{-1} region, with the amide II band of the NH$_2$ at 1666—1655and that of the NH group at 1575—1550 cm^{-1}. These assignments are supported by deuteration studies which show shifts for the amide II bands but not for ν_{CO} [164].

In symmetrical dialkyl ureas the 1575 cm^{-1} band vanishes whereas those at 1656 and 1610 cm^{-1} remain. In contrast, in unsymmetrical dialkyl ureas, it is the 1656 cm^{-1} band which vanishes whilst the others remain. Other symmetrical dialkyl ureas have been studied by Mido [165] with similar results. He quotes ν_{CO} in the 1640—1625 cm^{-1} range with little deuterium sensitivity, and the D-sensitive amide II band at 1580 cm^{-1}. Kutepov [166] *et al.* have studied symmetrical diaryl ureas where the carbonyl band appears close to 1640 cm^{-1}. Gompper *et al.* [115] have discussed the characteristic frequencies of the NH—CO—N— grouping in cyclic systems.

12.6(e). Compounds containing the group R—CO—NH—N—. These compounds appear to be little affected by the presence of the additional nitrogen atom, and the spectrum of diformylhydrazine, for example, does not differ very greatly from that of other secondary amides [80]. In cyclic systems, such as pyrazolones, however, it is much more difficult to identify the characteristic amide frequencies due to the multiplicity of bands which occur [119, 120].

12.7. Polypeptides and proteins

12.7(a). General. The possibility that infra-red methods would throw some light on the problem of the structures of proteins and similar materials was appreciated by a number of relatively early workers in this field [13–15, 36–40]. These workers were able to show that proteins exhibited a common pattern of bands, including those at 3300 cm^{-1} due to hydrogen-bonded NH groups, and the amide I and II bands in the 1600–1500 cm^{-1} region, whilst the remainder of the spectrum was largely determined by the nature of the component amino-acids. In addition, the weak band near 3080 cm^{-1} was noted by some workers and assigned tentatively to an additional NH absorption [40]. However, in recent years a stimulus has been given to this work by the development of polarised radiation techniques which enable the internal orientation of certain groupings to be studied. Coupled with advances in the methods of preparation of simple model polypeptides and of improved methods for the purification of proteins themselves, this work is giving useful results which have been reviewed by Sutherland [121] and by Elliott [122, 167], and by Fraser [103].

In addition, studies have been made on very simple peptides containing only a few amino-acid residues. In the simpler members the characteristic peptide pattern is dominated by the influence of the zwitterion structures which are undoubtedly present. For this reason they have been discussed separately below, although the results are nevertheless also directly relevant to the discussion which follows on the characteristic frequencies of larger polypeptides.

12.7(b). Simple peptides. The spectra of simple peptides have been investigated by a number of workers [25, 41, 123–127]. In dipeptides such as glycyl-glycine the influence of the zwitterion structure is greatest and the spectrum is more complex than in the higher members of the series. This compound is known from X-ray studies to exist in the zwitterion form with an open-chain structure [34]. It has NH absorptions at 3300 cm^{-1}, 3080 cm^{-1}, 1630 cm^{-1}, 1575 cm^{-1}, 1540 cm^{-1} and 1240 cm^{-1}, as indicated by the shifting of these bands on deuteration [25, 41]. The first two correspond to NH stretching absorptions, whilst the 1540 cm^{-1} band is the normal amide II band which may be an NH deformation. The 1630 cm^{-1} and 1575 cm^{-1} bands have parallels in the amino-acid I and II bands and may be associated with the NH_3^+ grouping. The carbonyl group of the peptide link absorbs normally at 1655 cm^{-1}, and there are strong bands at 1608 cm^{-1} and at 1400 cm^{-1} which probably correspond to the COO^- group. Edsall has found a similar 1400 cm^{-1} absorption in the Raman spectrum [42]. A weak band near 1680 cm^{-1} in the infra-red has been found by both the main groups of workers in this field, which is tentatively assigned to the CO of the un-ionised carboxyl group. Ascending through the series to hexa- and polyglycine the influence of the zwitterion structure becomes progressively less, so that the 1400 cm^{-1} band (which is the stronger of the two COO^{-} vibrations) eventually vanishes, whilst the 1680 cm^{-1} band is retained. Thompson *et al.* [25] report the latter as being strengthened in the higher members, whilst Blout and Linsley [41] say that it is diminished in intensity. This is probably a difference only in words, in that if the band is in fact due to the un-ionised COOH group, its absolute intensity will fall progressively as the chain length and the molecular weight increase, whilst its relative intensity per COOH group will increase due to the increasing proportion of un-ionised carboxyl groups. The advent of laser Raman spectroscopy has led to the study of many polypeptides in solution in water. These include polyglycines [168], polyalanines [169], and polyprolines [143] and many others. The vibrational spectra and their application to the determination of the configurations of polypeptides have been discussed in detail by Koenig [151]. The most prominent features under these conditions are of course the amide I and amide III bands.

In the infra-red none of the polyglycines shows absorption higher than 3330 cm^{-1}, indicating that all the NH groups are capable of hydrogen bonding in the crystal state. The 3300 cm^{-1} band showed a regular and stepwise increase in intensity relative to the CH$_2$ absorption with increases in molecular weight, with a tendency towards a limiting value for the absorption ratios as the number of CH groups approached that of the amide links [41]. The remaining characteristic absorptions of the higher peptides take on more and more the appearance of the typical protein spectrum as the chain length is increased and the influence of the zwitterion diminished. All the polyglycines absorbed at 1015 cm^{-1} ± 10 cm^{-1}, and this may be a characteristic of the diglycine structure [41], particularly as it is not found in glycyl-alanine or in any other polypeptides in which two glycine residues are not directly linked [127]. Polypeptides from single amino-acids tend to resemble proteins in their spectra as soon as the number of residues is three or more [109] and then exhibit the typical amide absorptions, such as the amide II, III and V bands referred to above [127]. Polyproline and polysarcosine [124] are, of course, somewhat different, as they lack NH links.

Mixed polypeptides have also been studied by both groups of workers [25, 41]. The overall pattern is similar to that of the polyglycines, but a few interesting differences have been noted. In particular, many of the glycine—leucine peptides show absorptions at 3400 cm^{-1} and 3520 cm^{-1} in addition to the 3300 cm^{-1} and 3080 cm^{-1} absorptions. This clearly indicates the presence of unbonded NH_2 groupings and suggests that the crystal packing of these materials is such that it is not possible for all the terminal NH_2 groups to enter into hydrogen bonding. Interesting results have also been obtained from a study of variations in the optical forms of the amino acids used in their preparation. When more than one asymmetric carbon is present, marked spectral changes follow in some cases. Thus the spectrum of *dl*-glycyl-alanine is identical with that of the *ld*-compound, but differs markedly from the *dd*- or *ll*-forms, which are themselves identical. Similarly, *lll*-trialanine is identical with the *ddd*-form, but shows distinct differences from a mixed isomer preparation [123]. These differences are centred mainly in the NH stretching and the amide I and II absorptions, and are connected with the steric differences introduced. Elliott [126] has also reported small differences between the active and meso forms of polypeptides such as polyglycine, polyalanine, etc. These are more pronounced in the α-folded forms than in the β modifications, but in the former they are still small and amount to shifts of about 9 cm^{-1} in the carbonyl absorption. A number of simple peptides have also been examined in heavy water solution by Lenormant *et al.* [62, 128], who have followed the change in the carbonyl and NH_2 absorption following alterations in the pH of the solution. Parallel studies on larger molecules have been made by Ehrlich and Sutherland [129, 130]. These have established that the band at 1550 cm^{-1} contains contributions from both the COO^- and the amide II absorptions. In consequence, the changes in the state of ionisation which occur on denaturation of such materials will affect this absorption independently of any accompanying configurational changes.

Acetylated compounds with no residual NH_2 or COOH groups have also been studied [25, 43, 131, 132]. The spectra in these cases are much more like those of the polyamides. In studies on compounds such as acetylglycine-*N*-methylamide, Mizushima and co-workers [8, 43, 63, 68] have reported the persistence of a band near 3330 cm^{-1}, in addition to the free NH absorption at 3450 cm^{-1}, in dilute solution in carbon tetrachloride. Further, more detailed, studies [44, 132, 137] have shown that both inter- and intra-molecular bonding can lead to absorption at this point, so the intrepretation of structure from this region is very difficult. In acetylproline -*N*-methylamide the molecules all assume a folded configuration in carbon tetrachloride, but in chloroform, in which the tendency to hydrogen bonding is reduced, a second configuration can arise, capable of intermolecular bonding. This tendency is increased in acetyl piperidine α-carboxylic acid *N*-methyl-amide [132], in which steric effects reduce further the tendency to intramolecular hydrogen bonds. In dilute solution in carbon tetra-chloride this material exhibits absorption at 3390 cm^{-1} assigned to intramolecular bonds, accompanied by another band at 3367 cm^{-1} due to intermolecular links. This evidence is indicative of the difficulty in assigning 3280 cm^{-1} absorption of proteins wholly to *trans*-intermolecular bonding.

12.7(c). The NH stretching band near 3330 cm^{-1}. The band centred near 3330 cm^{-1} is certainly due to a bonded NH structure, as shown by deuteration and polarisation studies. As with the simpler peptides, there has been much discussion as to whether this band can be regarded as arising wholly from a *trans*-type hydrogen bond between chains or whether intramolecular bonds are involved, or both. It has been shown that poly-γ-benzyl-α-glutamate does not absorb at 3450 cm^{-1} in dilute chloroform solution, indicating that all the NH groups are able to bond within the molecule itself. In contrast, polycarbobenzoxy-*dl*-lysine, which has NH and CO groups in both the main and the side-chains, does absorb at 3450 cm^{-1} under these conditions, indicating that the polypeptide chains are then effec-tively isolated [11, 49, 51]. The folding of the chains in solution may be different from that in the solid state, but it does afford evidence that in the α folded form the hydrogen bonds of the peptide links are intramolecular in this case. On the other hand, Shigorin *et al.* [133–135], on the basis of studies with poly-caprolactams and similar products, assign the 3300 cm^{-1} absorption wholly to intermolecular bonding, and associate the weaker absorp-tion near 3200 cm^{-1} with intramolecular association. As with the simpler peptides, it is probable that both types of association give rise to absorption near 3300 cm^{-1}. Thus both poly-γ-benzyl-*l*-glutamate [136] and poly-α-*l*-glutamic acid [137] can be prepared in two distinct modifications which differ in the directions of orientation of the NH group. One is probably a coiled α-form with intramolecular bonds, and the other is an open or random form. Both absorb close to 3300 cm^{-1}. Similarly, nylon and other large synthetic polymers in which intramolecular bonding is improbable also absorb near this point. The interpretation of changes in intensity in this band, such as those which follow denaturation [45], is therefore very difficult.

As mentioned earlier, the 3300 cm^{-1} absorption is accompanied [50] by a weaker absorption near 3200 cm^{-1}. This is assigned by Shigorin *et al.* [133, 135] to intramolecular association only, but their evidence for this would not appear to be conclusive in view of the data quoted above and of the fact that both α- and β-keratins absorb at this point [135].

12.7(d). The NH stretching band near 3080 cm^{-1}. Originally this band was assigned as being due to intramolecular hydrogen bonds of the NH group additional to the interchain bonds which give rise to the 3330 cm^{-1} absorption [6, 7]. Later work has, however, indicated that this band is an overtone of the amide II absorption in accordance with earlier suggestions by Badger [82].

12.7(e). The amide I and II bands. The amide I and II absorptions appear in all polypeptides and proteins. The majority of these materials absorb in the normal 1650 cm^{-1} and 1550 cm^{-1} regions, but minor differences occur amongst them which may to some extent reflect changes in the hydrogen bonding pattern. The cyclic peptide gramicidin [47] for example, absorbs at 1610 cm^{-1} and 1513 cm^{-1}. The influence of changes in the type of bonding on these absorption frequencies is very well brought out in studies on the folded and extended forms of the same materials, carried out in conjunction with work on dichroism and X-ray patterns [46, 52, 54]. Poly-*l*-glutamic benzyl ester shows the amide CO absorption as a single band at 1658 cm^{-1}, and X-ray evidence indicates that it exists wholly in the α-form. In low-molecular-weight preparations, however, it exists in a random form or in some uncoiled state, and the amide I absorption then occurs [136] near 1630 cm^{-1}. The counter-parts of these forms have been observed in solution [113]. The methyl ester, however, shows two bands at 1658 cm^{-1} and at 1628 cm^{-1}, of which the latter shows no orientation. When cast from formic acid, in what is believed to be the β form, only the 1629 cm^{-1} component is shown. Similarly, two amide II frequencies at 1527 cm^{-1} and 1550 cm^{-1} correspond to the β and α forms. Hurd et al. [69] have also been able to observe carbonyl shifts in synthetic polypeptides following differences in their method of preparation which they ascribe to the existence of α and β forms, and Krimm [138] has suggested that the differences in the carbonyl frequencies may arise from differences in the angles formed by the O . . . H−N− hydrogen bonds. This is, however, disputed by Cannon [78], who prefers to regard them as arising from different degrees of interaction of the $^-$OCN$^+$ dipoles.

The position of these bands may therefore give some indication of the molecular form, but in the solid state other factors also may be involved. For example, on raising the temperature of nylon, frequency shifts of the CO and NH bands occur which are very similar to those described for a change from the α to β form, although there is strong evidence that nylon retains the extended β form under these conditions [48]. Similarly, an absorption in the 1625 cm^{-1} region in chitin has been tentatively assigned to an amide CO group which is hydrogen bonded to an OH group [55]. The interpretation of the amide II frequencies is also fraught with considerable difficulties. Marked changes in these bands occur, for example, on deuteration as well as on change of configuration [128, 139, 140], whilst the presence of the ionised carboxyl absorption in the same region can also lead to confusion. In many proteins the amide II absorption is in part overlaid by the ionised carboxyl band, so changes which occur in this region with changes of state may only reflect the alteration in the COO$^-$ group on moving away from the iso-electric point [128, 129, 130, 137]. The interpretation of group frequencies in this region therefore calls for great caution.

Some useful data on proteins can also be obtained from the overtone region, especially as it them becomes possible to work in the presence of liquid water. These absorptions also show dichroism, but it must not necessarily be assumed that this will be always in the same direction as the parent fundamental, and many of the bands in this region are combination bands arising from more than one fundamental. Thus a band near 4600 cm^{-1} in many proteins was originally identified with the carbonyl absorption [53] as it shifts with alterations of the helical form and is present in polymethyl-methacrylate but absent in polyethylene. However, this has recently been shown by Hecht and Wood [141] to be a combination band of

$2\nu CO$ with the 1250 cm^{-1} absorption. These workers have also elucidated the origins of a number of other high-frequency absorptions of this type, and Frazer [142] has also discussed assignments in this region and the significance of the results of dichroic studies.

12.8. Infra-red dichroism in protein studies

The use of polarised infra-red for studying the orientation of specific groupings within a crystal or oriented film is a well established technique. It consists simply in passing polarised radiation through the sample in two directions at right angles and observing the intensity changes which occur in certain bands. In this way the orientation of individual groups in relation to the whole can be studied. This method, which was applied by Glatt et al. to proteins in 1947 [56, 57], has since been extensively used by others in this field, and a substantial body of evidence has been built up which indicates that differentiation between the folded (α) and extended (β) forms of proteins and polypeptides is possible in this way [46, 49–51, 59, 61, 125]. Useful reviews of this technique have recently been given by Elliott [122, 167] and by Bamford et al. [17]

Ambrose, Elliot and Temple [46] studied the NH stretching absorptions of oriented polyglutamic esters, α- and β-keratins, and oriented myosin and tropomyosin. In all cases in which the material is believed to exist in the folded α-form the NH stretching bands at both 3310 cm^{-1} and 3060 cm^{-1} showed parallel dichroism, whereas they were markedly perpendicular in feather keratin, which is the β form. Similar findings in respect of the polyglutamic ester where reported by Ambrose and Hanby [51]. This excludes the possibility that the dichroism is shown by side-chain NH groups, as these are absent from this material. This work was later extended to studies on the C=O stretching and NH deformation modes [49, 52], which, in addition to frequency shifts, were found to show changes in their direction of orientation on change of form. Polyglutamic ester, for example, in the α-form shows strong parallel dichroism. No orientation in the β-form could be obtained, but the CO absorption in a similar copolymer material showed a complete alteration from the parallel to perpendicular dichroism with a change of form.

The NH deformation modes behave similarly, and in this case the α-form has perpendicular dichroism and the β-form parallel dichroism. Similar data can also be obtained from the 4810 cm^{-1}

band, which arises from a combination of the NH stretch and amide II absorptions [56, 141]. The technique has also been applied to the study of simple secondary amides in crystals [33, 109].

This interpretation of the results in which only a single form of hydrogen bond is proposed for α-keratin and similar structures was criticised by Darmon and Sutherland [6] as an over-simplification, and they point out that the attachment of numerical values to dichroic ratios is hazardous in view of the discrepancy between the small dichroic ratios found for NH in α-keratin and nylon and the large ratio to be expected [6].

The need for caution in the interpretation of polarisation studies is indicated by the recent work of Fraser and Price [59]. They point out that the transition moment of the carbonyl stretching mode responsible for the amide I absorption will be displaced from the CO direction due to interaction with the C–N vibration, and to an orbital following effect. They have calculated the dichroism of an oriented polypeptide in which the chains have the configuration of Pauling and Corey [60] with a 3.7 residue helix, and have shown that if allowance is made for a displacement of the CO transition moment by 20° towards the NO, the dichroic value obtained is very close to that observed experimentally [54]. The objection to the 3.7 helix based on polarisation studies [61] cannot therefore be maintained.

12.9. Bibliography

1. Walter and Kubersky, *J. Mol. Structure*, 1972, **II**, 207.
2. Jones and Smith, *J. Mol. Structure*, 1968, **2**, 475.
3. Davies, *Discuss. Faraday Soc.*, 1950, **9**, 325.
4. Richards and Thompson, *J. Chem. Soc.*, 1947, 1248.
5. Randall, Fowler, Fuson and Dangl, *Infra-red Determination of Organic Structures* (Van Nostrand, 1949).
6. Darmon and Sutherland, *Nature*, 1949, **164**, 440.
7. Astbury, Dalgliesh, Darmon and Sutherland, *ibid.*, 1948, **162**, 596.
8. Mizushima, Shimanouchi and Tsuboi, *ibid.*, 1950, **166**, 406.
9. *The Chemistry of Penicillin.* (Princeton University Press, 1949), p. 390.
10. Darmon, *Discuss. Faraday Soc.*, 1950, **9**, 325.
11. Elliot and Ambrose, *Nature*, 1950, **165**, 921.
12. Hallam and Jones, *J. Mol. Structure*, 1968, **I**, 413, 425.
13. Buswell, Rodebush and Roy, *J. Amer. Chem. Soc.*, 1938, **60**, 2444.
14. Buswell, Downing and Rodebush, *ibid.*, 1940, **62**, 2759.
15. Buswell and Gore, *J. Phys. Chem.*, 1942, **46**, 575.
16. Tsuboi, *Bull. Chem. Soc. Japan*, 1949, **22**, 215.
17. Bamford, Brown, Elliott, Hanby and Trotter, *Proc. Roy. Soc.*, 1953, **141B**, 49.

18. Hallam and Jones, *J. Mol. Structure*, 1970, **5**, 1.
19. Sheehan and Bose, *J. Amer. Chem. Soc.*, 1950, **72**,5158.
20. *Idem, ibid.*, 1951, **73**, 1761.
21. Sheehan, Hill and Buhle, *ibid.*, p. 4373.
22. Sheehan and Corey, *ibid.*, p. 4756.
23. Sheehan and Laubach, *ibid.*, p. 4752.
24. Sheehan and Ryan, *ibid.*, p. 4367.
25. Thompson, Nicholson and Short, *Discuss. Faraday Soc.*, 1950, **9**, 222.
26. Cutmore and Hallam, *Spectrochim. Acta*, 1969, **25A**, 1767.
27. Lenormant, *Discuss. Faraday Soc.*, 1950, **9**, 319.
28. *Idem, Bull. Soc. chim.*, 1948, **15**, 33.
29. *Idem, Ann. Chim.*, 1950, **5**, 459.
30. Lenormant and Chouteau, *J. Physiol.*, 1949, **A203**, 41.
31. Chouteau and Lenormant, *Compt. rend. Acad. Sci. (Paris)*, 1951, **232**, 1479.
32. Skinner and Perkins, *J. Amer. Chem. Soc.*, 1950, **72**, 5569.
33. Sandeman, *Proc. Roy. Soc.*, 1955, **A232**, 105.
34. Hughes and Moore, *J. Amer. Chem. Soc.*, 1949, **71**, 2618.
35. Elliot, Ambrose and Temple, *J. Chem. Phys.*, 1948, **16**, 877.
36. Stair and Coblentz, *J. Res. Nat. Bur. Stand.*, 1935, **15**, 295.
37. Wright, *J. Biol. Chem.*, 1939, **127**, 137.
38. Bath and Ellis, *J. Phys. Chem.*, 1941, **45**, 204.
39. Lenormant, *Compt. rend. Acad. Sci. (Paris)*, 1945, **221**, 58.
40. Darmon and Sutherland, *J. Amer. Chem. Soc.*, 1947, **69**, 2074.
41. Blout and Linsley, *ibid.*, 1952, **74**, 1946.
42. Edsall, Otvos and Rich, *ibid.*, 1950, **72**, 474.
43. Mizushima, Shimanouchi, Tsuboi, Sugita, Kato and Kondo, *ibid.*, 1951, **73**, 1330.
44. Mizushima, Shimanouchi, Tsuboi and Souda, *ibid.*, 1952, **74**, 270.
45. Uzman and Blout, *Nature*, 1950, **166**, 862.
46. Ambrose, Elliot and Temple, *ibid.*, 1949, **163**, 859.
47. Klotz, Griswold and Gruen, *J. Amer. Chem. Soc.*, 1949, **71**, 1615.
48. Sutherland and Tanner, *Ohio Symposium on Molecular Spectroscopy*, 1951.
49. Elliot and Ambrose, *Discuss. Faraday Soc.*, 1950, **9**, 324.
50. Holliday, *ibid.*, p. 325.
51. Ambrose and Hanby, *Nature*, 1949, **163**, 483.
52. Hanby, Waley and Watson, *J. Chem. Soc.*, 1950, 3239.
53. Ambrose, Elliot and Temple, *Proc. Roy. Soc.*, 1951, **A206**, 192.
54. Ambrose and Elliot, *ibid.*, 1951, **A205**, 47.
55. Darmon and Rudall, *Discuss. Faraday Soc.*, 1950, **9**, 251.
56. Glatt and Ellis, *J. Chem. Phys.*, 1948, **16**, 551.
57. Glatt, *ibid.*, 1947, **15**, 880.
58. Pauling, Corey and Branson, *Proc. U.S. Nat. Acad. Sci.*, 1951, **37**, 205.
59. Fraser and Price, *Nature*, 1952, **170**, 490.
60. Pauling and Corey, *Proc. U.S. Nat. Acad. Sci.*, 1951, **37**, 235.
61. Bamford, Brown, Elliot, Hanby and Trotter, *Nature*, 1952, **169**, 357.
62. Lenormant and Chouteau, *Compt. rend. Acad. Sci. (Paris)*, 1952, **234**, 2057.
63. Mizushima, Shimanouchi, Tsuboi, Sugita, Kurosaki, Mataga and Souda, *J. Amer. Chem. Soc.*, 1952, **74**, 4639.
64. Wasserman, Precopio and Liu, *ibid.*, p. 4093.
65. Haszeldine, *Nature*, 1951, **168**, 1028.

66. Lacher, Olsen and Park, *J. Amer. Chem. Soc.*, 1952, **74**, 5578.
67. Yakel and Hughes, *ibid.*, 6302.
68. Mizushima, Shimanouchi, Tsuboi, Sugita, Kurosaki, Matoga and Souda, *ibid.*, 1953, **75**, 1863.
69. Hurd, Bauer and Klotz, *ibid.*, 624.
70. LeKlein and Plesman, *Spectrochim. Acta*, 1972, **28A**, 673.
71. Venkataramiah, Venkata and Chalapathi, *J. Mol. Spectroscopy*, 1964, **12**, 300.
72. Walter and Kubersky, *Spectrochim. Acta*, 1970, **23A**, 1158.
73. Worsham and Hobbs, *J. Amer. Chem. Soc.*,1954, **76**, 206.
74. Russell and Thompson, *Spectrochim. Acta*, 1956, **8**, 138.
75. Klemperer, Cronyn, Maki and Pimentel, *J. Amer. Chem. Soc.*, 1954, **76**, 5846.
76. Cannon, *Mikrochem. Acta*, 1955, 555.
77. Oshida, Ooshika and Miyasaka, *J. Phys. Soc. Japan*, 1955, **10**, 849.
78. Cannon, *J. Chem. Phys.*, 1956, **24**, 491.
79. Kessler and Sutherland, *ibid.*, 1953, **21**, 570.
80. Miyazawa, Shimanouchi and Mizushima, *ibid.*, 1956, **24**, 408.
81. Miyazawa, *J. Chem. Soc. Japan*, 1956, **77**, 321, 619.
82. Badger and Rubakava, *Proc. Nat. Acad. Sci. U.S.*, 1954, **40**, 12.
83. Cleverley, quoted in *Chemical Applications of Spectroscopy* (Interscience, New York, 1956), p. 515.
84. Desseyn, Jacob and Herman, *Spectrochim. Acta*, 1972, **28A**, 1329.
85. Mecke and Mecke, *Chem. Ber.*, 1956, **89**, 343.
86. Mizushima, *Structure of Molecules and Internal Rotation* (Academic Press, New York, 1954), p. 117.
87. Mizushima, Shimanouchi, Nagakura, Kuratani, Tsuboi, Baba and Fujioka, *J. Amer. Chem. Soc.*, 1950, **72**, 3490.
88. Letaw and Gropp, *J. Chem. Phys.*, 1953, **21**, 1621.
89. Evans, *ibid.*, 1954, **22**, 1228.
90. Davies, Evans and Lumley Jones, *Trans. Faraday Soc.*, 1955, **51**, 761.
91. Jones and Cleverly, quoted in *Chemical Applications of Spectroscopy* (Interscience, New York, 1956), p. 522.
92. Gierer, *Z. Naturforsch*, 1953, **8B**, 644, 654.
93. Bellamy and Williams, *J. Chem. Soc.*, 1957, 4294.
94. Mizushima, Shimanouchi, Ichishima, Miyazawa, Nakagawa and Araki, *J. Amer. Chem. Soc.*, 1956, **78**, 2038.
95. Robson and Reinhart, *J. Amer. Chem. Soc.*, 1955, **77**, 498.
96. White, *ibid.*, 6008.
97. Beer, Kessler and Sutherland, *J. Chem. Phys.*, 1958, **29**, 1097.
98. Ramirez and Paul, *J. Amer. Chem. Soc.*, 1955, **77**, 1035.
99. McLachlan and Nyquist, *Spectrochim. Acta*, 1964, **20**, 1397.
100. Bellamy, *Advances in Infrared Group Frequencies* (Methuen, London, 1968).
101. Puranik and Ramiah, *Nature*, 1961, **191**, 796.
102. Huisgen, Brade, Walz and Glogger, *Chem. Ber.*, 1957, **90**, 1437.
103. Frazer, *Analytical Methods in Protein Chemistry* **Vol.** 2 (Pergamon Press, New York, 1960).
104. Brügel, *Private communication*.
105. Schiedt, *Angew Chem.*, 1954, **66**, 609.
106. Miyazawa, Shimanouchi and Mizushima, *J. Chem. Phys.*, 1958, **29**, 611.
107. Brockmann and Musso, *Chem. Ber.*, 1956, **89**, 241.
108. Frazer and Price, *Proc. Roy. Soc.*, 1953, **141B**, 66.

109. Abbott and Ambrose, *ibid.*, 1956, **234A**, 247.
110. Mecke, Mecke and Luttringhaus, *Zeit. Naturforsch.*, 1955, **10B**, 367.
111. Becher and Griffel, *Naturweiss*, 1956, **43**, 467.
112. Becher, *Chem. Ber.*, 1956, **89**, 1593.
113. Doty, Holtzer, Bradbury and Blout, *J. Amer. Chem. Soc.*, 1954, **76**, 4493.
114. Huisgen and Walz, *Chem. Ber.*, 1956, **89**, 2616.
115. Gompper and Herlinger, *ibid.*, 2825.
116. Dijkstra and De Jonge, *Rec. trav. chim.*, 1956, **75**, 1173.
117. Boivin and Boivin, *Can. J. Chem.*, 1954, **32**, 561.
118. Boivin, Bridges and Boivin, *ibid.*, 242.
119. Gargon, Boivin, and Paquin, *ibid.*, 1953, **31**, 1025.
120. Gargon, Boivin, McDonald and Yaffe, *ibid.*, 1954, **32**, 823.
121. Sutherland, *Advances in Protein Chemistry*, 1952, **7**, 291.
122. Elliott, *J. App. Chem.*, 1956, **8**, 341.
123. Ellenbogen, *J. Amer. Chem. Soc.*, 1956, **78**, 363, 366, 369.
124. Berger, Kurtz and Katchalski, *ibid.*, 1954, **76**, 5552.
125. Elliott and Malcolm, *Trans. Faraday Soc.*, 1956, **52**, 528.
126. Elliott, *Proc. Roy. Soc.*, 1954, **A221**, 104.
127. Asai, Tsuboi, Shimanouchi and Mizushima, *J. Phys. Chem.*, 1954, **59**, 322.
128. Lenormant and Blout, *Nature*, 1953, **172**, 770.
129. Ehrlich and Sutherland, *ibid.*, 671.
130. *Idem, J. Amer. Chem. Soc.*, 1954, **76**, 5268.
131. Mizushima, Tsuboi, Shimanouchi, Sugita and Yoshimoto, *ibid.*, 1954, **76**, 2479.
132. Mizushima, Tsuboi, Shimanouchi and Asai, *ibid.*, 1954, **76**, 6003.
133. Mikhailov, Shigorin and Makar'eva, *Doklady. Acad. Nauk. S.S.S.R.*, 1952, **87**, 1009.
134. Shigorin, Mikhailov and Makar'eva, *ibid.*, 1954, **94**, 711.
135. Shigorin, Mikhailov and Klyuyeva, *Zhur. Fiz. Khim.*, 1956, **30**, 1591.
136. Blout and Asadourian, *J. Amer. Chem. Soc.*, 1955, **78**, 955.
137. Blout and Idelson, *ibid.*, 1956, **78**, 497.
138. Krimm, *J. Chem. Phys.*, 1955, **23**, 1371.
139. Lenormant and Blout, *Bull. Soc., Chim. Fr.*, 1954, 859.
140. Lenormant, *Trans. Faraday Soc.*, 1956, **52**, 549.
141. Hecht and Wood, *Proc. Roy. Soc.*, 1956, **A235**, 174.
142. Frazer, *J. Chem. Phys.*, 1956, **24**, 89.
143. Rippon, Koenig and Walton, *J. Amer. Chem. Soc.*, 1970, **92**, 7455.
144. Kniseley, Fassel, Farquhar and Grey, *Spectrochim. Acta*, 1962, **18**, 1217.
145. Suzuki, *Bull. Chem. Soc. Japan*, 1962, **35**, 540, 1279.
146. Brown, Regan, Schultz and Sternberg, *J. Phys. Chem.*, 1959, **63**, 1324.
147. Flett, *Spectrochim. Acta*, 1962, **18**, 1537.
148. Renema, Thesis, University of Utrecht, 1972.
149. Nyquist, *Spectrochim. Acta*, 1963, **19**, 509.
150. Hall and Zbinden, *J. Amer. Chem. Soc.*, 1958, **80**, 6248.
151. Koenig, *Raman Spectroscopy in Biological Polymers* Part D, *J. Polymer. Sci.*, 1972, **6**, 59.
152. Speziale and Freeman, *J. Amer. Chem. Soc.*, 1960, **82**, 903.
153. Thompson and Jameson, *Spectrochim. Acta*, 1958, **13**, 236.
154. Miyazawa, *J. Mol. Spectroscopy*, 1960, **4**, 155; 168.
155. Rao, Rao, Goel and Balasubramanian, *J. Chem. Soc.* (B), 1971, 3071.
156. Uno and Machida, *Bull. Chem. Soc. Japan*, 1961, **34**, 545; 551.
157. Abrahamovitch, *J. Chem. Soc.*, 1957, 1413.
158. Jansen and Kramer, *Can. J. Chem.*, 1971, **49**, 1011.

159. Pinchas and BenIshai, *J. Amer. Chem. Soc.*, 1957, **79**, 4099.
160. Katritzky and Jones, *J. Chem. Soc.*, 1960, 676.
161. Sato, *J. Org. Chem.*, 1961, **26**, 770.
162. Katritzky and Jones, *J. Chem. Soc.*, 1959, 2067.
163. Freedman, *J. Amer. Chem. Soc.*, 1960, **82**, 2454.
164. Jose, *Spectrochim. Acta*, 1969, **25A**, 111.
165. Mido, *Spectrochim. Acta*, 1972, **28A**, 1508.
166. Kutepov and Dubov, *J. Gen. Chem. (Moscow)*, 1960, **30**(92), 3448.
167. Elliott, *Advances in Spectroscopy*, **Vol. 1**, 1959, **1**, 213.
168. Smith, Walton and Koenig, *Biopolymers*, 1969, **8**, 29.
169. Koenig and Sutton, *Biopolymers*, 1969, **8**, 167.

13

Amino-Acids,
their Hydrochlorides and Salts,
and Amido-Acids

13.1. Introduction and table

Edsall's studies on the Raman spectra of amino-acids provided the first direct optical evidence [1, 2, 60] for the dipolar structure of these materials, and his findings have been fully confirmed by similar later studies [3–5]. The first infra-red studies were made by Freymann, Freymann and Rumpf [6] working in the overtone region on the NH stretching frequencies, and they also obtained evidence for the presence of a quaternary nitrogen atom in the neutral amino-acids. Further data in the fundamental region were obtained by a number of other workers, mostly in studies in the 3000 cm^{-1} region [7–10]. Wright [11, 12] was also amongst the early workers in this field, and he was the first to point out that the spectrum of the racemic form of cystine in the solid state differs from that of either of the pure optical isomers. This fact has been fully substantiated by later workers on other amino-acids [13, 14], and it appears to be true also of meso-tartaric acid [15]. Some fifty pairs of d- and l-forms of amino-acids have been studied by Koegel et al. [32]. As with the polypeptides examined by Ellenbogen [33], the individual d- and l-isomers have identical spectra, but these can differ appreciably from those of the racemate. When two asymmetric carbon atoms are present, as in normal and allo-forms, differences become possible between the spectra of the optical isomers. Brockmann and Musso [34], and Erhart and Hey [35] report similar findings.

More recently, interest in amino-acids has been stimulated by work on protein hydrolysates, and also by the studies undertaken in connection with the penicillin problem. The spectra of a large number of amino-acids have now been described by several groups of workers [17–23, 33, 34, 55–60], and a number of correlations have been defined by which amino-acids and their hydrochlorides can be identified. These are as follows:

(a) The NH stretching region. No bands in the normal $3500-3300$ cm^{-1} range are shown, but instead an absorption appears near 3070 cm^{-1} due to the NH_3+ group. This correlation holds also for hydrochlorides, but not for the N-substituted amino-acids such as sarcosine or proline in which only the NH_2^+ group is involved. The NH_2^+ group has its stretching frequency in the 2900 cm^{-1} region [57, 58]. With salt formation the NH_2 group reappears without a charge and normal NH stretching bands are shown, unless chelation occurs, as in metal glycinates.

(b) An absorption corresponding to the ionised carbonyl group appears in amino-acids of all types and in their salts. In the corresponding hydrochlorides this band disappears and is replaced by a typical carbonyl absorption. In α-amino-acid hydrochlorides this carbonyl absorption is displaced about 20 cm^{-1} towards higher frequencies under the influence of the NH_3^+ group.

(c) Two characteristic bands in the $3000-2000$ cm^{-1} region are shown by most but not all amino-acids, whilst their hydrochlorides show an almost continuous series of bands between 3030 cm^{-1} and 2500 cm^{-1}.

(d) In addition to the ionised carboxyl group absorption, two other bands appear in the $1600-1500$ cm^{-1} region which must be associated with the NH_3^+ group. NH_2^+ amino-acids show a single band in the region of 1565 cm^{-1} [57, 58]. Amino-acid hydrochlorides containing the NH_3^+ group also show two bands.

(e) In addition, many amino-acids absorb near 1300 cm^{-1} and many hydrochlorides near 2000 cm^{-1}.

These correlations form a comprehensive basis for the identification of amino-acids as a class, and this can be supplemented in a number of ways, particularly by the observation of the changes in

NH, $C\begin{smallmatrix} O \\ O^- \end{smallmatrix}$ and C=O frequencies which occur in hydrochlorides and salts. Studies in solution in water and in heavy water have also been valuable in this connection [24, 30, 56]. It is also possible to use antimony trichloride as a solvent for special purposes [36].

The presence of sulphur or other functional groups does not appear to affect the validity of these assignments [13, 18, 32]. Dicarboxylic mono-amino-acids behave as expected in showing a carbonyl absorption in addition to the ionised carboxyl and NH_3^+ absorptions [13], and ornithine [22] and other diamino-carboxylic acids show free NH_2 absorptions in addition to those of a typical amino-acid, but compounds of this complexity have not been very fully studied. The number of carbon atoms separating the NH_2 and COOH groups does not appear to affect the existence of these materials in dipolar form. Randall *et al.* [17] find ϵ-aminocaproic acid to be a typical amino-acid in this respect and Gaümann and Günthard have obtained similar results with other amino-acids with widely separated functional groups [29]. Lenormant [13] has also shown the persistence of the NH_3^+ absorptions in $NH_2(CH_2)_{10}COOH$ and similar compounds, and Despas *et al.* [37] report similar findings.

With aromatic acids the presence of the aromatic ring absorptions in the $1600-1500$ cm^{-1} region complicates the spectrum a good deal, but nevertheless β-phenylalanine, tyrosine and similar products appear to be essentially normal [13, 17]. Anthranilic acid, on the other hand, shows normal carboxyl and amine absorptions slightly modified by the internal hydrogen bond, and so does not behave as a typical amino-acid. The absence of the zwitterion form in this case may be associated with the hydrogen bond effect.

N-acyl substitution results in compounds which combine some of the characteristic features of amino-acids and of amides. Their spectra show a number of distinctive features, some of which are similar to those of amino-acids, so that they are more conveniently considered here than under amides. In common with amides, however, the positions of key bands in these compounds are prone to alteration following changes of state, and it is important that comparisons should be made only between materials examined in comparable states. The differences associated with the spectra of solid and liquid esters have been employed by Randall *et al.* [17] for their identification. These compounds do not exist as zwitterions, and therefore show NH and CO frequencies and the amide I and II bands. Like the amino-acids, they also show absorptions in the $3000-2000$ cm^{-1} region. Dipeptides and similar products also containing the NH_3^+ group are dealt with separately in Chapter 12.

Table 13

NH_3^+ *Vibrations*. (NH_2 Amino-acids and their hydrochlorides. Not in salts)

NH_3^+ stretching frequencies	$3130-3030$ cm^{-1} (m.)
NH_3^+ deformation frequencies	Amino-acid 1, $1640-1610$ cm^{-1} (w.). (Hydrochlorides down to 1590 cm^{-1})
	Amino-acid 2, $1550-1485$ cm^{-1} (m.)

NH *Vibrations*. Amido-acids, $3390-3260$ cm^{-1}. (Esters lower) (m.)

COOH *Frequencies*

Ionic carboxyl (all types of amino-acids and salts but not hydrochlorides), $1600-1560$ cm^{-1} (s.)
Normal acid carbonyl (un-ionised). Hydrochlorides
$\quad$ $1754-1720$ cm^{-1} α-amino-acids (s.)
$\quad$ $1724-1695$ cm^{-1} α-amido-acids (s.)
$\quad$ $1730-1700$ cm^{-1} β, γ, and lower amino-acids (s.)

The Region $3000-2000$ cm^{-1}

Amino-acids, $2760-2530$ cm^{-1} (w.) $\left.\right\}$ Present in most, but not all cases
$\quad\quad\quad\quad$ $2140-2080$ cm^{-1} (w.)
Amino-acid hydrochlorides $\quad$ A series of almost continuous bands, $3030-2500$ cm^{-1} (w.)
Amido-acids, $2640-2360$ cm^{-1} $\left.\right\}$ Present in most, but not all
$\quad\quad\quad\quad$ $1945-1835$ cm^{-1} $\quad$ cases (w.)

Other Correlations (Amino-acids). A band near 2000 cm^{-1} is shown by most amino-acid hydrochlorides (w.)
A band near 1300 cm^{-1} is shown by most amino-acids and their derivatives (m.)

Other Correlations (Amido-acids)

Amide I (CO) α-acids, $1620-1600$ cm^{-1}. Other acids, $1650-1620$ cm^{-1} (s.)
Amide II, $1570-1500$ cm^{-1} (s.)

The correlations to be discussed are summarised in Table 13. In all cases the ranges quoted relate to materials examined in the solid state as this is the condition in which these materials are usually studied.

13.2. NH_3^+ Stretching vibrations

No absorption in the usual NH stretching region $3500-3300$ cm^{-1} is shown by any amino-acid or hydrochloride [10, 11, 12, 17, 18, 19]. Instead a single band appears in the $3130-3030$ cm^{-1} region. Both Freymann and Freymann [10] and Thompson *et al.* [19] have reported an absorption near 3070 cm^{-1} in a number of amino-acids which they have assigned to the NH_3^+ stretching mode, and although Randall *et al.* [17] have not commented on this correlation, their data have been re-examined by Fuson *et al.* [18], who found that all the thirteen amino-acids they examined absorbed in the 3130–3030 cm^{-1} range. Fuson *et al.* [18] have also given data on an additional five amino-acids which they found to absorb in this region. Finally, Koegel *et al.* [32] observe this band in the very extensive series of amino-acids they have studied. It is almost certainly due to the asymmetric NH_3^+ stretching mode, and the corresponding symmetric frequency is probably amongst the bands found between 3000 and 2000 cm^{-1}.

Hydrochlorides can be expected to absorb similarly. The simple amine hydrochlorides of butylamine [37] and methylamine [38] absorb here, and Thompson *et al.* [19] find a 3100 cm^{-1} band in glycine ester hydrochloride, whereas the sodium salt of phenyl-glycine with an uncharged NH_2 group shows a normal amine absorption near 3370 cm^{-1}. Gore *et al.* [24] have confirmed the presence of a band near 3000 cm^{-1} in a solution of glycine in heavy water containing DCl. NH_3^+ stretching frequencies also occur in co-ordination compounds such as the cobalt amines, and these have been studied by a number of workers [39–45]. However, in these cases the charge on the nitrogen atom is substantially smaller, and the asymmetric and symmetric frequencies occur near 3300 cm^{-1} and 3150 cm^{-1} respectively. This correlation will of course apply only to amino-acids which are capable of taking up the NH_3^+ structure. *N*-substituted amino-acids such as *N*-phenylglycine, sarco-sine or proline can only have the NH_2^+ group, which absorbs at lower frequencies. The limited number of such compounds studied precludes any precise correlation but this group probably absorbs near 2900 cm^{-1}. In the cases of hydrochlorides of secondary amines [38, 46] absorption from the NH_2^+ stretching mode may be expected somewhere near 2700 cm^{-1}. The NH^+ frequency may well be lower still, and Lord and Merrifield [31] have reported values of 2425 cm^{-1} for the hydrochlorides of a few tertiary amines examined in the solid state.

13.3 NH_3^+ Deformation vibrations. 'amino-acid bands' I and II

As indicated in Table 13, all amino-acids capable of possessing the NH_3^+ structure, and their hydrochlorides, show at least one, and probably two, characteristic bands in the $1600-1500$ cm^{-1} region in addition to any ionic carboxyl absorption [13, 17–19, 55, 56, 60]. The first of these, in the $1660-1610$ cm^{-1} range, is often weak, and in some cases appears only as a shoulder on the main ionic carboxyl absorption. The second characteristic band is usually more intense, but the absolute intensity appears to be variable [17]. It appears in the overall range $1550-1485$ cm^{-1}. The origin of these bands is indicated by their absence in salts and in N-substituted amino-acids [13, 17] and by their presence in the hydrochlorides Lenormant [13] attributes the second band to the NH$_3$$^+$ deformation and does not attempt to assign the first, whereas Randall *et al.* [17] assign the first to NH$_3$$^+$ and do not assign the second. It is now well established by deuteration studies [55] and by a normal coordinate treatment of glycine [28] that the weak higher frequency band is due to a degenerate NH$_3$$^+$ mode and that the stronger lower frequency band is due to the symmetric NH$_3$$^+$ deformation. In primary amine hydrochlorides [38] and in complex amines [41, 44, 47–49], however, the NH$_3$$^+$ structure shows stronger splitting.

'*Amino-acid I.*' This band, the degenerate NH$_3$$^+$ deformation, is intrinsically weak, so that whilst some authors report its presence, [17, 19, 24, 55, 56], others [20, 32] fail to find it, presumably because of the use of thinner films. The band ranges between 1640 and 1600 cm^{-1} but is most frequently found near 1600 cm^{-1}. It is well seen in the studies of Shiflin *et al.* [57] in which glycine and other amino-acids are studied in a range of different pH conditions, both in water and in discs. In solutions at high pH the NH$_3$$^+$ group is discharged and normal NH$_2$ bands appear. With NH$_2$$^+$ deformations, the number of compounds studied is too few for certainty, but the deformation frequency appears to occur near 1565 cm^{-1}.

Hydrochlorides with the NH$_3$$^+$ structure show a similar absorption. Randall *et al.* [17] quote a range of $1610-1590$ cm^{-1} for the compounds they have examined, whilst Lenormant's [13] values are in general agreement. The detection of this absorption is considerably simplified in this case by the suppression of the ionic carboxyl absorption in the same region. Tsuboi *et al.* [55] give 1565 cm^{-1} for this frequency in methylamine hydrochloride but this is a solution value which may be lowered by hydrogen bonding.

'*Amino-acid II.*' There is more general agreement on the occurrence of this band, which is usually considerably stronger than the first. It is reported as being present in the range given, in all the NH_3^+ acids and hydrochlorides examined by Randall *et al.* [17], Fuson *et al.* [18], Thompson *et al.* [19], Klotz and Gruen [20] and other workers [22, 32, 34, 55, 56, 59]. It is absent from the spectra of salts [13, 55, 56], and this supports the assignment to NH_3^+ deformations. As indicated above, however, its intensity is variable to some extent, although it is usually strong. Koegel *et al.* [32] finds this band close to 1515 cm^{-1} in almost all the fifty compounds they have studied. However, it is absent from the spectrum of *iso*valine and shows doubling in some other amino-acids, such as $\alpha\beta$-diamino-propionic and -butyric acids. In hydroxy-acids the band shows an interesting progression towards higher frequencies as the OH group is brought successively nearer to the NH_3^+. In hexahomoserine it is at 1504 cm^{-1}, rising to 1522 cm^{-1} in pentahomoserine. In homo-serine itself it appears at 1538 cm^{-1}, and in serine it vanishes, being presumably superimposed [34] on the COO$^-$ absorption at 1600 cm^{-1}. The band is most clearly seen in the spectra of secondary amine hydrochlorides when this absorption occurs [46] near 1600 cm^{-1}. The disappearance of the band on salt formation provides a convenient method of confirmation of identifications based upon it.

13.4. COOH Vibrations

13.4(a). The ionised carboxyl group. The position of the absorption bands of the ionised carboxyl group are fairly well established (Chapter 10.5) following the work of Lecomte [25], and Raman studies on both normal acid salts and amino-acids, by Edsall [1–5, 60]. Two absorptions corresponding to the asymmetric and symmetric modes occur near 1550 cm^{-1} and 1410 cm^{-1}. The former is usually more easy to identify in infra-red spectra, as many other absorptions can occur in the same region as the second band.

Amino-acids follow completely this expected pattern in that the neutral acids and their salts all show absorptions near 1600 cm^{-1} which can be assigned to the ionic carboxyl absorption [13, 17–22, 32, 34, 50, 55, 56]. All the various workers in this field are agreed on this assignment, which is supported by the large number of amino-acid spectra they have examined. Thompson *et al.* [19] have proposed the narrow range 1600–1590 cm^{-1}, but this is obviously too restricted in the light of the results of other workers. Fuson *et al.* [18] follow Randall *et al.* [17] in proposing the range 1600–1560 cm^{-1}, and this is sufficiently wide to cover all the acids examined by these groups except sarcosine [17], which absorbs at 1616 cm^{-1}. The more usual position of the asymmetric band, which is the more readily identifiable in the infra-red is 1600 cm^{-1} in α—amino-acids, but there is a progressive fall in frequency as the NH$_3$$^+$ group is moved further and further away from the ionised carboxyl group, and it ultimately reaches a value of about 1560 cm^{-1} in long chain acids with the polar groups at the ends.

This absorption vanishes on formation of the quaternary hydrochloride in which the ionisation of the carboxyl group is suppressed. The spectral changes following the change of the ionic carboxyl to the non-ionic form have been elegantly demonstrated by Gore *et al.* [24] working with glycine in water and in heavy water solution. In neutral solution the ionic carboxyl absorption is shown (1610 cm^{-1} and 1400 cm^{-1}) which is replaced by the carbonyl absorption at 1740 cm^{-1} on the addition of DCl. On the addition of NaOD the original ionic absorptions reappear at approximately the original positions (1590 cm^{-1}, 1400 cm^{-1}). Pearson and Slifkin [56] have also studied a number of amino-acids in solution at a range of five different pH values, and they also comment on the replacement of the COO$^-$ absorptions by the normal acid carbonyl at 1735 cm^{-1} in solutions of low pH.

This correlation is applicable to all types of amino-acids, regardless of whether or not the amine group has the NH$_2$$^+$ or NH$_3$$^+$ structure. A similar effect is reported in pyridine betaine [26], although in this case the absorption occurs as high as 1652 cm^{-1}.

With non-crystalline preparations it is sometimes possible for both the ionised and un-ionised carboxyl absorptions to appear [21], and this is a source of possible confusion in work of this kind. The effect is not, however, common.

The second absorption arising from the ionic carboxyl group is less easy to identify in the infra-red, where its intensity is appreciably lower than the 1600 cm^{-1} band, but it appears clearly in many Raman spectra [5]. Although many amino-acids do in fact absorb in this region, the assignment of any particular individual band to this mode is extremely difficult. Fuson *et al.* [18] have found bands near 1400 cm^{-1} and 1406 cm^{-1} in a series of aryl cysteines, one of which, they suggest, may be associated with the ionised carbonyl absorption, and Koegel *et al.* [32] have also identified this absorption as occurring at 1408 cm^{-1} in the vast majority of the amino-acids they have studied. Ehrlich and Sutherland [50] have also identified a band at 1410 cm^{-1} in *N*-methyl succinamic acid with the COO$^-$ symmetric mode. This band is well characterised in the Raman spectra where it is always strong and easily seen. In α-amino acids it appears in the 1420 to 1400 cm^{-1} region. In solutions of low pH it vanishes and is replaced by the normal 1250 cm^{-1} band of acids.

13.4(b). The un-ionised carboxyl absorption. The normal carbonyl absorption of the COOH group appears only in hydrochlorides in the amino-acid series, except in the case of dicarboxylic mono-amino-acids in which both the ionic and non-ionic forms appear.

Randall *et al.* [17] have commented upon the abnormally high-frequency range which they find for this absorption (1754–1724 cm^{-1}), which overlaps to some extent the frequency range of normal esters. However, Edsall [5] has pointed out that in Raman work, abnormal frequencies in the 1750–1740 cm^{-1} range occur only in the α-amino-acids, and that the remainder absorb normally in the 1730–1700 cm^{-1} range. Consideration of the spectral data of Randall *et al.* [17] and of Lenormant [13] confirm that this is true also for the infra-red spectra. The great majority of the compounds examined by Randall *et al.* [17], for example, are α-amino-acids absorbing in the higher frequency range, whereas δ-amino-*n*-valeric acid hydrochloride is shown as absorbing at 1701 cm^{-1}.

This general effect of an elevation of the frequency by about 20 cm^{-1} in α-amino-acids is not surprising, and must reflect the influence of the NH_3^+ Cl^- grouping on the C=O bond length of the adjacent carbonyl group. A similar effect is shown by halogen substitution in acids and ketones, whereas β-halogen substituents are sufficiently far removed from the carbonyl group to be without effect on its frequency.

This correlation should also be applicable to the hydrochlorides of NH_2^+ type acids, but, as in previous cases, not much information is available. Sarcosine hydrochloride again appears to be slightly abnormal in showing its carbonyl absorption at 1757 cm^{-1}, but both proline and *N*-phenylglycine hydrochloride absorb at 1730 cm^{-1}. Histidine hydrochloride absorbs at 1706 cm^{-1} and ornithine dihydrochloride [13] at 1739 cm^{-1}. As far as the limited data available allow, therefore, it seems reasonable to assume that the correlation is obeyed by this type of compound.

13.5. The region 3000–2000 cm^{-1}

Thompson *et al.* [19] have drawn attention to the presence of a weak, but definite band near 2100 cm^{-1} in the spectra of the simple amino-acids which is absent from the spectrum of phenylglycine sodium salt or from p-aminophenylacetic acid. This suggests its association with the zwitterion structure. Randall *et al.* [17] had earlier noted a similar effect and drawn attention to the presence of two bands in this region in the spectra of most but not all amino-acids. Koegel *et al.* [32] find a band at 2138 cm^{-1} in all the α-amino-acids they have studied, including many which are branched at the α-carbon atom. The band is obviously not a fundamental and it must be associated in some way with the zwitterion structure. Liefer and Lippincott [61] have identified this band as a combination band associated with the COO$^-$ group. It appears near 2130 cm^{-1} in all α-amino-acids, but is displaced in others. This is almost certainly due to the fact that the antisymmetric COO$^-$ stretch is itself displaced when the NH$_3$$^+$ group is moved further away than the α-carbon atom.

Hydrochlorides of amino-acids, along with other types, exhibit an almost continuous series of band of moderate intensity between 3030 cm^{-1} and 2500 cm^{-1} [17]. The series is highly characteristic and is readily identified. No interpretation of the great multiplicity of bands has been attempted in detail, but like the somewhat related series in carboxylic acid dimers it is probable that it originates in overtone and combination bands superimposed on a very strong and very broad NH band.

13.6 Other correlations (amino-acids)

The other likely possibility for the identification of amino-acids would be the C–N stretching vibration. This has been put by Suzuki [28] at 1034 cm^{-1} in glycine and Tsuboi *et al.* [55] have given specific assignments for a range of amino acids, which they have substantiated by N^{15} measurements. They quote a range of 1040 to 1000 cm^{-1} but unfortunately the intensity is only moderate in the infra-red and there are frequent cases of band shifts and of doubling through Fermi-resonance interactions with overtones of lower lying fundamentals. This vibration is therefore of very little use for identification, especially as the band is not particularly strong in the Raman spectrum.

13.7. Amino-acid salts

Amino-acids can, of course, form basic and acidic salts, and the frequencies arising in hydrochlorides have already been discussed above. Salt formation through the carboxyl group removes the zwitterion character, and the amino-group then shows the normal NH_2 absorptions of a primary amine. However, in α-amino-acids a different situation arises, as the amino-group can co-ordinate to many metals by donation of its lone-pair electrons. Compounds of this type are of considerable biological importance, and they have been studied by many workers [51–54]. The metal glycinates are compounds of this type, but their spectra are complicated by the presence of additional bands due to water of crystallisation. However, they exhibit typical NH_2 stretching frequencies near 3250 cm^{-1} and 3130 cm^{-1}. These are higher than the corresponding hydrochloride frequencies, which is consistent with the concept of lone-pair donation by the nitrogen atom. The ionised carbonyl frequencies appear at the usual positions. This has been interpreted as indicating a fully ionic bond between the oxygen and metal atoms, although this implies a degree of equivalence between the two oxygen atoms of the carboxyl group which it is difficylt to reconcile with the square coplanar configuration assigned to many of these chelates, such as copper. Rosenberg [54] has carried out low-temperature studies on chelates of glycine and leucine when a sharpening of the absorption bands takes place. In the leucine series, which is not hydrated, the NH_2 stretching frequencies rise to near 3350 cm^{-1} and 3250 cm^{-1}; but the COO^- absorption remains near 1600 cm^{-1}. It is also possible to differentiate between such compounds as the *cis*- and *trans*-forms of platinum glycine complexes by taking advantage of the minor differences which occur in the NH_2 absorptions in the two cases.

13.8. Amido-acids

13.8(a). NH vibrations. Fuson *et al.* [18] have analysed the results of Randall *et al.* [17] in addition to their own and have shown that in solid compounds the NH stretching vibration falls in the range 3390–3260 cm^{-1}. This covers fifteen α-amido-acids studied by Randall *et al.* and a further ten samples examined by themselves. On formation of esters in which the hydrogen bonding possibilities are reduced, this frequency increases, and is then closer to the normal range. Micheel and Schleppinghoff [16] also find values of 3350 cm^{-1} in a series of *N*-acetyl amino-acids.

13.8(b). The region 3000–2000 cm^{-1}. Like the amino-acids, the amido-acids show in many cases broad and rather weak absorption bands in this region. An analysis by Fuson *et al.* [18] of the published data indicates that a band in the range 2640–2360 cm^{-1} occurs in twenty-three out of twenty-six cases, whilst another in the range 1945–1835 cm^{-1} occurs in sixteen out of twenty-six cases. These bands, which are sometimes multiple, are probably combination bands of some sort, and the fact that they do not invariably occur reduces their value. However, it is important to remember in the interpretation of the spectra of unknowns that amido-acids, in addition to amino-acids and hydrochlorides, may show absorptions in this region. Micheel *et al.* [16] also comment on the presence of a band between 2600 and 2500 cm^{-1} in all their materials. This is probably a characteristic carboxylic acid frequency and corresponds to the absorption near 2650 cm^{-1} which is found in dimeric carboxylic acids (Chapter 10).

13.8(c). The carboxyl group absorption. The carboxyl absorption of these materials in the solid state is normal [17, 18] and falls in the range 1724–1695 cm^{-1}, except for hippuric acid, which absorbs at 1740 cm^{-1}, and N-chloroacetyl dl-serine, which absorbs at 1730 cm^{-1}. Randall *et al.* [17] suggest that this band occurs at frequencies above 1695 cm^{-1} for α-acids, and below this for others, but the number of amido-acids other than α-acids they have studied is not large. Of these several absorb at 1695 cm^{-1}, whilst others, such as ε-benzamido caproic acid, absorb at 1698 cm^{-1}. It is doubtful, therefore, whether this fine distinction can be sustained. Solution results in shifts of 30–60 cm^{-1} towards higher frequency in dilute solutions [18].

13.8(d). The amide I and II absorptions. The location of the amide I band at or below 1620 cm^{-1} is regarded by Randall *et al.* [17] as being characteristic of the α-amido-acids, as others absorb on the high-frequency side. In general this is well supported by their own findings (1620–1600 cm^{-1} in twelve out of fifteen cases) and by those of Fuson *et al.* [18], and Micheel and Schleppinghoff [16]. However, this rule cannot be applied inflexibly. Fuson *et al.* find dl-acetyl cystine to absorb as low as 1587 cm^{-1}, whilst dl-phenacetyl-β-phenylalanine [17] absorbs as high as 1665 cm^{-1}. In solution this band again shows appreciable shifts to higher frequencies (1700–1680 cm^{-1}).

The amide II band is essentially normal and falls in the range 1570–1500 cm^{-1} for the twenty-five compounds studied by the two main groups of workers [17, 18]. The shifts on solution are appreciably smaller than in the first case and are in the opposite direction. The frequency range found is approximately 1530–1510 cm^{-1} and the values of Fuson *et al.* [18] are in agreement with this. Micheel *et al.* [16] do not comment on this absorption in their compounds.

13.8(e). Other correlations. The presence of a band at or near 1225 cm^{-1} in many amido-acids and also in amino-acid hydrochlorides has been suggested as a possible aid in their identification [18, 27]. This is a COOH band, probably originating in a C–O stretching or OH deformation mode and its significance in identification work has been discussed in the section dealing with carboxylic acids.

13.9. Bibliography

1. Edsall, *J. Chem. Phys.*, 1936, **4**, 1.
2. *Idem, ibid.*, 1937, **5**, 225, 508.
3. Edsall and Scheinberg, *J. Chem. Phys.*, 1940, **8**, 520.
4. Edsall, *J. Amer. Chem. Soc.*, 1943, **65**, 1767.
5. Edsall, Otvos and Rich, *ibid.*, 1950, **72**, 474.
6. Freymann, Freymann and Rumpf, *J. Phys. Radium*, 1936, **7**, 30.
7. Heintz, *Compt. rend. Acad. Sci. (Paris)*, 1935, **201**, 1478.
8. Lenormant, *ibid.*, 1946, **222**, 1432.
9. Duval and Lecomte, *Bull. Soc. Chim. France*, 1943, **10**, 187.
10. Freymann and Freymann, *Proc. Indian Acad. Sci.*, 1938, **8**, 301.
11. Wright, *J. Biol. Chem.*, 1937, **120**, 641.
12. *Idem, ibid.*, 1939, **127**, 137.
13. Lenormant, *J. Chim. Phys.*, 1946, **43**, 327.
14. Sutherland, *Discuss. Faraday Soc.*, 1950, **9**, 319.
15. Lecomte, unpublished work quoted in Reference 13 above.
16. Micheel and Schleppinghoff, *Chem. Ber.*, 1955, **88**, 763.
17. Randall, Fowler, Fuson and Dangl, *Infra-red Determination of Organic Structures* (Van Nostrand, 1949).
18. Fuson, Josien and Powell, *J. Amer. Chem. Soc.*, 1952, **74**, 1.
19. Thompson, Nicholson and Short, *Discuss. Faraday Soc.*, 1950, **9**, 222.
20. Klotz and Gruen, *J. Phys. Colloid. Chem.*, 1948, **52**, 961.
21. *The Chemistry of Penicillin* (Princeton University Press, 1949), p. 407.
22. Larsson, *Acta Chem. Scand.*, 1950, **4**, 27.
23. Kuratani, *J. Chem. Soc. Japan.*, 1949, **70**, 453.
24. Gore, Barnes and Petersen, *Analyt. Chem.*, 1949, **21**, 382.
25. Duval, Lecomte and Douvillé, *Annals de Physique*, 1942, **17**, 5.
26. Rasmussen and Brattain, *J. Amer. Chem. Soc.*, 1949, **71**, 1073.
27. Josien and Fuson, *Compt. rend. Acad. Sci. (Paris)*, 1951, **232**, 2016.
28. Suzuki, Shimanouchi and Tsuboi, *Spectrochim. Acta*, 1963, **19**, 1195.
29. Gäumann and Günthard, *Helv. Chim. Acta*, 1952, **35**, 53.
30. Lenormant, *J. Chim. Phys.*, 1952, **49**, 635.
31. Lord and Merrifield, *J. Chem. Phys.*, 1953, **21**, 166.

32. Koegel, Greenstein, Winitz, Birnbaum and McCallum, *J. Amer. Chem. Soc.*, 1955, **77**, 5708.
33. Ellenbogen, *ibid.*, 1956, **78**, 363, 366, 369.
34. Brockmann and Musso, *Chem. Ber.*, 1956, **89**, 241.
35. Erhart and Hey, *ibid.*, 2124.
36. Lacher, Croy, Kianpour and Park, *J. Phys. Chem.*, 1954, **58**, 206.
37. Despas, Khaladji and Vergoz, *Bull. Soc. Chim. Fr.,* 1953, 1105.
38. Bellanato and Barcelo, *Anales. Fiz. Quim.*, 1956, **52B**, 469.
39. Powell and Sheppard, *J. Chem. Soc.,* 1956, 3108.
40. Chatt, Duncanson and Venanzi, *ibid.*, 1956, 2712.
41. Hill and Rosenberg, *J. Chem. Phys.*, 1956, **24**, 1219.
42. Beattie and Tyrrell, *J. Chem. Soc.*, 1956, 2849.
43. Caglioti, Silvestrom, Sartori and Scrocco, *Ricerca. Sci.*, 1956, **26**, 1743.
44. Pentland, Lane and Quagliano, *J. Amer. Chem. Soc.*, 1956, **78**, 887.
45. Fujita, Nakamoto and Kobayashi, *ibid.*, 3295.
46. Heacock and Marion, *Can. J. Chem.*, 1956, **34**, 1782.
47. Merritt and Wiberley, *J. Phys. Chem.*, 1955, **59**, 55.
48. Barrow, Kreuger and Basolo, *J. Inorg. Nuclear Chem.*, 1956, **2**, 340.
49. Mizushima, Nakagawa and Quagliano, *J. Chem. Phys.*, 1956, **25**, 1367.
50. Ehrlich and Sutherland, *J. Amer. Chem. Soc.*, 1954, **76**, 5268.
51. Sweeney, Curran and Quagliano, *ibid.*, 1955, **77**, 5508.
52. Sen, Mizushima, Curran and Quagliano, *ibid.*, 211.
53. Svatos, Curran and Quagliano, *ibid.*, 6159.
54. Rosenberg, *Acta Chem. Scand.*, 1956, **10**, 840.
55. Tsuboi, Takenishi and Nakamura, *Spectrochim. Acta*, 1963, **19**, 271.
56. Pearson and Slifkin, *Spectrochim. Acta*, 1972, **28A**, 2403.
57. Slifkin, Smith and Walmsley, *Spectrochim. Acta*, 1969, **25A**, 1479.
58. Herlinger, Wenhold and Veachlong, *J. Amer. Chem. Soc.*, 1970, **92**, 6481.
59. Kirschenbaum and Park, *Spectrochim. Acta*, 1961, **17**, 785.
60. Edsall, *J. Amer. Chem. Soc.*, 1943, **65**, 1767.
61. Liefer and Lippincott, *J. Amer. Chem. Soc.*, 1957, **79**, 5098.

14

Amines and Imines

14.1. Introduction and table

Primary amines can be identified by the presence of two absorption bands in the NH stretching region arising from the symmetric and asymmetric vibrations of the hydrogen atoms. In some cases a third band is shown in this region due to hydrogen bonding effects. Hydrogen bonding results in a shift towards lower frequencies in all cases, but the bonds are considerably weaker than those of OH groups, so that the bands are sharper and are not shifted to anything like the same extent. Unfortunately there is a good deal of overlapping between the OH and NH vibrations in this region, so that differentiation is not always possible. Frequency shifts of the NH stretching vibration occur also in structures such as amines and hydrochlorides in which the amine group is charged (NH_3^+). These cases are considered separately.

Primary amines also show an NH_2 deformation absorption near 1650 cm^{-1}, and, like the stretching mode, this is subject to frequency shifts on hydrogen bonding. Deuteration is a useful aid in recognising these absorptions. In addition to these, a characteristic band can often be identified in aromatic amines, which is probably a C—N stretching absorption. There is also an absorption in the low-frequency region corresponding to the external deformation of the NH_2 group, but its position in relation to the structural environment has not be adequately defined.

Secondary amines show only a single NH stretching band in the $3500-3200 \text{ cm}^{-1}$ region, which is also subject to small changes on hydrogen bonding, and imines absorb in very much the same region.

The NH deformation in these compounds is too weak to be of any real value for identification purposes, although there appears to be an enhancement of the intensity in a limited number of cases. As with primary amines, C—N linkages of aromatic compounds can be identified, whereas the corresponding aliphatic materials show weaker absorptions which are more variable in position.

Tertiary amines are extremely difficult to identify spectroscopically. The C—N stretching band in aromatics can be identified in some cases, but there is no satisfactory correlation for aliphatic materials. The possibility of identification of the CH_3 —N group is also discussed.

The correlations discussed are listed in Table 14 below.

Table 14

NH *Stretching Absorptions*

Primary amines.	Dialkyl 3398–3381 cm^{-1}, 3344–3324 cm^{-1} (both w.)
	Diaryl 3509–3460 cm^{-1}, 3416–3382 cm^{-1} (both m.)
Secondary amines.	Dialkyl 3360–3310 cm^{-1} (w.)
	Alkyl/Aryl near 3450 cm^{-1} (m.)
Hydrazines $RNNH_2$.	3427–3354 cm^{-1} and 3327–3139 cm^{-1} (m.)
Heterocyclic NH.	3490–3430 cm^{-1} (m.)
Imines.	3350–3320 cm^{-1} (m.)

All the above intensify, broaden and move to lower frequencies on association.

NH *Deformation Frequencies*

| Primary amines. | 1650–1590 cm^{-1} (ms.) |
| Secondary amines. | 1650–1550 cm^{-1} (medium aryl, weak alkyl) |

C—N *Stretching Frequencies*

Primary amines.	Primary α-carbon atom 1079 $\mp$ 11 cm^{-1} (m.)
	Secondary α-carbon atom 1040 $\mp$ 3 cm^{-1} (m.)
	Tertiary α-carbon atom 1030 $\mp$ 8 cm^{-1} (m.)
Secondary amines.	Primary α-carbon atom 1139 $\mp$ 7 cm^{-1} (m.)
	Secondary α-carbon atom 1181 $\mp$ 10 cm^{-1} (m.)
Alkyl amines.	1150–1020 cm^{-1} (m.)
Aryl amines.	Primary 1340–1250 cm^{-1} (s.)
	Secondary 1350–1280 cm^{-1} (s.)
	Tertiary 1360–1310 cm^{-1} (s.)

14.2. NH Stretching vibrations

14.2(a). General. The occurrence of an absorption band near 3500 cm^{-1} in primary amines was first noted by Coblentz [1], and the fact that this was of general occurrence and was to be associated with the N–H stretching vibration was established by Bell [2–5], and by Ellis [6–7], who examined a considerable number of compounds and were able to show that the absorption was present in all primary and secondary amines, but absent from tertiary materials. The precise positions of these bands in primary and secondary amines in solution have since been established with considerable precision. In some cases such as aromatic primary amines it is possible to forecast the frequencies reasonably well.

14.2(b). Primary amines. Primary amines in dilute solution in non-polar solvents give two absorption bands in the region 3500–3300 cm^{-1}. The first of these, which is due to the asymmetric stretching mode, is usually found near 3500 cm^{-1}, and the second, which arises from the corresponding symmetrical mode, near 3400 cm^{-1}. Both of these are subject to small changes with alteration of the polarity of the solvent, and to rather larger changes in concentrated solutions in which intermolecular association can occur. Intramolecular bonding also lowers these frequencies, as will be seen from the discussion of hydrogen bonding effects below.

In alkyl amines in dilute solution these bands appear [8, 9, 14, 15, 19, 43, 46] in the narrow ranges 3398–3381 and 3344–3324 cm^{-1}. In anilines [10, 13, 20, 33, 47, 48] and naphthylamines [36] the range is somewhat wider, being 3509–3460 and 3416–3382 cm^{-1}. The frequency shifts in this series can be related to the electron-donating or -withdrawing properties of the ring substituents [20, 33, 47]. The higher frequencies correspond to electron-withdrawing substituents and the lower to electron donors.

The surprising rise in both frequencies which occurs on passing from an alkyl to an aryl amine has been discussed by Mason [33] who has also explained the increased separation of the bands. It results from the changes which occur in the nitrogen orbitals of the NH bonds as a result of interaction with the aromatic ring. This imparts more *s* character to these orbitals with the result that the bonds shorten and the frequencies rise. The bond angle is also increased by this interaction with the result that the coupling increases and the bands become more widely separated. This same phenomenon occurs to an even more marked extent in amides where the frequencies are still higher, and the band separation is increased to about 115 cm^{-1} as compared with about 65 cm^{-1} in alkyl amines. Mason [33] has shown that it is possible to calculate from the frequencies both the s character of the orbitals and the HNH bond angles.

Data is also available on aminoanthraquinones [16], naphthylamines [17], and aminopyridines [50–52]. Although made at lower resolution, these are in good agreement with the ranges quoted.

A substantial amount of high-resolution data is also available on metallic co-ordination compounds in which a primary amine acts as an electron donor. Compounds such as (L, am PtCl$_2$), where L is one of a series of ligands and am is a primary or secondary amine, have been studied in solution by Chatt, Duncanson and Venanzi [53, 54].

Bellamy and Williams [55] have reviewed the high-resolution data available. They point out that as both the asymmetric and symmetric stretching frequencies depend basically upon the same force constant, they must be directly related to each other. Over a range of sixty-four primary amines for which reliable data are available, they find that the equation $\nu_{sym} = 345.53 + 0.876\, \nu_{as}$ is obeyed in all cases with a standard deviation of 4.8 cm^{-1}. This is a valuable diagnostic relationship for free NH$_2$ frequencies, particularly as it is applicable also to co-ordination compounds in which these frequencies are lowered by as much as 200 cm^{-1}. However, as they point out, the relationship must be expected to fail in situations in which one NH link is bonded and the other is not, as in the dimeric complexes described by Chatt $et\ al.$ [54] or in compounds such an anthranilic acid. In some solid co-ordination compounds, however, the equality between the NH links is restored and the relation is again obeyed. This therefore offers a sensitive method for the detection of non-equivalent NH links in primary amines. This approach has been refined for the more specific case of the aniline series by Kreuger [93] who gives the equation $\nu_s = 1023 \mp 0.682\, \nu_{as}$ which applies with a precision of ∓ 1. Whetsel $et\ al.$ [94] have devised a similarly high precision equation for the corresponding overtone bands.

Intensity studies on the NH stretching bands of primary amines have been made by Califano and Moccia [86] and by Kreuger and Thompson [20].

Vampiri [56] has made the interesting observation that the total absorption intensity of primary amines in the 3600–3300 cm^{-1} region is approximately double that of the corresponding secondary amines. There is a good relationship between the Hammett σ values of the substituents and the observed intensities [20]. The symmetric mode is usually the more intense of the two and is the more sensitive to changes in the nature of the substituents.

Hydrazine and its derivatives have been studied by several authors [18, 57, 91, 92]. The results for substituted hydrazines R_2NNH_2 are interesting [92] in that there is a major change as compared with the alkyl amines. The observed frequency ranges are 3427–3354 and 3327–3139 cm^{-1}. It will be seen that the symmetric vibration is at a much lower frequency than in alkyl amines and that the band separation has increased to about 180 cm^{-1}. However, unlike the cases discussed above this does not originate in any widening of the amine bond angle. Partial deuteration experiments have shown very clearly that the two NH bonds in this series are not equivalent. This is attributed to an effect similar to that which is responsible for the low symmetric stretching frequencies of OCH_3 groups. One of the amino hydrogen atoms lies in a *trans* position to the lone pair of the second nitrogen atom and the NH bond is weakened by a non-bonding orbital interaction. The other hydrogen atom is unaffected. In $(C_2H_5)_2NNH_2$ the bands appear at 3362 and 3149 cm^{-1} but in $(C_2H_5)_2NNHD$, the bands corresponding to the two possible arrangements of the hydrogen atoms are at 3351 and 3145 cm^{-1}. It follows that the coupling is in fact extremely small and that in this series the bands correspond more nearly to individual separate NH bonds than to antisymmetric and symmetric modes.

14.2(c). Secondary amines and imines. Secondary amines shown only a single NH stretching absorption in dilute solution, although a second band at low frequencies is sometimes shown at higher concentrations when hydrogen bonding effects occur.

The most detailed studies of secondary amines are due to Russell and Thompson [59], who have examined both the intensity and frequency of this band in a wide range of compounds. Both were found to be very sensitive to the nature of the surrounding structure. In aliphatic secondary amines the frequency falls in the range 3360–3310 cm^{-1} with low intensity. In alkylaryl amines the frequency rises sharply to near 3450 cm^{-1} and the intensity increases by a factor of 50. In pyrrole, indole and similar compounds the frequency is as high as 3490 cm^{-1} and the intensity is again much enhanced. In this series, at least, the observed intensity would seem to have considerable diagnostic value. These results have been fully confirmed by other workers [19, 20, 33, 74, 95], and it seems that the behaviour of the NH stretching band of secondary amines is precisely parallel to that found in the primary compounds. In methyl-aniline for example the NH stretch is almost exactly at the mean position of the two bands of aniline itself, and correlations of Kreuger and Thompson [20] between the frequencies or intensities of primary amines and the Hammett σ values prove to be equally applicable to the corresponding secondary anilines.

A very substantial amount of data is available on NH stretching frequencies in heterocyclic ring systems. Indoles have been studied by Fuson et al. [13], by Witkop [35, 41, 42] and by Mirone and Vampiri [45]. Indole itself absorbs at 3491 cm^{-1} but values as low as 3371 cm^{-1} are quoted for some derivatives. However it is doubtful if this is wholly due to substituent effects as not all cases were studied in dilute solution and it is more probable that self-association is responsible for much of this shift.

Pyrroles have been studied by Fuson et al. [13] and by Jones and Moritz [49]. They find an overall range of 3495 to 3431 cm^{-1} and the latter authors have developed an equation which allows the calculation of this frequency in pyrroles variously substituted with methyl or carbethoxy groups in different positions. Marion et al. [34] quote a range of 3480–3440 cm^{-1} for related systems in alkaloids.

Other heterocyclic systems studied include the phenazines [61], triazoles and imidazoles [89] and aziridenes [88]. In the last named the NH bands are more typical of alkyl amines and absorption is between 3360 and 3340 cm^{-1} in the vapour. In the liquid state self-association lowers this to 3250–3240 cm^{-1} for alkyl aziridenes. When acyl groups are substituted in the ring two NH bands appear corresponding to two different invertomers. Marked variations in both frequency and intensity have been noted by Barr and Haszeldine [60] in NH groups with fluorine substituents at the α-carbon atom. The NH frequency of bis-2-2-2-trifluoro-dimethylamine, for example, is 3460 cm^{-1}, and the extinction coefficient is about 50 times greater than that of dimethylamine itself.

The imino group has a relatively narrow frequency range as is to be expected from a structure in which the NH bonds are reasonably well insulated from the rest of the molecule. Mathis et al. [90] quote values of 3346 cm^{-1} for ten benzimidic esters and find slightly lower values for phenylacetamidic esters, which show twin bands between 3325–3326 and 3322–3326 cm^{-1}. However in $(C_6H_5)_2C=NH$ the frequency falls to 3265 cm^{-1}, presumably because of the conjugation effect [97]. In addition, a large number of aminopyrimidines have been examined by Brownlie [23] and by Short and Thompson [24], but whilst it is possible that many of these exist in the imido-form, the complex tautomerism which these compounds can undergo makes any correlation difficult.

14.3. Hydrogen bonding in amines

Intermolecular and/or intramolecular hydrogen bonding effects are shown by most amines under suitable conditions. As with the hydroxyl group, no attempt will be made to review all the work — especially in the overtone region — which has been carried out in this connection, as a number of excellent reviews already exist [25, 26, 62], but a general indication of the conditions under which frequency shifts may be expected, and of the extent to which they are likely to occur, is given below.

14.3(a). Intermolecular hydrogen bonds. In the condensed phase hydrogen bonding occurs with amines causing a small fall in the frequency, which is usually considerably less than 100 cm^{-1}. Fuson *et al.* [13] have demonstrated that intermolecular bonding of the type originally suggested by Gordy [27] and by Thompson and Harris [28] does occur in pyrrole and related compounds. Typical shifts in the vNH frequency region observed by Fuson *et al.* [13] on passing from liquids or highly concentrated solutions to very dilute solutions were approximately 80–90 cm^{-1} in pyrrole and indole, 70 cm^{-1} for carbazole and 30 cm^{-1} for diphenylamine. The exact difference in any one case depends, of course, on the solvent and the concentration, as the shift in the bonded absorption is slightly less in the more dilute solution, whilst the free NH position varies somewhat with the polarity of the solvent [13]. The changes which take place in the intensity and position of the NH band in pyrrole in different solvents at various concentrations have been studied by several groups of workers [63–68]. Aniline is a much weaker proton donor, and although there is a frequency shift on passing from the vapour to the liquid, the hydrogen bonds formed are weak and only small further shifts occur in more basic solvents.

NH groups are also capable of interacting with alcohols [29] and with ketonic groups [30], although the shifts are again small. Sutherland [30] has indicated that, in general, association with ketonic groups gives absorption in the range 3320–3240 cm^{-1} and with other nitrogen atoms within the range 3300–3150 cm^{-1}. The possibility of the simultaneous occurrence of absorptions corresponding to both the free and bonded NH absorption should not be overlooked. Thus Flett has noted that solid β-naphthylamine and 2-aminoanthraquinone show three absorption peaks in the NH region [16], and we have found a similar effect with *ortho-* and *meta*-chloranilines, *p*-phenylenediamine, *p*-nitraniline and other amines. Pyrrole and other secondary amines at suitable concentrations show two absorptions [13]. There is a danger of confusion in the second case, with the two free NH bands of primary amines, but this can be avoided by studies at other concentrations. In any case, the broadening of the low-frequency band is usually a sufficient indication that a hydrogen bond is involved.

14.3(b). Intramolecular bonds. In alcohols one of the commonest cases of intramolecular bonding arises with conjugated systems containing the group $\underset{\substack{\| \\ O}}{C}-\underset{\substack{\| \\ OH}}{C}=C,$ such as salicylic acid and acetyl-acetone. This finding is repeated in amines in which a similar conjugated system favours the formation of internal bonds. In such cases, however, shifts observed in the NH frequencies are not usually large, indicating a considerably weaker bond, although the shifts in the carbonyl absorption are a good deal greater than would be expected from this. For example, in methyl *N*-methylanthranilate the carbonyl frequency of 1685 cm^{-1} indicates the presence of a strong hydrogen bond, whilst the NH frequency of 3361 cm^{-1} is not appreciably different from its usual position in intermolecular bonding [31]. However, the fact that such chelation is occurring is indicated by the carbonyl frequency of methyl N-*N*-dimethyl-anthranilate which reverts to the normal ester value of 1730 cm^{-1}. Hathway and Flett [17] have obtained somewhat similar findings for a series of nitronaphthylamines. In 2-nitro-1-naphthylamine in dilute solution the asymmetrical NH frequency is raised from 3486 cm^{-1} in α-naphthylamine to 3528 cm^{-1}, whilst the symmetrical NH frequency falls at the same time from 3412 cm^{-1} to 3378 cm^{-1}. A smaller fall is observed in the case of 1-nitro-2-naphthylamine. In both these cases alteration in the N=O frequency near 1340 cm^{-1}

afforded confirmation of chelation. However, bonding was found to be weak or absent in 3-nitro-2-naphthylamine and 8-nitro-1-naphthylamine in which the NH frequencies were virtually the same as in the naphthylamines themselves. The former case is attributed to the reduced double-bond character of the 2 : 3-double bond of naphthalene, whilst in the second case the absence of the conjugated double bond apparently prevents the formation of a chelated bond, despite the fact that the amino- and nitro-groups are actually closer together in the compound than in the 1-nitro-2-amino-material. Chelation also occurs in 1-aminoacridine, in 8-aminoquinoline [40] and in certain amino-azo compounds [75], but the frequency changes are small.

Bellamy and Pace [96] have used intramolecular bonds of NH_2 groups to explore the changes in coupling which result from inequalities in the NH bonds. By taking a series of intramolecularly-bonded amines in which only one hydrogen atom was associated, and studying these in different polar solvents which afforded a bonding site for the other, they obtained a wide range of systems with different relative NH bond strengths. These differences were measured by partial deuteration techniques, and a comparison with the original compounds then afforded a measure of the coupling. They were able to show that coupling would vanish in amines in which one bond was different from the other by more than 250 cm^{-1}. At this point the two bands correspond to free and bonded NH bonds.

Cromwell *et al.* [32] have found that β-amino-αβ-unsaturated ketones show chelation effects, but in this case a considerable shift of the NH frequency is reported, as no NH bands could be observed in the materials examined, and it was concluded that they were coincident with the CH stretching bands near 3000 cm^{-1}. The position is further complicated in these cases by their observation that a smaller but still appreciable carbonyl shift occurs in four compounds in which the amino-group is fully substituted and unable to form hydrogen bonds. The suggestion has been made that in these

$$O^- \qquad \overset{\diagdown \diagup}{N^+}$$
$$\mid \qquad \qquad \parallel$$

cases the structure $R-C=C-C-R$ may contribute appreciably to the ground state, and it is possible that a contribution from similar resonant forms rather than hydrogen bonding accounts also for the small shifts observed in the NH frequencies in the cases quoted earlier, in which the NH shifts are so much smaller than would be expected in relation to the shifts in the carbonyl absorption. Flett [16] has suggested an explanation along these lines to account for the small NH shifts observed in 1-amino- and 1 : 4-diamino-anthraquinones, and in this case the suggestion is supported by the fact that 2-amino-anthraquinone exhibits the same phenomenon of carbonyl shift. However, there is a possibility of ambiguity in the assignment of the C=O frequencies in these cases, as the NH_2 deformation is to be expected in the same region.

14.4. NH Deformation vibrations

14.4(a). Primary amines. The situation in respect of our knowledge of NH deformations has recently been discussed by Sutherland [30]. There is general agreement that the NH_2 internal deformation mode occurs in the range 1650–1590 cm^{-1} in simple amines, but this is based on an extremely limited number of compounds, such as methylamine [8, 9, 46], aniline [10–12], hydrazine [18] and substituted hydrazines [36], and a small number of long-chain amines [69]. This band is of course related to the amide II band of primary amides and it occurs in the same frequency region. It is strong in the infra-red but is very weak in the Raman spectra. The appearance of a band at 1650 cm^{-1} in most amino-substituted pyrimidines is regarded as evidence for their existence in the amino-rather than the imino-form.

In the absence of more precise data it is not possible to relate movements of this band within the given frequency range with any structural features. As a deformation mode this absorption moves to higher frequencies on bonding [37], but the shifts are often not sufficient to take the absorption out of the overall range quoted. Primary amides have been studied particularly in this connection, and in a typical case the NH deformation of hexoamide is given as 1635 cm^{-1} in the solid state, falling to 1595 cm^{-1} in very dilute solution [38]. In the case of primary amines where the hydrogen bonds will be weaker than in amides, the frequency shifts are correspondingly smaller but they are of course in the same direction.

Primary amines should also show absorptions at longer wavelengths due to the NH_2 wagging mode. Stewart [19] discusses the position of this band in about twenty primary amines. The overall range is 840–760 cm^{-1} but this is capable of subdivision into narrower ranges depending on the degree of substitution at the α-carbon atom. However in liquid systems this band is readily identified by its considerable breadth.

14.4(b). Secondary amines. The NH deformation absorpti·
is a strong to medium intensity band in primary amines,
extremely weak in secondary aliphatic amines, so that it f
cannot be detected at all with the film thicknesses
employed. With secondary aromatic amines the position is confused
to some extent by the presence of the ring C=C stretching
absorptions in the same region, coupled with the fact that in certain
cases these become intensified when a nitrogen atom is directly
attached to the ring [31].

However, whilst some aromatics examined as thin mulls seem to
show no additional band in this region, others, such as phenyl-β-
naphthylamine and methyl-N-methylanthranilate [31], do show
some differences from the disubstituted amines. Hadži and
Skrbljak [70] have recently reinvestigated this question using
deuteration techniques. Their results show clearly that there is
indeed an additional weak absorption in the 1510 cm^{-1} region in
aromatic secondary amines. This is an NH deformation mode which
is coupled with the lower-frequency C–N stretching mode. However,
in some instances, such as methyl aniline, it is superimposed on the
ring vibrations and is difficult to identify. In thioamides the band is
clearly marked and has much greater intensity [71–73].

In aliphatic secondary amines this band is too weak to be detected
readily. In cyclic bases, for example, only the normal aromatic
absorptions are found in this region unless salt formation
occurs [58]. Similarly, Barr and Haszeldine [74] could not identify
this band in dimethyl or diethylamine, although a band near
1500 cm^{-1} was traced in some fluorinated derivatives. Imines with
the structure –C=NH are similar to other secondary amides in that
any NH deformation absorption shown is extremely weak, but of
course they also show an absorption in the same region corresponding to the terminal C=N linkage.

14.5. C–N Stretching vibrations

14.5(a). Aromatic amines. Colthup [21] has given the following
correlation for aromatic amines. A strong band appears as follows:

Primary, $1340–1250 \text{ cm}^{-1}$; Secondary, $1350–1280 \text{ cm}^{-1}$;
Tertiary, $1360–1310 \text{ cm}^{-1}$.

These are based on his own work, and no publications have dealt in
detail with these particular correlations. However, its persistence in

the same approximate frequency range in all three types indicates that these must be C—N stretching vibrations the frequency of which is largely determined by the aromatic character of the carbon atom. This interpretation is also that of Barnes *et al.* [44], who put forward this correlation in more general terms.

The intensity of these bands appears to be rather variable, some being relatively strong and others of not more than medium intensity. This may also be associated with some differences in substitution or other structural features. Lieber *et al.* [22] have also found these bands in a series of thirteen tetrazoles of various types. More recently Hadži [70] has studied the effects of deuteration upon this absorption in aniline derivatives. The band near 1260 cm^{-1} moves to near 1350 cm^{-1}. This identifies it with a C—N stretching mode, which in the hydrogen compounds is lowered in frequency by coupling with the NH deformation frequency.

14.5(b). Aliphatic amines. Colthup [21] has also indicated correlations for aliphatic amines with absorptions in the 1220—1020 cm^{-1} range. This correlation has been confirmed and extended by Stewart [19] by measurements on numbers of primary and secondary amines. He proposed much more limited sub-divisions depending on the degree of substitution at the α-carbon atom. Secondary amines with a primary α-carbon atom absorb at 1139 $\mp$ 7 cm^{-1}, and this rises to 1181 $\mp$ 10 when the α-carbon atom is a secondary one.

Primary amines show three ranges, 1079 $\mp$ 11 cm^{-1} for primary α-carbons, 1040 $\mp$ 3 cm^{-1} for secondary α-carbons and 1030 $\mp$ 8 cm^{-1} for the tertiary cases. These bands are reported to be strong in the infra-red, but some caution is needed in their use for diagnosis. However the bands always appear clearly and are well recognisable in the Raman spectra which provides a valuable confirmation.

Tertiary amines are difficult to identify in the infra-red. The most characteristic bands are the antisymmetric and symmetric C—N stretching modes and these can best be identified in the Raman spectra where the former occurs in the range 1070—1050 cm^{-1} and the latter near 830 cm^{-1}.

In unsaturated tertiary amines it is possible to differentiate between $\alpha\beta$- and $\beta\gamma$-unsaturation by observing the frequency changes which occur in the 1600—1700 cm^{-1} region on salt formation [87].

14.6. The CH$_3$—N group

The deformation mode of the CH$_3$ group attached to a nitrogen atom could be expected to give an absorption band of reasonably characteristic frequency similar to that of the CH$_3$—C absorption which occurs at 1375 cm^{-1}. In this case, however, alterations in the substituents on the nitrogen atom might have a greater influence on the frequency. This vibration in simple molecules appears to occur in the 1460—1430 cm^{-1} region. Examination of the spectra of a range of more complex materials containing this group indicates that whilst many of them show absorption near 1430 cm^{-1}, in others the band, if it occurs at all, is indistinguishable from the main CH deformation vibration. The band is not strong, and the usefulness of the correlation is further diminished by the fact that a considerable proportion of other nitrogen-containing compounds appear also to show weak bands in the region 1430—1400 cm^{-1}. Dimethylamino-compounds such as dimethylamine and bisdimethylamino-fluorophosphine oxide do not show any behaviour comparable with that of *iso*propyl derivatives in showing any splitting of this band into two components of equal intensity. It must therefore be concluded that, despite the constant frequency shown by the CH$_3$ deformation mode when attached to carbon, silicon and a number of other elements, the corresponding NCH$_3$ vibration is too weak in intensity or too variable in position to afford any useful correlation. However, the NCH$_3$ group can be identified by the characteristically low CH stretching frequency which results from the interaction of the nitrogen lone pair with a *trans* hydrogen atom. This correlation has already been discussed in Chapter 2.

14.7. The groups NH$_3$$^+$, NH$_2$$^+$ and NH$^+$

Amine derivatives with positively charged nitrogen atoms commonly occur in salts, in co-ordination compounds involving ammonia and in zwitterion structures. The latter have been discussed in the previous chapter, and will not be reconsidered here.

14.7(a). NH$_3$$^+$ absorptions. These occur in complex amines and in primary amine hydrochlorides. In the former series they have been studied by many workers [53, 54, 76—83]. The charge on the nitrogen atom is appreciably smaller than in hydrochlorides, so that the frequencies are nearer to those of normal amines. A few such compounds studied in solution show stretching frequencies near 3380 cm^{-1} and 3280 cm^{-1}. However, most studies have been made on solids in which intermolecular bonding effects lower the range to $3350-3150 \text{ cm}^{-1}$ and sometimes lead to a multiplicity of bands. The NH_3 deformation occurs near 1600 cm^{-1}, and the NH_3 rocking frequency near 800 cm^{-1}. These have been confirmed by deuteration studies.

Hydrochlorides of simple primary amines have been little studied, but appear similar to those of amino-acids. Methylamine hydrochloride, for example, absorbs at 3075 cm^{-1} and at 2972 cm^{-1}, and has an NH deformation band [84] at 1617 cm^{-1}. In hydroxylamine hydrochloride the NH bonds are more strongly associated and are assigned [84] at 2955 cm^{-1} and 2667 cm^{-1}.

14.7(b). NH_2^+ absorptions. A large number of co-ordination compounds containing aniline derivatives, methylamine and other amines have been studied by Chatt *et al.* [53] in the high-frequency region. In dilute solution these behave like normal primary amines in that the relationship of Bellamy and Williams [55] is obeyed, but the influence of the electron donation is shown by the substantial reductions which take place in the NH stretching frequencies. These occur about 200 cm^{-1} lower than the same bands in aniline. Further reductions accompany hydrogen bonding in concentrated solutions and in solids. The NH_2 deformations of compounds of this type have been studied by Mizushima *et al.* [81], and are more nearly normal.

The hydrochlorides of secondary bases have been examined by Heacock and Marion [58], who found a complex series of absorptions between 2800 cm^{-1} and 2000 cm^{-1}. This was especially marked in aromatics, in which three or four bands recurred between $2780-2600 \text{ cm}^{-1}$, in addition to others at lower frequencies. On deuteration these bands become attenuated, indicating that an NH motion is involved. Out of seventeen salts studied, fifteen had a band in the range $2760-2690 \text{ cm}^{-1}$, all absorbed between 1620 cm^{-1} and 1560 cm^{-1} (NH_2 deformation), and all absorbed near 800 cm^{-1} (NH_2 rock). Barcello *et al.* [84] report similar findings for dimethylamine hydrochloride.

14.7(c). NH^+ absorptions. In trimethylamine hydrochloride the NH^+ stretching band [84] is at 2735 cm^{-1}. In salts of pyridine, indolene and Shiff's bases the group $C=NH^+$ absorbs at $2500-2325 \text{ cm}^{-1}$.

However, when no other hetero-atom is in the vicinity of the NH^+ group, a second band appears between 2200 cm^{-1} and 1800 cm^{-1}. Witkop [85] has discussed the diagnostic applications of this effect.

14.8. Bibliography

1. Coblentz, *Investigations of Infra-red Spectra* (Carnegie Institute, 1905).
2. Bell, *J. Amer. Chem. Soc.*, 1925, **47**, 2192.
3. *Idem. ibid.*, p. 3039.
4. *Idem. ibid.*, 1926, **48**, 813.
5. *Idem. ibid.*, p. 818.
6. Ellis, *J. Amer. Chem. Soc.*, 1927, **49**, 347.
7. *Idem. ibid.*, 1928, **50**, 685.
8. Cleaves and Plyler, *J. Chem. Phys.*, 1939, **7**, 563.
9. Bailey, Carson and Daly, *Proc. Roy. Soc.*, 1939, **A173**, 339.
10. Buswell, Downing and Rodebush, *J. Amer. Chem. Soc.*, 1939, **61**, 3252.
11. Gordy, *ibid.*, 1937, **59**, 464.
12. Williams, Hofstadter and Herman, *J. Chem. Phys.*, 1939, **7**, 802.
13. Fuson, Josien, Powell and Utterback, *ibid.*, 1952, **20**, 145.
14. Segal, *App. Spectroscopy*, 1961, **15**, 112, 148.
15. Baldwin, *Spectrochim. Acta*, 1962, **18**, 1455.
16. Flett, *J. Chem. Soc.*, 1948, 1441.
17. Hathway and Flett, *Trans. Faraday Soc.*, 1949, **45**, 818.
18. Giguère and Liu, *J. Chem. Phys.*, 1952, **20**, 136.
19. Stewart, *J. Chem. Phys.*, 1959, **30**, 1259.
20. Kreuger and Thompson, *Proc. Roy. Soc.*, 1959, **A250**, 22; 1957, **A243**, 143.
21. Colthup, *J. Opt. Soc. Amer.*, 1950, **40**, 397.
22. Lieber, Levering and Patterson, *Analyt. Chem.*, 1951, **23**, 1594.
23. Brownlie, *J. Chem. Soc.*, 1950, 3062.
24. Short and Thompson, *ibid.*, 1952, 168.
25. Pauling, *The Nature of the Chemical Bond* (Oxford University Press, 1950).
26. Hunter, Price and Martin, *Report on the Symposium on the Hydrogen Bond* (Institute of Chemistry, 1950).
27. Gordy, *J. Chem. Phys.*, 1939, **7**, 167.
28. Thompson and Harris, *J. Chem. Soc.*, 1944, 301.
29. Baker, Davies and Gaunt, *ibid.*, 1949, 24.
30. Sutherland, *Discuss. Faraday Soc.*, 1950, **9**, 274.
31. Rasmussen and Brattain, *J. Amer. Chem. Soc.*, 1949, **71**, 1073.
32. Cromwell, Miller, Johnson, Frank and Wallace, *ibid.*, p. 3337.
33. Mason, *J. Chem. Soc.*, 1958, 3619; 1961, 22.
34. Marion, Ramsay and Jones, *J. Amer. Chem. Soc.*, 1951, **73**, 305.
35. Witkop, *ibid.*, 1950, **72**, 614.
36. Bryson, *J. Amer. Chem. Soc.*, 1960, **82**, 4862.
37. Thompson, Nicholson and Short, *Discuss. Faraday Soc.*, 1950, **9**, 222.
38. Richards and Thompson, *J. Chem. Soc.*, 1947, 1248.
39. Witkop and Patrick, *J. Amer. Chem. Soc.*, 1951, **73**, 713.
40. Short, *J. Chem. Soc.*, 1952, 4584.
41. Witkop and Patrick, *J. Amer. Chem. Soc.*, 1951, **73**, 1558.
42. *Idem. ibid.*, p. 2188.
43. Orville Thomas, Parsons and Ogden, *J. Chem. Soc.*, 1968, 1048.

44. Barnes, Gore, Stafford and V. Zandt Williams, *Analyt. Chem.*, 1948, **20**, 402.
45. Mirone and Vampiri, *Atti accad. Nazl. Lincei Rend. Classe Sci. fis. mat. e. Nat.*, 1952, **12**, 405.
46. Barcello and Bellanato, *Spectrochim. Acta*, 1956, **8**, 27.
47. Califano and Moccia, *Gazz. Chim.*, 1956, **86**, 1014.
48. Richtering, *Z. Phys. Chem.*, 1956, **9**, 393.
49. Jones and Moritz, *Spectrochim. Acta*, 1965, **21**, 295.
50. Angyal and Werner, *J. Chem. Soc.*, 1952, 2911.
51. Shigorin, Danyushevskii and Gold'farb, *Izvest, Akad. Nauk. S.S.S.R. Otdel khim Nauk.*, 1956, 120.
52. Costa, Blasina and Sartori, *Z. Phys. Chem.*, 1956, **7**, 123.
53. Chatt, Duncanson and Venanzi, *J. Chem. Soc.*, 1955, 4461.
54. *Idem, ibid.*, 1956, 2712.
55. Bellamy and Williams, *Spectrochim. Acta*, 1957, **9**, 341.
56. Vampiri, *Gazz. Chim.*, 1954, **84**, 1087.
57. Shull, Wood, Aston and Rank, *J. Chem. Phys.*, 1954, **22**, 1191.
58. Heacock and Marion, *Can. J. Chem.*, 1956, **34**, 1782.
59. Russell and Thompson, *J. Chem. Soc.*, 1955, 483.
60. Barr and Haszeldine, *ibid.*, 1955, 4169.
61. Stammer and Taurin, *Spectrochim. Acta*, 1967, **19**, 1625.
62. Hunter, *Progress in Stereochemistry*, Vol. 1 (Butterworth, London, 1954), p. 224.
63. Josien and Fuson, *J. Chem. Phys.*, 1954, **22**, 1169.
64. Fuson and Josien, *J. Phys. Radium*, 1954, **15**, 652.
65. Josien and Fuson, *J. Chem. Phys.*, 1954, **22**, 1264.
66. Mirone and Fabbri, *Gazz. Chim.*, 1956, **86**, 1079.
67. Tuomikoski, *Mikrochem. Acta*, 1955, 505.
68. *Idem, J. Phys. Radium*, 1955, **16**, 347.
69. Despas, Khaladji and Vergoz, *Bull. Soc. Chim. Fr.*, 1953, 1105.
70. Hadži and Skrbljak, *J. Chem. Soc.*, 1957, 843.
71. Hadži, *ibid.*, 847.
72. Mecke and Mecke, *Chem. Ber.*, 1956, **89**, 343.
73. Mecke, Mecke and Luttinghaus, *Zeit. Naturforsch.*, 1955, **10B**, 367.
74. Salimov and Tatevskii, *Doklady Akad. Nauk. S.S.S.R.*, 1957, **112**, 890.
75. Bagratishvili, *ibid.*, 1954, **96**, 753.
76. Powell and Sheppard, *J. Chem. Soc.*, 1956, 3108.
77. Duval, Duval and Lecomte, *Compt. Rend. Acad. Sci. (Paris)*, 1947, **224**, 1632.
78. Beattie and Tyrell, *J. Chem. Soc.*, 1956, 2849.
79. Hill and Rosenberg, *J. Chem. Phys.*, 1956, **24**, 1219.
80. Mizushima, Nakagawa and Quagliano, *ibid.*, 1956, **25**, 1367.
81. Mizushima, Nakagawa and Sweeney, *ibid.*, 1956, **25**, 1006.
82. Svatos, Curran and Quagliano, *J. Amer. Chem. Soc.*, 1955, **77**, 6159.
83. Barrow, Kreuger and Basolo, *J. Inorg. Nuclear Chem.*, 1956, **2**, 340.
84. Bellanato and Barcello, *Anales. real Soc. Espan. fis. y. quim. Madrid*, 1956, **52B**, 469.
85. Witkop, *Experimentia*, 1954, **10**, 420.
86. Califano and Moccia, *Gazz. Chim.*, 1957, **87**, 58.
87. Leonard and Gash, *J. Amer. Chem. Soc.*, 1954, **76**, 2781.
88. Mathis, Mathis, Imberlin and Lattes, *Spectrochim. Acta*, 1974, **30A**, 741.
89. Mathis, Baccar, Kateka Bon, N'Gondo M'Pondo, *J. Mol. Structure*, 1971, **7**, 381.

90. Mathis, Baccar, Barrens and Mathis, *J. Mol. Structure*, 1971, **7**, 355.
91. Blair and Gardner, *J. Chem. Soc.* (**C**), 1970, 2707.
92. Hadži, Jan and Ocvirk, *Spectrochim. Acta*, 1969, **25A**, 97.
93. Kreuger, *Nature*, 1962, **194**, 1077.
94. Whetsel, Robertson and Krell, *Analyt. Chem.*, 1958, **30**, 1598.
95. Moritz, *Spectrochim. Acta*, 1960, **16**, 1176.
96. Bellamy and Pace, *Spectrochim. Acta*, 1972, **28A**, 1869.
97. Perrier, Datin and Lebas, *Spectrochim. Acta*, 1969, **25A**, 169.

15

Unsaturated Nitrogen Compounds

15.1. Introduction and table

The data available on unsaturated nitrogen-containing compounds are extremely variable. The position of the C≡N stretching vibration in nitriles has been extensively studied and is clearly defined; the influence of conjugation or of halogen substitution is also known. Similarly, azides can be recognised without undue difficulty. The double-bond vibrations, C=N and N=N, are reasonably well defined. The approximate positions of the absorptions due to each are known, but little is known of factors influencing the position or intensity. In conjugated cyclic compounds the C=N vibration appears to be subject to frequency shifts which are much greater than any shown by the corresponding C=C vibrations, so that considerable caution is required in the interpretation of the spectra of compounds containing such structures. The correlations discussed are listed in Table 15.

15.2. The —C≡N stretching vibration

15.2(a). General. Infra-red absorption from nitriles is to be expected in the triple-bond region 2300–2000 cm^{-1}. Early studies on isolated molecules confirmed that this was so [1–8] and the correlation was put on a firm foundation by Kitson and Griffiths [9] who studied the spectra of some seventy nitriles. These workers found that in thirty-four saturated mono- and di-nitriles (excluding malononitrile) the C≡N frequency lies between 2260 cm^{-1} and 2240 cm^{-1}, and in a further seventeen unsaturated but not conjugated nitriles the

Table 15

C≡N *Stretching Vibrations*	
Saturated alkyl nitriles	$2260-2240$ cm^{-1} (s.)
Aryl nitriles	$2240-2220$ cm^{-1} (s.)
$\alpha\beta$-unsaturated alkyl nitriles	$2235-2215$ cm^{-1} (s.)
Isocyanates	$2275-2240$ cm^{-1} (s.)
C=N *Stretching Vibrations*	
Open-chain compounds	$1690-1590$ cm^{-1} (v.)
Open-chain $\alpha\beta$-unsaturated compounds	$1660-1590$ cm^{-1} (v.)
Conjugated cyclic compounds	$1660-1480$ cm^{-1} (v.)
N=N *Stretching Vibrations*	$1630-1575$ cm^{-1} (v.)
$-$N=C=N$-$	$2155-2130$ cm^{-1} (s.)
N=N=N *Stretching Vibrations* (Azides)	$\left\{\begin{array}{l} 1340-1180 \text{ cm}^{-1} \text{ (w.)} \\ 2160-2120 \text{ cm}^{-1} \text{ (s.)} \end{array}\right.$

frequencies were within the same range. Twelve further nitriles in which the C≡N was conjugated with a double bond (excluding fumaronitrile) have a strong band between 2232 cm^{-1} and 2218 cm^{-1}, and four compounds with nitrile groups of both types exhibit both absorptions. Similarly with aromatic compounds, those in which the C≡N group was directly attached to the ring (twelve compounds) absorbed in the range 2240–2221 cm^{-1}, whilst two others, in which the C≡N group was separated from the ring by methylene groups, showed the normal frequencies of 2254 and 2248 cm^{-1}. As is to be expected, the nature of other ring substituents affects the frequency to some extent, but the range given includes the absorption shown by compounds with either electron repelling or electron attracting substituents. These results have been supplemented by further work by Felton and Orr [35] and by Skinner and Thompson [36], who have also studied numerous nitriles with similar results.

These findings have been fully substantiated by later workers [10, 14, 16, 20, 22, 37, 38, 82], and some refinements of the correlations have been developed. Thus, in aromatic nitriles [10, 16] the overall frequency shifts with changes in the substituents are very small, ranging from 2222 cm^{-1} for α-naphthyl nitrile to 2240 cm^{-1} for *meta*-nitro benzonitrile, but the shifts do reflect the Hammett σ values of the substituents so that the frequency can be predicted with reasonable confidence if these are known. This is also true of the intensities where the wider spread of values makes this more useful.

With alkyl nitriles the overall range is so small as to make any differentiation difficult. However there is a regular frequency fall as one passes along the series from acetonitrile (2247 cm^{-1}) to *tert*-butyl nitrile (2236 cm^{-1}), indicating a slight frequency fall for each successive substitution at the α-carbon atom. Halogens substituted at this point have a different effect. The first chlorine raises the frequency by 11 cm^{-1} but further substitution reduces the frequency to nearer the original value.

From this it is clear that the C≡N group can be identified with reasonable certainty in the infra-red provided the intensity is sufficient. It is usually possible to differentiate between saturated and conjugated nitriles, but there is insufficient difference between alkenyl and aryl nitriles to enable any distinction to be made from the C≡N frequency. As regards malononitrile and fumaronitrile mentioned above as being the two exceptions to these correlations, they are both cases in which some interaction between the C≡N groups is to be expected, and a similar discrepancy exists in the case of the C=O frequencies of the corresponding carboxylic acids. They are not therefore significant in so far as the general applicability of these correlations is concerned. Another apparent exception is diazoguanidine cyanide, which is reported by Lieber *et al.* [13] to absorb at 2126 cm^{-1}. This large shift is much greater than would be expected from the −N=N− conjugation.

The above correlations, of course, are not applicable to compounds of the type −N≡C which are *iso*-nitriles. The number of such compounds examined is very few, but the triple bond of this type [11] probably absorbs in the range $2180-2120 \text{ cm}^{-1}$. McBride and Beachell [29] have compared the spectra of a limited number of nitriles and *iso*-nitriles. The former absorb in the ranges given, depending upon the nature of the substituents, but the alkyl *iso* nitriles absorbed between $2183-2144 \text{ cm}^{-1}$ and the aryl derivatives near 2145 cm^{-1}. More recent work by Ugi and Meyer [47] on the infra-red spectra of 18 *iso*-nitriles suggests that the overall range may be somewhat smaller, as might be expected from the parallel with the nitriles. They suggest $2146-2134 \text{ cm}^{-1}$ for alkyl *iso*-nitriles and $2125-2109 \text{ cm}^{-1}$ for aromatic derivatives. The difference between these narrow ranges and those of McBride *et al.* [29] may well be due to the fact that the latter have taken account of vapour phase data as well as results from solution. In the vapour phase methyl *iso*cyanide [39] absorbs at 2166 cm^{-1}. They also observed a new band near 1592 cm^{-1} in the spectra of the *iso*-nitriles which is absent from those of the nitriles. There is, however, no band in this region in methyl *iso*-cyanide which has been studied by Williams [39].

15.2(b). Intensities of C≡N bands. The intensity of the nitrile absorption band has been studied by Kitson and Griffith [9], and others [12, 35, 36, 40], and in great detail by later workers [10, 20, 22, 37]. The data have been reviewed by Brown [20]. There is a marked variation in intensity in various types of nitrile, so that the band varies from very strong to undetectable. The band is usually strong in alkyl nitriles which contain only C, H and N atoms, and there is a steady increase in intensity as the carbon chain lengthens from C_1 to C_5, thereafter it remains constant. However there are very wide variations with the solvents used, and also large substituent effects. The presence of an α-halogen atom can reduce the intensity by an order of magnitude, whilst a directly attached nitrogen atom can raise it by as much as two orders. The introduction of an oxygenated group into the molecule results in a 'quenching' of the C≡N absorption intensity, and its effect is greater when the oxygen-containing group is attached to the same carbon atom as the nitrile. In acetone cyanhydrin, for example, the C≡N intensity is about a third of the normal and in *dl*- and *meso*-diacetyl cyanhydrin the nitrile band was weak but detectable. In the corresponding acetoxy-compounds it could not be detected at all.

With αβ-unsaturated nitriles, which have been studied by Heilmann *et al.* [38], the intensities are usually greater than in the saturated compounds, but show very considerable variations with substitution at the double bond. The replacement of the hydrogen atoms of the vinyl group of vinyl nitrile by methyl groups enhances the C≡N intensity by 2–3 times. Such large changes from what must be very small differences in the polarity of the nitrile group are very remarkable.

Aromatic nitriles show similar large intensity variations [10, 16, 20]. In passing from a strongly electron attracting substituent in *para*-nitrobenzonitrile to a strongly donating substituent in the *para*-dimethylamino derivative, there is a frequency change of 17 cm^{-1} but an intensity change of 25 times. Again there is a good linear relationship between the intensity changes and the Hammett σ values of the substituents.

The ability of the C≡N band to appear to vanish from the spectrum due to a large reduction in intensity from causes which are largely unpredictable, is of course a limitation to its use in diagnosis. However, it is fortunate that in the Raman spectra there is no such erratic behaviour and the band is always strong [14]. In the alkyl series, some intensity variations do occur but these are small so that the band can always be recognised without difficulty. In aromatic nitriles the bands are even more clearly marked, as their Raman intensities are usually 2 or 3 times those of their alkyl counterparts.

15.3. *iso*Cyanates and carbodiimides

15.3(a). *iso*Cyanates. This class of compound has been studied by Davison [42], by Hoyer [41], and by Ham and Willis [55]. Davison studied ten *iso*cyanates and found a very intense absorption at $2269 \text{ cm}^{-1} \pm 6 \text{ cm}^{-1}$. This is unaltered by changes of state and by conjugation. The intensities of this band in two aromatic *iso*cyanates were assessed as being over 100 times that of the corresponding band in alkyl cyanides which absorb in the same region, so that differentiation should be possible in this way. These findings have been fully confirmed by Hoyer [41]. More than forty *iso*cyanates have been studied by this worker, who finds both alkyl and aryl compounds to absorb in the range $2274–2242 \text{ cm}^{-1}$, with the majority near 2270 cm^{-1}. No clear distinction was traceable between conjugated and non-conjugated compounds, but the high intensity of this band was confirmed. In 1-cyano-3-*iso*cyanato-propane, for example, the N=C=O absorption at 2272 cm^{-1} could be clearly differentiated from the much weaker nitrile absorption on the low-frequency side. Phenyl *iso*cyanate is anomalous in showing a doublet at 2278 cm^{-1} and 2260 cm^{-1}. However Ham and Willis [55] point out that this is exceptional, and that this band is usually a singlet and does not show the multiple peaks of some other cumulative bond systems such as the *iso*thiocyanates. The 2270^{-1} absorption of *iso*cyanates corresponds to the asymmetric stretching mode of the N=C=O group, and a weaker symmetric frequency is to be expected. This has been identified at 1377 cm^{-1} in methyl *iso*cyanate by Eyster and Gillette [25]. If a positive identification could be made it would make the differentiation of *iso*cyanates from nitriles much simpler. Unfortunately this second band is too weak to be recognised [42] with any certainty in the infra-red, but it is of course very intense in the Raman spectrum. In all cases other than the methyl derivative the Raman spectrum shows a strong band in the 1450–1400 cm^{-1} region. Differentiation between nitriles and isocyanates are therefore possible through the use of the Raman spectra, and perhaps also by making use of the very high intensity of the isocyanate band in the infra-red.

15.3(b). Carbodiimides. Compounds of this class contain the structure R–N=C=N–R and have been little studied. Khorana [43, 45] has studied a very limited number of examples and noted absorption at 2150 cm^{-1}. A more detailed study of both frequency and intensity has been made by Meakins and Moss [44]. Nine carbodiimides were examined by these authors, and all showed a very intense absorption in the range 2152–2128 cm^{-1} accompanied in three instances by a much weaker combination band at lower frequencies. Little frequency or intensity differences was found between alkyl and aryl derivatives. The intensity of the band was assessed by a variety of methods, and the extinction coefficient ϵ^a was of the order of 1400. This is some 2½ times as strong as a normal ketonic carbonyl absorption. Mogul [60] has reviewed the infra-red and Raman spectra of carbodiimides. The findings confirm those of Meakins, and others in the infra-red, but add the additional observation that the band doubles when an aromatic ring is directly attached to the diimide group. In the Raman spectra the symmetric vibration is of course the more intense and this occurs near 1460 cm^{-1}.

15.4. The —C=N— stretching vibration

15.4(a). General. A good deal of work has been done on the position of the —C=N— stretching absorption in various classes of compounds, and the situation has been reviewed in detail by Fabian, Legrand and Poirier [46]. The band remains a difficult one to identify, however, owing to the considerable changes in intensity which follow changes in its environment. Also, the information available on the effects of conjugation in ring systems is often conflicting and indecisive. This is partly due to the fact that in many of the compounds studied the conjugation is with C=C links and the frequencies of the two are so close that, in ring systems at least, it is doubtful whether either can be regarded as retaining its individual character. In ring systems such as the thiazoles, as with pyrrole itself, it is more useful to consider the characteristic ring breathing vibrations than to seek to identify individual C=N vibrations. In conjugated, nitrogen-containing, five-membered ring systems, as in pyridine and similar aromatic compounds the individuality of the C=N absorption is completely lost, and systems of this type will be considered separately in the section on heterocyclic aromatic materials (Chapter 16).

In acyclic compounds and cyclic materials without internal conjugation, Barnes *et al.* [15] assigned the C=N absorption to the 1650 cm^{-1} region, and this assignment has been confirmed by Fabian *et al.* [46], who assign the overall range of 1674–1665 cm^{-1} to non-conjugated compounds of the type R–CH=N–R, with a slight lowering in some cases when one of the carbon substituents carries a branched chain. More recent work covering oximes [67], imines [61, 71–73], hydrazones [68] and semicarbazides [69, 70] has confirmed this general conclusion but has led to some extension of the frequency ranges. With carbon atoms attached at all points to the C=N bond the frequency lies between 1665 and 1645 cm^{-1} with some reduction when one of the substituents is aromatic. The effect of aromatic conjugation is variable and depends in part on the point of attachment. The band moves to about 1630 cm^{-1} when the ring is attached to the nitrogen atom, but only to 1640 cm^{-1} when it is attached to the double-bond carbon. However exceptions occur so that the behaviour of the C=N frequency on conjugation can be erratic. Thus, there is a fall of 40 cm^{-1} on passing from alkyl ketoximes to acetophenone oxime.

Unlike the C=C frequency, that for C=N shows some sensitivity to phase changes, so that different values may be found in the solution as compared with the solid. Oximes for example absorb between 1685 and 1660 cm^{-1} if unassociated but fall to 1660–1640 cm^{-1} in the condensed phase. In solution, acetoxime, for example, absorbs at 1675 cm^{-1}, *cyclo*hexanone oxime at 1669 cm^{-1} and *cyclo*pentanone oxime[12] at 1684 cm^{-1}. The slight rise on passing from a six-membered ring to a five-membered ring recalls the similar rise in the C=O frequency of *cyclo*pentanone over that of *cyclo*hexanone, and is due to ring strain. Formaldoxime and acetone oxime have been studied in some detail by Califano and Lüttke [48], whilst Palm and Werbin [49] and Hadži and Premu [67] have studied a range of geometric isomers. Oximes are unaffected by the geometric form, insofar as the frequencies are concerned, although there are some indications of differences in the relative intensities [67]. Aldoximes and ketoximes both absorb in the same frequency range, and both show larger than usual frequency shifts from aromatic conjugation. Benzaldoxime exists in two different crystal forms which absorb at 1645 and 1635 cm^{-1}. In the Raman spectra the C=N band is very much stronger than in the infra-red and is always easy to identify.

Imines also appear to be very similar. Picard and Polly [50] quote 1640−1633 cm^{-1} for non-conjugated compounds. However, this probably refers to samples in the condensed phase, and higher values would be expected in solution, when they would probably fall within the 1680−1660 cm^{-1} range. Baguley and Elvidge [51] quote a solid-state frequency of 1664 cm^{-1} for an exocyclic =NH bond attached to a five-membered ring, which would be expected to be raised a little above the normal value. Mathis *et al.* [71−73] give a slightly wider range for a series of benzimidic and phenylacetimidic esters in the solid state (1659−1635 cm^{-1}). Conjugation again has a smaller effect than in olefins and in their series it reduces the frequency by 10−20 cm^{-1}. However in diphenyl ketimine, the frequency falls as low as 1595 cm^{-1} in the liquid state [61]. When one or more NH groups are attached to the carbon atom of the C=N link, as in guanidines and related materials, it becomes more difficult to identify the C=N links with any certainty. In general, the frequencies appear to be slightly higher than usual. Lieber, Levering and Patterson [13] quote an overall range of 1689−1657 cm^{-1} for compounds of this type, but Pickard and Polly find a wide variation between 1718 and 1590 cm^{-1}. Other data on these compounds by Mathis *et al.* [71−73] support the wide frequency range, with the conjugated systems having lower values. Raman data, where the band can be more readily identified, have shown that *N*-mono-substituted amidines show two bands due to *syn* and *anti* forms. With acyl groups on the carbon and nitrogen atoms these appear between 1658 and 1632 cm^{-1}. With *N*-di-substituted amidines only one band is shown due to the *syn* form. This is in the 1633−1617 cm^{-1} range for aryl substituted derivatives. Non-conjugated C=N links in oxazolones also occur [17] within the range 1683−1668 cm^{-1}. Provided therefore that the C=N link is not conjugated and carries no charge on the nitrogen atom, the stretching frequency in unstrained compounds of all types may be expected in the overall range 1680−1650 cm^{-1} with somewhat lower values in the solid state. Departures from this will be found, however, particularly in the guanidines and in strained-ring systems. In pyrazolones, for example [52], the $C=N$ frequency can rise to 1700 cm^{-1}. However, when the nitrogen atom of the $C=N$ bond is so substituted that it is able to take on a more polar character the characteristic frequency alters considerably. Goulden [30] quotes limits of 1659−1510 cm^{-1} for a series of compounds of this type, and the metallic chelates of some oximes have also been found to show considerable variations in their $C=N$ frequencies [31]. The direction of frequency shift in such cases depends in part on the nature of the charge on the nitrogen atom. In $N^+=C$ compounds, for example, the frequencies are usually raised a little above the normal, and Leonard et al. [53, 54] have made use of this in the differentiation of $\alpha\beta$-unsaturated tertiary amines by

observing the high $C=\overset{+}{N}\diagup_{\diagdown}$ frequencies which occur when salt

formation allows tautomerism to take place. However the direct attachment of chlorine to the nitrogen atom lowers the frequency to about 1600 cm^{-1}.

The infra-red and Raman spectra of hydrazones have been reviewed by Kitaev et al. [68]. Compounds with the $C=N-NH_2$ structure absorb near 1650 cm^{-1} but this is reduced to 1640−1610 cm^{-1} by substitution on the terminal amino group. However it appears to make little difference whether the terminal substituents are alkyl or aryl groups.

Semicarbazones and thiosemicarbazones have been reviewed by Raevskii et al. [69, 70]. The former absorb near 1645 cm^{-1} with a fall to 1610 cm^{-1} for aryl derivatives. The latter absorb in the range 1680−1612 cm^{-1}.

In general the influence of conjugation on the $C=N$ frequency is very similar to the case of the $C=C$ frequency, although the shifts are a little less. Acetylazine [19] with two $C=N$ bands in conjugation absorbs at 1664 cm^{-1}, but with crotonaldehyde azine [19] the interpretation is more complex, as there are two $C=C$ groups as well as two $C=N$ links. However, here also the strongest band is at 1650 cm^{-1}. Similarly, α-furylazine absorbs at 1635 cm^{-1}. When aromatic rings form the conjugating system, the effect is usually small. The $C=N$ frequency in solution then falls in the overall range [45] 1660−1630 cm^{-1}, with slightly lower values for solids and associated compounds such as the aromatic aldoximes. The presence of a second aromatic ring on the other side of the $C=N$ link leads to a concentration of the observed frequencies near the lower end of this range. The influence of aryl conjugation on the C=N absorption in oxazolones has also been extensively studied in connection with the penicillin programme. The general finding [17, 18] was that non-conjugated oxazolones absorbed in the range $1683-1668$ cm^{-1} and that this was only lowered to $1657-1641$ cm^{-1} by aromatic-type conjugation.

15.4(b). The intensity of C=N absorptions. The intensity of the C=N— absorption varies widely with the nature of the attached group. Cross and Rolfe [12] have noted that in oximes it is extremely weak, but even then varies by a factor of two in non-conjugated compounds. On the other hand, Goulden [30] noted that in imino-thioethers the —C=N— absorption was relatively strong, whilst in $N:N'$dimethylbenziamide it is extremely strong [56]. Fabian and Legrand [56] have compared the extinction coefficients of the —C=N— absorptions in various systems. In a compound such as N-propylidene propylamine the ϵ value in carbon tetrachloride is 140. Replacement of one or both alkyl groups by aromatic rings raises this value slightly to 180. On the other hand, oximes such as *cyclo*hexanone oxime have an ϵ value of 20 under these conditions, falling to less than 8.5 in the oxime of acetophenone.

The substitution of a sulphur atom on to the carbon of the double bond raises the extinction coefficient to 218 (conjugated 270), whilst nitrogen has an even greater effect. Values ranging from $\epsilon = 365$ to 880 are quoted for variously substituted benzamidines. Oxygen attached to the double bond at the carbon atom also has a very marked effect, and methyl N-phenyl benzimidate has an ϵ value of 670.

Whilst extinction coefficients have not the precision of intensity measurements based on band areas, the relative orders of magnitude are clearly so widely different that intensities measured in this way will give useful information as to the immediate environment of the —C=N— bond.

15.5. The —N=N— stretching vibration

The N=N linkage has only very weak absorptions in the infra-red and if the molecule is in the *trans* form and is symmetrically substituted the band is forbidden. This band is therefore better identified in the Raman spectrum where it is strong. Herzberg [21] has identified the strong band at 1575 cm^{-1} in the Raman spectrum of azomethane as belonging to the N=N vibration. A band in the 1550 cm^{-1} region is also shown in the Raman spectra of esters of azo dicarboxylic acids but is absent in the infra-red. Other examples of this band near 1550 cm^{-1} occur in acyl alkyl diimides [74], diacyl diimides [75] and 1-pyrazoline [76].

The N=N frequency of aromatic azo compounds is also difficult to identify in the infra-red. The frequencies are lower in these cases and are not only weak but are overlaid by other aromatic bands. LeFevre *et al.* [34, 57] studied many aromatic azo compounds and made a very tentative identification of the N=N band in the infra-red at 1406 ∓ 14 cm^{-1}, but Dolinsky and Jones [58] were unable to find any specific absorptions from this group. The problem has been resolved by Kubler *et al.* [77] and by Hacher [79] using Raman spectra. This shows that *trans*-azobenzene absorbs at 1442 cm^{-1} and the *cis* form at 1511 cm^{-1}. Derivatives of azobenzene and of azonaphthalene absorb in the $1450-1380$ cm^{-1} range, but are difficult to observe in the infra-red [78, 79].

Dolinsky and Jones [58] make the interesting observation that *ortho*-hydroxy or aminoazo-compounds were abnormal and different from the *p*-substituted compounds, which led them to conclude that zwitterion structures involving the $-N=\overset{+}{N}H-$ structure were involved. Hadži [59] had investigated this further using deuteration techniques. He found that complex mixtures of tautomers involving the hydrazones are formed in this way, whilst compounds such as 4-phenylazo-1-naphthol and the *o*-methyl derivative of 1- and 4-phenylazophenol exist as true $-N=N-$ compounds. He was, however, unable to identify any characteristic $-N=N-$ frequency in any of these compounds.

There is also a possibility that a characteristic frequency for the skeleton $-C-N=N-C$ may exist, as Tetlow [23] has shown that both *cis*- and *trans*-azobenzenes show absorption at 927 cm^{-1} which is absent from hydrazobenzene, and that this band shows a Christiansen filter effect only in the *trans*-form, which may indicate its association with a skeletal group along the direction of maximum polarisability.

15.6. The diazo group

Diazomethane is a linear molecule with the structure $CH_2=N=\overset{+}{N}$ and absorbs at 2101 cm^{-1}. The assignment of this to the stretching frequency is supported by fundamental studies on two diazo-cyanides [32, 33]. Diazonium salts have been investigated by Aroney *et al.* [62] and by Whetsel *et al.* [63]. Altogether a total of over fifty aryl diazonium salts have been studied by these groups. Aroney *et al.* [62] find the characteristic $-CN_2^+$ frequency at 2261 cm^{-1} with a range of about ± 20 cm^{-1}. Whetsel [63] *et al.* confirm these figures but extend the range slightly. These latter workers studied the influence of both the anion and the diazonium cation on this frequency. Changes in the former produce only negligible shifts, but ring substitution by groups with strong electron-attracting properties has a marked effect. Diazonium salts of *p*-nitroaniline, for example, absorb at 2294 cm^{-1}, whereas the corresponding *p*-diethylamino-derivatives absorb at 2151 cm^{-1}. In the extreme case of diazophenol the frequency falls to 2110 cm^{-1}. However, Le Fevre *et al.* [64] have recently shown that both diazophenols and diazonaphthols exist in fact as quinone diazides, so the frequency fall due to the increased resonance is readily understood on this basis. The range found for compounds of this type was 2173–2014 cm^{-1}, depending upon the nature and positions of other ring substituents. Only a limited amount of data is available on alkyl diazo compounds. These absorb in the 2050–2035 cm^{-1} range when mono-substituted, and the frequency falls to 2030–2000 cm^{-1} on di-substitution [80].

The high frequency of this absorption is typical of $X=X=X$ structures, and it cannot therefore always be readily differentiated from *iso*cyanates and similar materials. However, the recognition of this band in the spectrum of the naturally occurring azaserine has led to its successful synthesis [65].

15.7. Azides

Azides can be readily identified by the strong $N=N=N$ asymmetric stretching absorption which occurs with great consistency close to 2130 cm^{-1}. The corresponding symmetric vibration is at considerably lower frequency, and is not only much weaker, but is also much more variable in position, so that it is of relatively little use for analytical purposes.

The spectrum of hydrazoic acid as been examined by Eyster [24], who has assigned the band at 2141 cm^{-1} to the $N=N=N$ asymmetric vibration and that at 1269 cm^{-1} to the corresponding symmetric mode. Methyl azide has also been fully studied [25], and the corresponding bands in this case are at 2141 cm^{-1} and 1351 cm^{-1}.

A systematic study of azides as a group has been made by Sheinker and Syrkin [26], who examined the Raman spectra of sodium azide and six other materials. They found a strong band in all cases in the range $2167-2080$ cm^{-1} and a second weaker one in the range $1343-1177$ cm^{-1}. In accordance with the earlier work these were assigned to the asymmetric and symmetric vibrations respectively. It should be noted, however, that some inorganic azides have a simple linear structure so that the symmetric frequency is not found in the infra-red. In ammonium azide crystals, for example, only the 2050 cm^{-1} band is shown and the symmetric band is absent [66].

Further confirmation of these findings has come from Boyer [27] and also from Lieber, Levering and Patterson [13]. Boyer studied the addition of hydrazoic acid to conjugated systems and examined the infra-red spectra of the products. Ten azides in all were examined, and all showed strong absorptions near 2128 cm^{-1}. The frequency did not appear to be particularly sensitive to environmental changes, and 1-phenyl-1-azido-2-nitroethane, for example, absorbed at the same frequency as triazoacetone.

Lieber *et al.* [13, 81] have given data on a number of azides, and their results are in general agreement with those of Sheinker and Syrkin [26]. Thus azido ethers absorb near 2120 cm^{-1} and most organic azides absorb in the $2120-2080$ cm^{-1} range. However, inorganic azides absorb at rather lower frequencies near 2030 cm^{-1} in most instances. In the Raman spectra it is of course the symmetrical stretching frequency which is the intense band and this occurs near 1250 cm^{-1} the Raman spectra of inorganic azides have been examined by Kahovec and Kohlrausch [28]. Their results agree with the infra-red data in showing only a small frequency range for a variety of cations.

15.8. Bibliography

1. Reitz and Sabathy, *Monatsh.*, 1938, **71**, 100.
2. *Idem, ibid.,* p. 131.
3. Reitz and Skrabel, *ibid.,* 1937, **70**, 398.
4. Snyder and Eliel, *J. Amer. Chem. Soc.*, 1948, **70**, 1857.
5. Marvel, Brace, Miller and Johnson, *ibid.,* 1949, **71**, 34.
6. Marvel and Brace, *ibid.,* p. 37.
7. Holmstedt and Larsson, *Acta Chem. Scand.,* 1951, **5**, 1179.
8. Richardson and Bright-Wilson, *J. Chem. Phys.,* 1950, **18**, 155.
9. Kitson and Griffith, *Analyt. Chem.,* 1952, **24**, 334.
10. Jesson and Thompson, *Spectrochim. Acta,* 1958, **13**, 217.
11. Mellon, *Analytical Spectroscopy* (Wiley, 1950).
12. Cross and Rolfe, *Trans. Faraday Soc.,* 1951, **47**, 354.

13. Lieber, Levering and Patterson, *Analyt. Chem.,* 1951, **23**, 1594.
14. Jesson and Thompson, *Proc. Roy. Soc.,* 1962, **268A**, 68.
15. Barnes, Gore, Liddel and Van Zandt Williams, *Infra-red Spectroscopy* (Reinhold, 1944).
16. Brown, *J. Amer. Chem. Soc.,* 1958, **80**, 794.
17. Rasmussen and Brattain, *The Chemistry of Penicillin* (Princeton University Press, 1949), p. 400.
18. Thompson, *ibid.,* p. 387.
19. Blout, Fields and Karplus, *J. Amer. Chem. Soc.,* 1948, **70**, 194.
20. Brown, *Chem. Reviews,* 1958, **58**, 581.
21. Herzberg, *Infra-red and Raman Spectra of Polyatomic Molecules* (Van Nostrand, 1945), p. 357.
22. Caldow and Thompson, *Proc. Roy. Soc.,* 1960, **254A**, 1.
23. Tetlow, *Research,* 1950, 3, 187.
24. Eyster, *J. Chem. Phys.,* 1940, 8, 135.
25. Eyster and Gillette, *J. Chem. Phys.,* 1940, 8, 369.
26. Sheĭnker and Syrkin, *Izvest. Akad. Nauk. S.S.S.R. Ser. Fiz.,* 1950, **14**, 478.
27. Boyer, *J. Amer. Chem. Soc.,* 1951, **73**, 5248.
28. Kahovec and Kohlrausch, *Mh. Chem.,* 1947, **77**, 180.
29. McBride and Beachell, *J. Amer. Chem. Soc.,* 1952, **74**, 5247.
30. Goulden, *J. Chem. Soc.,* 1953, 997.
31. Duyckaerts, *Bull. Soc. Roy. Sci. Liége,* 1952, **21**, 196.
32. Sheppard and Sutherland, *J. Chem. Soc.,* 1947, 453.
33. Anderson, Le Fèvre and Savage, *ibid.,* 443.
34. Le Fèvre, O'Dwyer and Werner, *Chem. and Ind.,* 1953, 378.
35. Felton and Orr, *J. Chem. Soc.,* 1955, 2170.
36. Skinner and Thompson, *ibid.,* 1955, 487.
37. Caldow, Cunliffe Jones and Thompson, *Proc. Roy. Soc.,* 1960, **254A**, 17.
38. Heilmann, Bonnier and Arnaud, *Compt. Rend. Acad. Sci. Paris,* 1959, **248**, 3578.
39. Williams, *J. Chem. Phys.,* 1956, **25**, 656.
40. Sensi and Gallo, *Gazz. Chim.,* 1955, **85**, 224, 235.
41. Hoyer, *Chem. Ber.,* 1956, **89**, 2677.
42. Davison, *J. Chem. Soc.,* 1953, 3712.
43. Khorana, *Can. J. Chem.,* 1954, **32**, 261.
44. Meakins and Moss, *J. Chem. Soc.,* 1957, 993.
45. Khorana, *Chem. Reviews,* 1953, **53**, 145.
46. Fabian, Legrand and Poirier, *Bull. Soc. Chim. Fr.,* 1956, 1499.
47. Ugi and Meyer, *Chem. Ber.,* 1960, **93**, 239.
48. Califano and Lüttke, *Zeit. Phys. Chem.,* 1956, **6**, 83.
49. Palm and Werbin, *Can. J. Chem.,* 1953, **31**, 1004.
50. Pickard and Polly, *J. Amer. Chem. Soc.,* 1954, **76**, 5169.
51. Baguley and Elvidge, *J. Chem. Soc.,* 1957, 709.
52. Gagon, Boivin, McDonald and Yaffe, *Can. J. Chem.,* 1954, **32**, 823.
53. Leonard and Gash, *J. Amer. Chem. Soc.,* 1951, **76**, 2781.
54. Leonard, Thomas and Gash, *ibid.,* 1955, **77**, 1552.
55. Ham and Willis, *Spectrochim. Acta,* 1960, **16**, 279.
56. Fabian and Legrand, *Bull. Soc. Chim. Fr.,* 1956, 1461.
57. Le Fèvre, O'Dwyer and Werner, *Austral. J. Chem.,* 1953, **6**, 341.
58. Dolinsky and Jones, *J. Assoc. Off. Agric. Chem.,* 1954, **37**, 197.
59. Hadži, *J. Chem. Soc.,* 1956, 2143.
60. Mogul, *Nuclear Sci. Abstracts,* 1967, **21**, 47014.
61. Perrier, Datin and Lebas, *Spectrochim. Acta,* 1969, **25A**, 169.

62. Aroney, Le Févre and Werner, *J. Chem. Soc.,* 1955, 276.
63. Whetsel, Hawkins and Johnson, *J. Amer. Chem. Soc.,* 1956, **78**, 3360.
64. Le Fèvre, Sousa and Werner, *J. Chem. Soc.,* 1954, 4686.
65. Fusari, Frohardt, Ryder, Haskell, Johanessen, Elder and Bartz, *J. Amer. Chem. Soc.,* 1954, **76**, 2878.
66. Dows, Whittle and Pimentel, *J. Chem. Phys.,* 1955, **23**, 1475.
67. Hadži and Premu, *Spectrochim. Acta,* 1967, **23A**, 35.
68. Kitaev, Buzykin and Troepol'skaya, *Russian Chem. Rev.,* 1970, **39**, 441.
69. Raevskii, Shagidullin and Kitaev, *Bull. Acad. Sci. USSR, Chem. Ser.,* 1966, 200.
70. Raevskii, Shagidullin and Kitaev, *Dokl. Acad. Sci. USSR,* 1966, **170**, 853.
71. Mathis, Baccar, Barrens and Mathis, *J. Mol. Structure,* 1971, **7**, 355.
72. Baccar, Mathis, Seccher and Mathis, *J. Mol. Structure,* 1971, **7**, 369.
73. Mathis, Baccar, Katcka Bon and N'Gondo M'Pondo, *J. Mol. Structure,* 1971, **7**, 381.
74. Lynch, Maclachlan and Siu, *Can. J. Chem.,* 1971, **49**, 1598.
75. Crawford, Mishla and Dummel, *J. Amer. Chem. Soc.,* 1966, **88**, 3959.
76. Durig, Karriker and Harris, *J. Chem. Phys.,* 1970, **52**, 6096.
77. Kubler, Luttke and Weckherlin, *Zeit. Electrochem.,* 1960, **64**, 650.
78. Morgan, *J. Chem. Soc.,* 1961, 2151.
79. Hacker, *Spectrochim. Acta,* 1965, **21**, 1989.
80. Yates, Shapiro, Yoda and Fugger, *J. Amer. Chem. Soc.,* 1957, **79**, 5756.
81. Lieber and Thomas, *App. Spectroscopy,* 1961, **51**, 144.
82. Hildago, *Compt. Rend. Acad. Sci. Paris,* 1959, **249**, 395.

16

Heterocyclic Aromatic Compounds

16.1. Introduction and table

There is a reasonably close analogy between the ring vibrations of benzene and those of pyridine and quinoline, but there are considerable differences in the hydrogen deformation vibrations. However, the out-of-plane hydrogen deformation vibrations appear to be like those of benzene compounds containing an additional substituent. An α-mono-substituted pyridine, therefore, behaves like an *ortho*-di-substituted aromatic compound in this respect. Substituted pyridines have been extensively studied in both the Raman and infra-red and correlations are now available which enable the substitution pattern to be recognised. Some more limited data are available on the diazines, pyrazine and pyridazines.

The pyrimidines and purines have been more extensively studied on account of their intrinsic interest as important biological compounds. The compounds studied, however, have been primarily the hydroxy- and amino-derivatives and there was a good deal of argument as to whether these existed as non-aromatic tautomers. This has now been resolved. It is clear that the hydroxy compounds do indeed exist in the keto form, but that in the amino compounds the aromatic character is retained. When this is the case the frequencies are very parallel to those of benzene. More limited data are also available on triazines and tetrazines.

Heterocyclic rings which have two double bonds in a five-membered ring are all aromatic in character and show typical ring-breathing modes and hydrogen deformations. All those molecules for which a detailed vibrational analysis is available show one very characteristic feature. This is that at least one and often all the CH stretching bands appear above 3100 cm^{-1}. This is sufficiently different from the six-membered ring aromatics as to be diagnostic. Unfortunately these bands are not strong in the infra-red and the intensity is further reduced when an oxygen atom is present, so that this characteristic is not always immediately obvious.

The infra-red data on heterocyclics generally has been very fully covered in two reviews by Katritzky and Ambler [42] and Katritzky and Taylor [43]. These are very comprehensive and between them they provide over 1600 references. In view of this and of the very valuable supplementary reviews of the Raman spectra given by Dollish *et al.* [44] no attempt at a full coverage of the literature is attempted in this chapter.

The correlations discussed are listed in Table 16.

Table 16

Six-Membered Ring Heterocyclics

Pyridines and Quinolines

CH stretching bands	3090–3060 and 3050–3010 cm^{-1} (w.)
Ring breathing bands	1615–1585 cm^{-1} (m.)
	1588–1560 cm^{-1} (m.)
	1520–1465 cm^{-1} (m.)
	1438–1410 cm^{-1} (m.)
	(Two further bands in fused rings.)
CH deformations 2-substitution	790–740 cm^{-1} (799–781 alkyl groups) (s.)
3-substitution	820–770 cm^{-1} (810–789 alkyl groups) (s.)
4-substitution	850–790 cm^{-1} (820–794 alkyl groups) (s.)

A characteristic pattern of overtone bands occurs in the range 2000–1650 cm^{-1}.

Pyrimidines

CH stretching bands	3090–3060 and 3050–3010 cm^{-1} (w.)
Ring breathing bands	1590–1553 cm^{-1} (m.)
	1574–1541 cm^{-1} (m.)
	1508–1409 cm^{-1} (m.)
	1469–1328 cm^{-1} (m.)
CH deformation	860–795 cm^{-1}

Five-Membered Ring Heterocyclics

CH stretching frequencies	3150–3050 cm^{-1}, one to three bands; one of which is usually above 3100 cm^{-1}.
Ring breathing bands	Multiple bands 1600–1400 cm^{-1} which vary systematically with the compounds.
CH deformations	900–700 cm^{-1} strong, vary with compound and with its substitution.

16.2. Pyridines and quinolines

16.2(a). General. The nitrogen atom of pyridine is isoelectronic with the CH group, and as there is little difference in mass or in the bond strengths to adjacent atoms, it is to be expected that the majority of the fundamental frequencies will be very close to those of benzene, and that differences will only arise in relation to vibrations which involve collective motions of the hydrogen atoms. Klein and Turkevitch [1] were the first to show that this is so and this was fully confirmed by early studies on alkylated pyridines and on deuterated derivatives [1–5, 22, 23, 26].

More recent studies by Cook and Church [41], Groenewege [45] and by Katritzky and his coworkers [46–48] have provided a substantial basis for the assignments of the bands of pyridine derivatives. Apart from the usual ring and hydrogen deformation modes it is interesting to find that the pyridines show a series of overtone bands between $2000-1650 \text{ cm}^{-1}$ which are parallel to those of substituted benzenes and which are also indicative of the substitution pattern [41, 49]. None of these patterns is the same as any of those of the benzene series but their occurrence may well cause some confusion if it is not suspected that a pyridine ring is present. Other heteroaromatics also give overtone bands in this range, with yet different patterns.

16.2(b). CH stretching vibrations. The data available confirm the expectation that the C–H stretching modes of pyridines and quinolines will be essentially similar to those of benzene. Pyridine and the picolines all show CH absorptions in the range $3070-3020 \text{ cm}^{-1}$, [1, 4, 5, 6] which appear as a series of multiple absorptions under high resolution [25, 27]. The positions and numbers of these vary somewhat with the substitution involved, but the observed frequencies are in each case very close to those of the benzene homologues. In most cases in the infra-red two bands are observed between $3090-3060$ and $3050-3010 \text{ cm}^{-1}$ just as in benzene. In substituted pyridines the intensity of the CH bands is related to the inductive properties of the substituents, and follows a parabolic relationship with σ, going through a minimum at σ, 0.7. This is due to the reversal of the CH bond polarity which takes place with electron-accepting substituents [50].

16.2(c). Ring breathing vibrations. Pyridines have four ring breathing vibrations in the region $1610-1400 \text{ cm}^{-1}$ and are very similar to the benzenes. The highest frequency band of the group falls between 1615–1585 cm^{-1} with occasional examples of lower values with heavy substituents. There is some small dependence on the nature and position of the substituents. In mono-substituted derivatives the frequency falls from 1615 cm^{-1} to 1585 cm^{-1} as the substituent changes from an electron donor to an acceptor.

The second band of this series falls between 1588–1560 cm^{-1}. Unlike the parallel benzene band it is usually of medium intensity and its appearance is not therefore necessarily related to conjugation. The separation of this pair of bands is about 40 cm^{-1} for 4-alkyl pyridines but about 20 cm^{-1} for the 2- and 3-compounds [41].

The other pair of bands fall between 1520–1465 cm^{-1} and 1438–1410 cm^{-1}. The former is usually more intense than the other. The frequency of the 1500 cm^{-1} band tends to rise with an increasing degree of substitution.

These bands arise from similar fundamentals to those which are responsible for the ring breathing modes of benzene. It is therefore to be expected that they will show considerable intensity variations which are due to the electronic effects of the substituents. As with benzenes it is found that the intensity changes can be predicted from a consideration of the magnitude of the change in dipole moment which a given group will produce.

With quinolines, isoquinolines and related compounds, the pattern is that of pyridine with the bands of the other aromatic ring superimposed. Complex patterns are then found. Quinolines and isoquinolines show six bands between 1625 and 1430 cm^{-1} with two others between 1400 and 1370 cm^{-1}. Katritzky and Taylor [43] list these bands for several other polycyclic azines.

16.2(d). Ring vibrations and hydrogen deformations. Assignment of pyridine and quinoline substitution. In the 1300 cm^{-1} region, pyridine and its derivatives show a series of four bands due to in plane hydrogen bending modes and a ring vibration near 1000 cm^{-1}. These bands are all weak in the infra-red, as they are in benzene, and they do not correlate well with the changes in the substituents. In the Raman spectra these bands are stronger, particularly that near 1000 cm^{-1} which is very intense. This falls at 1030 cm^{-1} in pyridine, at 1030–1000 cm^{-1} in 3-substituted compounds and at 1000–985 cm^{-1} in 2- or 4-substituted derivatives. ·

The remaining characteristic region is that between 900 cm^{-1} and 700 cm^{-1} in which CH deformations occur. It has been shown in the case of benzene substitution that the strongest band in this region originates in the out-of-plane vibrations of the unsubstituted hydrogen atoms of the ring, and that the principal factor determining the frequency is the number of such free hydrogen atoms which are adjacent to one another. A similar effect can be expected in pyridine derivatives, so that pyridine with five free hydrogen atoms would be similar in this region to a mono-substituted benzene, whilst for example γ-picoline, which has two pairs of free hydrogen atoms, should be similar to a *para*-di-substituted benzene. This expectation is realised in practice. Thus pyridine [2] absorbs at 750 cm^{-1}, α-picoline [3] at 755 cm^{-1}, β-picoline at 790 cm^{-1} and γ-picoline at 800 cm^{-1}, corresponding to mono-, *ortho, meta-* and *para*-substituted benzenes respectively. A series of eight variously substituted ethyl pyridines and nineteen substituted methyl pyridines have also been shown to follow this correlation with slightly wider divergencies [26]. Cook *et al.* [41] report similar findings. and these have been fully confirmed by later workers. Katritzky and Taylor [43] summarise the available data as follows: 2-substitution: 794–781 cm^{-1} for alkyl groups but 780–740 cm^{-1} for others; 3-substitution 810–789 cm^{-1} for alkyl, and 820–770 cm^{-1} overall; 4-substitution 820–794 cm^{-1} for alkyl and 850–790 cm^{-1} for others. Multiple substitution gives the bands appropriate for the numbers of adjacent hydrogen atoms. In all cases a second band appears at lower frequency corresponding to a ring deformation. A similar mode is responsible for the 695 cm^{-1} band in mono-substituted benzenes. This band is less diagnostic and can occur over a wider frequency range; usually it is in the 740–730 cm^{-1} range but this falls to 715–710 cm^{-1} in 3-substituted derivatives.

In the quinolines the CH out-of-plane deformation correlation still appears to hold good if each ring is considered separately. Thus both 2 : 6- and 2 : 7-dimethylquinolines have only two adjacent free hydrogen atoms in each of the two rings. Both compounds absorb very strongly at 831 cm^{-1} and 835 cm^{-1}, respectively (corresponding to *para*-substituted aromatics). In addition, both show a second weaker band in the 900–850 cm^{-1} region which might possibly be associated with the vibration of the remaining isolated ring hydrogen atom.

2 : 3- and 2 : 4-Dimethylquinolines contain four free hydrogen atoms in the carbocyclic ring and only one in the heterocyclic ring. These show their strongest bands in this region at 755 cm^{-1} and 758 cm^{-1} (corresponding to *ortho*-substitution), respectively. They

also both absorb in the 900—850 cm^{-1} range. *iso*Quinoline is similar in showing its strongest band in this region at 745 cm^{-1}, which could be assigned to the four hydrogens of the carbocyclic ring. In this case a second strong band occurs at 829 cm^{-1} which could arise from the two adjacent free hydrogen atoms of the heterocyclic ring, whilst absorption at 864 cm^{-1} may be associated with the single free hydrogen atom.

The apparent ease with which the quinoline derivatives comply with these correlations is surprising. However, it should be remembered that, as in the case of the aromatic compounds, substitution by strongly polar groupings will materially alter these CH frequencies.

16.2(e). Hydroxy and amino pyridines and quinolines. The constitution of compounds such as 2-hydroxypyridines are of interest from the point of view of the possible tautomeric pyridone structures. Dipole-moment evidence provides strong support for the view that the pyridone form is favoured exclusively, and this has been fully substantiated by infra-red studies by a number of workers [28, 31]. Thus not only 2-hydroxy- but also 4-hydroxypyridine exists in this form, as do many related quinoline derivatives.

With aminopyridines, on the other hand, there is little or no evidence of imine formation. 2-Aminopyridine [32] in dilute solution absorbs at 3500 cm^{-1} and 3410 cm^{-1}, and is thus very similar to aniline; 4-aminopyridine is similar [33]. Bellamy and Williams [34] have shown that these, and the corresponding frequencies of other heterocyclic amines [35], follow the relationship which connects the asymmetric and symmetric stretching frequencies of primary amines so that there can be little doubt that this is the preferred structure.

16.3. Pyrimidines and purines

16.3(a). General. A good deal of work has been carried out on pyrimidines owing to their biological importance as components of nucleic acids. However, owing to the many tautomeric forms in which it is possible to write the structures of some substituted pyrimidines—including some in which the aromatic character of the ring is completely lost—early work was mainly directed to the elucidation of the actual structures present.

The first publications of spectra of pyrimidines are due to Blout and Fields [9—11], who characterised the spectra of a number of nucleic acids, purines and pyrimidines, and showed that it was probable that uracil existed in the keto form. This particular finding was supported by Lacher, Campion and Park [12], who examined uracil, 5-chlorouracil and thymine in the overtone region. Blout and Mellors [13] also explored the possibilities of the direct application of infra-red spectra to tissue samples. Brownlie, Sutherland and Todd [14] used infra-red methods to demonstrate the occurrence of hydrogen bonding in certain glycosidaminopyrimidines, but their observations were confined to the carbonyl absorption region. The spectra of a number of pyrimidines and purines have also been collected by Randall *et al.* [15], whilst spectra in antimony trichloride have been given by Lacher *et al.* [36].

More recently a very large number of substituted pyrimidines have been studied by many workers [16–18, 22, 37, 38], and it is now well established that hydroxy derivatives do not exist as such, but in the non-aromatic form. However amino pyrimidines and other derivatives are fully aromatic and their spectra are closely related to those of benzene and pyridine [42–44, 61, 62].

16.3(b). CH Stretching vibrations. By analogy with pyridine and benzene, C–H stretching vibrations in pyrimidine and its derivatives will give rise to absorptions near 3050 cm^{-1}. Pyrimidine itself has four bands between 3100 and 3000 cm^{-1} and in the infra-red, most substituted compounds show at least two bands in this range. However, in tri-substituted pyrimidines with only one free ring hydrogen atom the band is very weak, and no absorption is to be expected from this cause in the tetra-substituted materials.

16.3(c). Ring breathing modes. Pyrimidines have the usual four bands in the $1600–1400 \text{ cm}^{-1}$ range. Pyrimidine itself has only three bands at 1570, 1467 and 1402 cm^{-1}, but that at 1570 cm^{-1} is degenerate and this is split in substituted compounds. The overall ranges are 1590–1553, 1574–1541, 1508–1409, and 1469–1328 cm^{-1} These ranges can be narrowed for individual classes such as 2-substituted compounds.

16.3(d). Ring vibrations etc. Short and Thompson [18] have drawn attention to the fact that a series of chloro-substituted and ethoxy-substituted pyrimidines which they have examined all absorb near 990 cm^{-1} and near 810 cm^{-1}. The former was assigned to a ring vibration, and the latter thought to be due to this also, although the possibility that it is a CH deformation mode was not excluded. Bands near these frequencies are illustrated in the great majority, but not all of the spectra of the thirty-nine amino- and hydroxy-derivatives given in their paper. Later studies have confirmed these absorptions as characteristic of the pyrimidine ring. The 990 cm^{-1} band is indeed a ring mode and appears within a few cm^{-1}, of this value in lightly-substituted derivatives. Even with tri-substitution the overall range is only $1017-964 \text{ cm}^{-1}$ [44].

The 810 cm^{-1} band is now known to be a hydrogen deformation and its position is therefore more sensitive to the position of substituents. In 2-substituted pyrimidines it is between 818 and 799 cm^{-1} but values up to 860 cm^{-1} can be reached with more complex substitution.

16.4. Other aromatic six-membered ring heterocycles

Assignments for pyridazine and pyrazine were suggested by Ito *et al.* [40], who noted that the latter differed from other heteroaromatics in having no infra-red absorptions above 1480 cm^{-1}. However, this is a consequence of its symmetry and the usual four bands in the $1600-1400 \text{ cm}^{-1}$ range are shown in the Raman spectrum. Several bands also appear in the infra-red when the symmetry is lost by substitution. Alkyl pyrazines show three or four bands in this range. Pyridazines have been studied as 3 : 6-substituted compounds and show two bands between $1600-1540 \text{ cm}^{-1}$ and $1430-1395 \text{ cm}^{-1}$ [63]. Both series show medium to weak bands between 1300 and 1000 cm^{-1} from in plane CH bends and ring modes. These are better seen in the Raman spectra [44]. Below 1000 cm^{-1} the out of plane CH bends appear. Pyridazine has its as a band at 760 cm^{-1} which would correspond to four adjacent hydrogens, and pyrazine at 804 cm^{-1}.

s-Triazines have been studied by Lancaster *et al.* [7]. All symmetrically substituted derivatives show only two ring stretching bands near 1560 and 1410 cm^{-1}. In asymmetrical compounds the degeneracy is split but is still small, so that two bands are still usual. All *s*-triazines absorb in the 800 cm^{-1} region. The position of the band depends upon the resonance characteristics of the substituents and the overall range is $817-736 \text{ cm}^{-1}$.

The infra-red spectrum of tetrazine has been discussed by Spencer *et al.* [8] but not much is known of its derivatives. The CH stretch is rather higher than usual (3090 cm^{-1}) and the ring stretches are at

1521 and 1418 cm^{-1}. A ring breathing band at 1017 cm^{-1} is strong in the Raman but not in the infra-red.

The spectra of polycyclic heteroaromatics are complex and cannot be covered adequately in the space available here. Katritzky and Taylor [43] summarise the characteristic frequencies of many such systems such as quinoxalines, naphthyridines, pyridopyridazines and pyrimidotriazines.

16.5. Aromatic five-membered ring heterocycles

16.5(a). One heteroatom. Pyrrole, furan and thiophene. These compounds have the same fundamental modes and in most cases corresponding bands can be identified at essentially similar frequencies in all three. Exceptions only arise for those modes which involve any substantial movement of the heavy sulphur atom of thiophene. The spectra are closely related to the aromatic heterocycles already discussed. The CH stretching frequencies are high and one if not all of them always appears above 3100 cm^{-1}, but the intensities are variable and the Raman spectra are more useful in this region. These bands have been studied in the infra-red by Lebas and Josien [19] and by Joeckle *et al.* [20]. Where these bands can be identified they can give useful structural information. Thus, Sobolev *et al.* [54] working in the Raman, showed that whereas furan has bands at 3153, 3122 and 3092 cm^{-1}, the last of these vanishes with substitution at the 2-position and that the first also vanishes in 2 : 5-substitution. The Raman data on pyrroles is fully reviewed by Jones [55], and on this general group by Dollish *et al.* [44].

Ring stretching bands occur in the 1600–1400 cm^{-1} range. They are more sensitive than the corresponding bands of benzene to changes in the substituents so that wide frequency ranges occur. The intensities also vary widely with the substituents. Katritzky [42, 43] has summarised the available data, and gives the frequency ranges for various substitution patterns. As an example of the variations, the highest frequency band in this region in thiophene is at 1591 cm^{-1}. This moves to 1523 $\mp$ 9 cm^{-1} on 2-substitution. Furans also show a band in the 1600 cm^{-1} which is rather more stable in position, but is not strong. Pyrroles have their highest frequency fundamental between 1550 and 1500 cm^{-1} but show an overtone near 1600 cm^{-1} which can be confused with the ring modes. The second band of furans is between 1512 and 1435 cm^{-1}, of thiophenes between 1435 and 1400 cm^{-1}, and of pyrroles between 1485 and

1439 cm^{-1}. The third band lies between $1405-1336 \text{ cm}^{-1}$ (furans), $1255-1210 \text{ cm}^{-1}$ (thiophenes) and $1418-1384 \text{ cm}^{-1}$ (pyrroles). The in plane CH bends lie between 1300 and 1000 cm^{-1} and are best characterised in the Raman spectra.

The out of plane CH bends and ring modes are responsible for a number of strong bands between 1000 and 700 cm^{-1}. These vary widely in position with substitution changes in the pyrroles and thiophenes, but much less so with the furans. Katritzky [42, 43] lists the positions of each of the bands in this region and gives the frequency ranges for various substitution patterns.

16.5(b). Two heteroatoms in five-membered rings. 1 : 2-positions.
This series comprises the pyrazoles, the isoxazoles, the *iso*thiazoles and dithiols. The high CH stretching frequencies are again a characteristic, and three fundamental ring stretches occur between 1560 and 1340 cm^{-1}. In many cases more bands are found due, presumably to overtones. Thus pyrazole absorbs at 1530, 1446 and 1394 cm^{-1}, but its derivatives show four bands including one between 1580 and 1600 cm^{-1}. Isoxazole is similar in having an overtone at $1666-1610 \text{ cm}^{-1}$ whereas the highest fundamental of the parent is 1560 cm^{-1}. In *iso*thiazole the highest fundamental is at 1489 cm^{-1}. In view of this complexity it is therefore necessary to compare the observed bands very carefully with the extensive data relating to their shifts with substitution given in the reviews of Katritzky [42, 43] and Dollish [44], and in the papers on isoxazoles by Slack and Wooldridge [56], and on pyrazoles [57] by Zerbi *et al.* In the Raman spectra a very strong band due to ring breathing modes occurs at $1300-1000 \text{ cm}^{-1}$ in isoxazoles, near 980 cm^{-1} in *iso*thiazoles, and between 1040 and 1000 cm^{-1} in pyrazoles. It is usually weak in the infra-red. The out of plane hydrogen bends give rise to a number of strong bands in the infra-red between $900-700 \text{ cm}^{-1}$. The strongest of these is usually in the 750 cm^{-1} region but is sensitive to substitution changes. The positions of all bands in this range are summarised by Katritzky [43].

16.5(c). Five-membered rings with heteroatoms at the 1 : 3-positions. Imidazoles, oxazoles and thiazoles, again show high CH stretching bands and the ring stretching modes are again complicated by both the appearance of overtones and of the wide frequency shifts which can result from substituent changes. Katritzky [43] lists the band positions but this region is better characterised in the Ramans spectra where the bands are stronger. The Raman spectra also show the very strong ring band at 1060 cm^{-1} in imidazoles, 1040 cm^{-1} in thiazoles and 1095 cm^{-1} in oxazoles. This band is usually weak in the infra-red but is strong in thiazole. A number of strong bands are shown by all this series between 900 and 700 cm^{-1}. Again the frequency positions vary considerably with substitution.

16.5(d). More than two heteroatoms. Triazoles have been discussed by Otting [39] and by Borello [58] and Curtin [59], thiadiazoles and triazoles by Rao [60] who also deals with tetrazoles. The data on these and on oxadiazoles are summarised by Katritzky [42, 43]. The same features as before are found, i.e. high CH stretching frequencies, multiple ring breathing bands of variable position and strong CH bends at lower frequencies.

16.5. Bibliography

1. Kline and Turkevitch, *J. Chem. Phys.*, 1944, **12**, 300.
2. Cannon and Sutherland, *Spectrochimica Acta*, 1951, **4**, 373.
3. Freiser and Glowacki, *J. Amer. chem. Soc.*, 1948, **70**, 2575.
4. Roberts and Szwarc, *J. Chem. Phys.*, 1948, **16**, 981.
5. Marion, Ramsay and Jones, *J. Amer. Chem. Soc.*, 1951, **73**, 305.
6. Coulson, Hales and Herington, *J. Chem. Soc.*, 1951, 2125.
7. Lancaster, Stamm and Colthup, *Spectrochim. Acta*, 1961, **17**, 155.
8. Spencer, Cross and Wiberg, *J. Chem. Phys.*, 1961, **35**, 1939.
9. Blout and Fields, *Science*, 1948, **107**, 252.
10. *Idem. J. Biol. Chem.*, 1949, **178**, 335.
11. *Idem. J. Amer. Chem. Soc.*, 1950, **72**, 479.
12. Lacher, Campion and Park, *Science*, 1949, **110**, 300.
13. Blout and Mellors, *ibid.*, p. 137.
14. Brownlie, Sutherland and Todd, *J. Chem. Soc.*, 1948, 2265.
15. Randall, Fowler, Fuson and Dangl, *Infra-red Determination of Organic Structures* (Van Nostrand, 1949).
16. Brownlie, *J. Chem. Soc.*, 1950, 3062.
17. Thompson, Nicholson and Short, *Discuss. Faraday Soc.*, 1950, **9**, 222.
18. Short and Thompson, *J. Chem. Soc.*, 1952, 168.
19. Lebas and Josien, *Bull. Soc. Chim. France*, 1957, 250.
20. Joeckle, Lemperle and Heckle, *Ziet. Naturforsch.*, 1967, **22A**, 395.
21. Corrsin, Fox and Lord, *J. Chem. Phys.*, 1953, **21**, 1170.
22. Brown and Short, *J. Chem. Soc.*, 1953, 331.
23. Anderson, Bak, Brodersen and Rastrup-Andersen, *J. Chem. Phys.*, 1955, **23**, 1047.
24. Dimroth and Lenke, *Chem. Ber.*, 1956, **89**, 2608.
25. Finnegan, Henry and Olsen, *J. Amer. Chem. Soc.*, 1955, **77**, 4420.
26. Shindo and Ikekawa, *Pharm. Bull. Japan*, 1956, **4**, 192.
27. Tallent and Siewers, *Analyt. Chem.*, 1956, **28**, 953.
28. Gibson, Kynaston and Lindsey, *J. Chem. Soc.*, 1955, 4340.

29. Ramirez and Paul, *J. Amer. Chem. Soc.*, 1955, **77**, 1035.
30. Scheinker and Resinkow, *Doklady, Akad. Nauk. S.S.S.R.*, 1955, **102**, 109.
31. Scheinker and Pomerantsev, *Zhur. Fiz. Khim.*, 1956, **30**, 79.
32. Shigorin, Danyushevskii and Gold'farb, *Izvest. Akad. Nauk. S.S.S.R., Otdel Khim. Nauk.*, 1956, 120.
33. Costa, Blasina and Sartori, *Z. Phys. Chem.*, 1956, **7**, 123.
34. Bellamy and Williams, *Spectrochim. Acta*, 1957, **9**, 341.
35. Angyal and Werner, *J. Chem. Soc.*, 1952, 2911.
36. Lacher, Bitner, Emery, Sefel and Park, *J. Phys. Chem.*, 1955, **59**, 615.
37. Montgomery, *J. Amer. Chem. Soc.*, 1956, **78**, 1928.
38. Tanner, *Spectrochim. Acta*, 1956, **8**, 9.
39. Otting, *Chem. Ber.*, 1956, **89**, 2887.
40. Ito, Shimada, Kuraishi and Mizushima, *J. Chem. Phys.*, 1956, **25**, 597.
41. Cook and Church, *J. Phys. Chem.*, 1957, **61**, 458.
42. Katritzky and Ambler, in *Physical Methods of Heterocyclic Chemistry.* Ed. Katritzky. Vol. II (Academic Press, New York, 1963).
43. Katritzky and Taylor, in *Physical Methods in Heterocyclic Chemistry.* Ed. Katritzky. Vol. 4 (Academic Press, New York, 1971).
44. Dollish, Fateley and Bentley, *Characteristic Raman Frequencies of Organic Compounds* (Wiley and Sons, New York, 1974).
45. Groenewege, *Spectrochim. Acta*, 1958, **II**, 579.
46. Katritzky and Hands, *J. Chem. Soc.*, 1958, 2202.
47. Katritzky, Hands and Jones, *J. Chem. Soc.*, 1958, 3165.
48. Katritzky and Gardner, *J. Chem. Soc.*, 1958, 2198.
49. Podall, *Analyt. Chem.*, 1957, **29**, 1423.
50. Schmid and Joeckle, *Spectrochim. Acta*, 1966, **22**, 1645.
51. Goya, Takanishi and Okano, *J. Pharm. Soc. Japan*, 1966, **86**, 952.
52. LaFaix and Lebas, *Spectrochim. Acta*, 1970, **26A**, 1243.
53. Salisbury, Ryan and Mason, *J. Heterocyclic Chem.*, 1967, **4**, 431.
54. Sobolev, Aleksanyan, Karakhanov, Bel'skii, and Ovodova, *J. Struct. Chem., USSR*, 1963, **4**, 330.
55. Jones, *Advances in Heterocyclic Chemistry* (Academic Press, New York, 1970), Vol. 11, p. 443.
56. Slack and Wooldridge, *Advances in Heterocyclic Chemistry*, Vol. 4 (Academic Press, New York, 1965), p. 107.
57. Zerbi and Alberti, *Spectrochim. Acta*, 1962, **18**, 407; 1963, **19**, 1261.
58. Borello and Zecchina, *Spectrochim. Acta*, 1963, **19**, 1703.
59. Curtin and Alexandrou, *Tetrahedron*, 1963, **19**, 1679.
60. Rao and Venkataraghaven, *Can. J. Chem.*, 1964, **42**, 43.

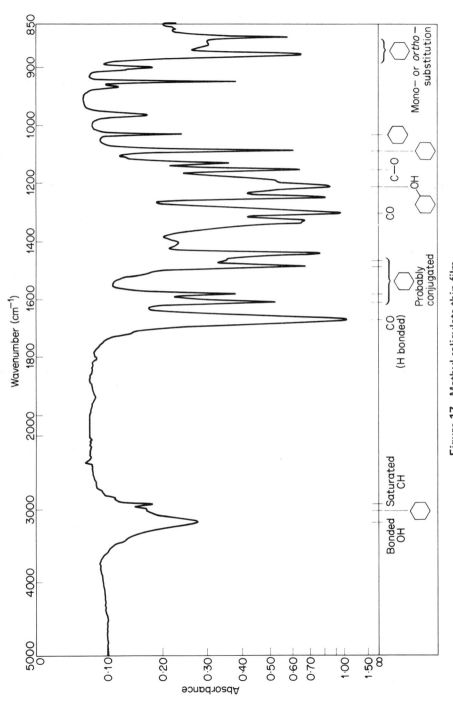

Figure 17. Methyl salicylate thin film

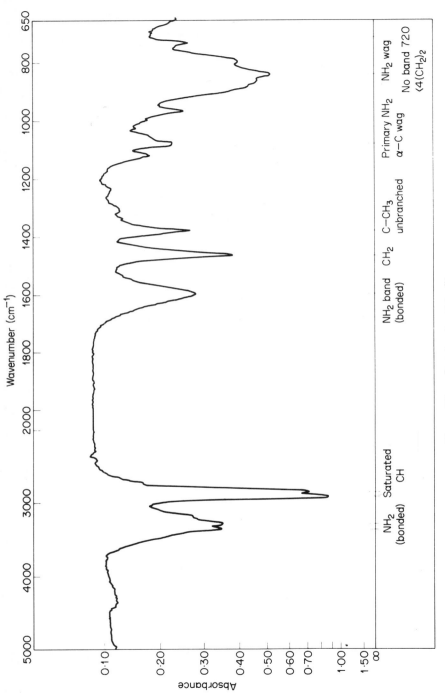

Figure 18. *n*-Butylamine thin film

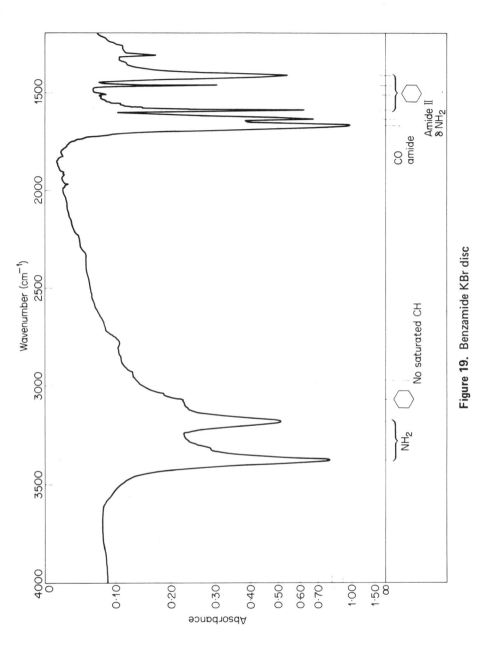

Figure 19. Benzamide KBr disc

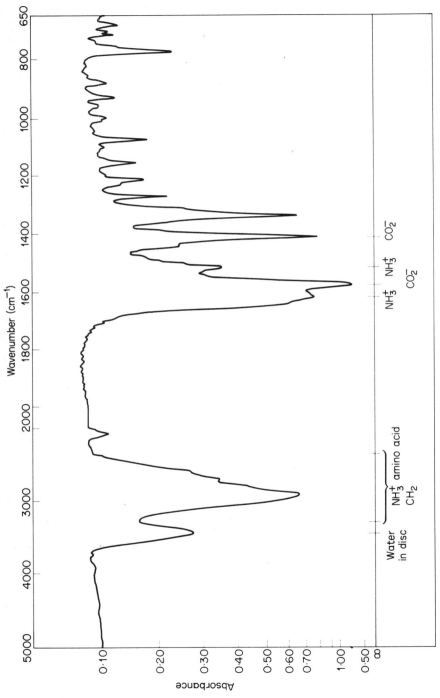

Figure 20. Methionine KBr disc

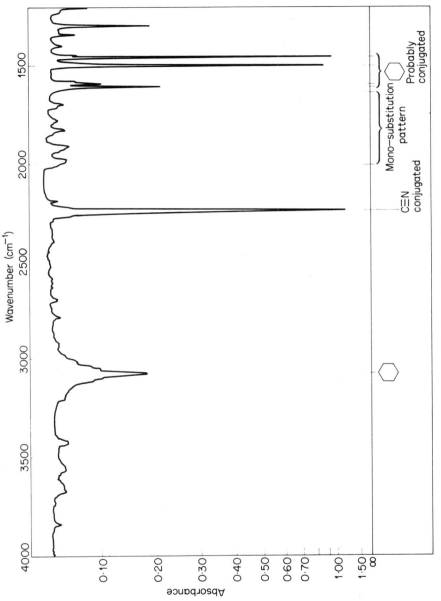

Figure 21. Benzonitrile thin film

325

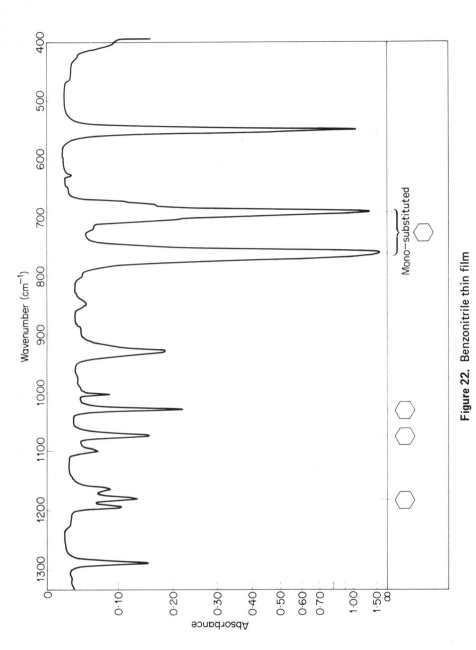

Figure 22. Benzonitrile thin film

326

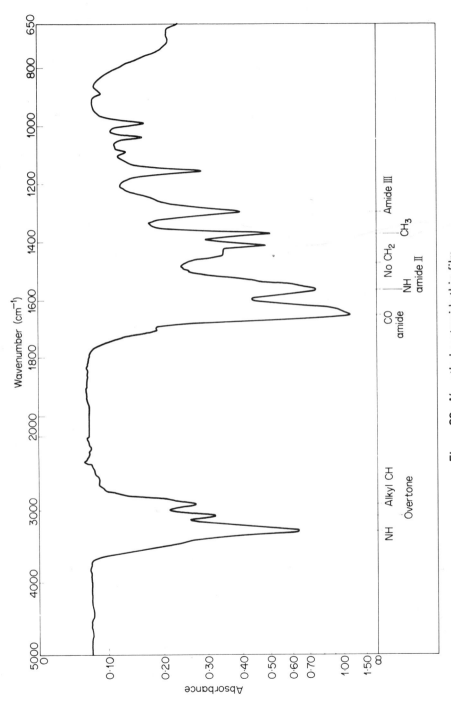

Figure 23. *N*-methyl acetamide thin film

327

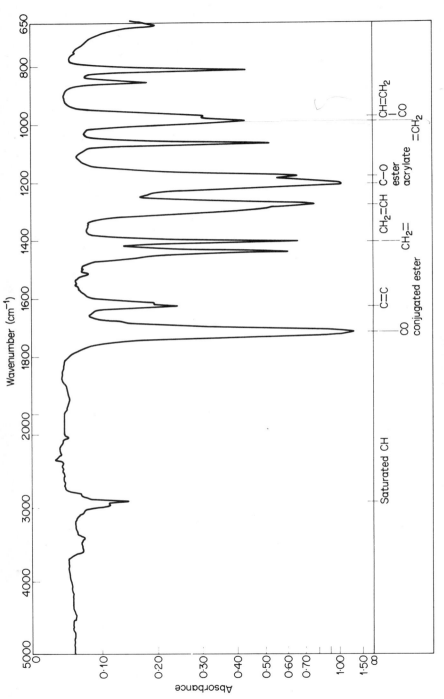

Figure 24. Methyl acrylate thin film

328

Part Four | Vibrations involving other elements; groups of inorganic origin

17

Nitro- and Nitroso-Compounds, Nitrates and Nitrites

17.1. Introduction and table

In this chapter compounds containing the grouping $R-NO_2$, $R-NO$, or NO_3^- are considered, together with the amine oxides. In compounds such as the oximes, in which the $N-O$ link has a purely single bond character, the frequency which arises [34] is variable in position and the vibration is probably coupled. In these circumstances this particular nitrogen-oxygen frequency has not been considered further.

Compounds containing the $R-NO_2$ group are covalent nitro-compounds, covalent nitrates and nitramines. Each of these three classes exhibits two extremely strong absorption bands in the $1650-1500$ cm^{-1} and $1350-1250$ cm^{-1} regions corresponding to the asymmetric and symmetric stretching valence vibrations of the NO_2 group. The frequencies at which these bands occur are sensitive to changes in the nature of the R group, so that, although there is some overlapping, it is often possible to differentiate between the three main classes. In addition, there is some evidence of characteristic absorptions occurring in the low-frequency ranges due to skeletal or deformation modes.

The valence vibration bands are not materially affected by changes of state, except in a few cases in which hydrogen bonding on to the NO_2 group can occur. The shifts involved are not in any case sufficient to take the absorption out of the ranges quoted. In the low-frequency region, however, very considerable changes often occur with nitro-compounds following changes of state or of crystal form, so that considerably less reliance can be placed upon any results from this region.

With compounds containing the R—N=O group only a single band in the 1600 cm^{-1} region would be expected, but in nitrites this is almost always double, due to the co-existence of *cis*- and *trans*-forms. By a study of the relative intensities of these two bands it is often possible to obtain some data on the nature of the hydrocarbon residue R in R—O—N=O structures. In compounds in which the nitroso-group is attached directly to carbon or nitrogen, the position of the N=O absorption has been more difficult to determine, partly owing to the ease of dimerisation of certain compounds of this class. However, the appropriate group frequencies now appear to be reasonably well defined. In nitrites two additional useful frequencies are found in the low-frequency region, but this is not so in the case of nitroso-compounds.

Ionic nitrates have a strong band centred near 1380 cm^{-1} and a medium intensity band near 800 cm^{-1} which enable them to be characterised.

The correlations discussed are listed in Table 17 below.

Table 17

RNO$_2$	CNO$_2$	N=O	stretching bands.	
		Alkyl	Primary α-carbon	1554 $\mp$ 6 and 1382 $\mp$ 6 cm^{-1} (vs)
			Secondary α-carbon	1550 $\mp$ 3 and 1370 $\mp$ 4 cm^{-1} (vs)
			Tertiary α-carbon	1539 $\mp$ 5 and 1349 $\mp$ 5 cm^{-1} (vs)
			1,1-dinitro	1575 $\mp$ 12 and 1332 $\mp$ 5 cm^{-1} (vs)
			1.1.1-trinitro	1600 $\mp$ 3 and 1302 $\mp$ 5 cm^{-1} (vs)
		Aryl C—N	1555—1487 and stretching band	1357—1318 cm^{-1} (both vs)
			Alkyl and aryl	875—830 cm^{-1} (ms)
	ONO$_2$	N=O	stretching bands	1640—1620 and 1285—1270 cm^{-1} (vs)
	NNO$_2$	N=O	stretching bands	1630—1550 and 1300—1250 cm^{-1} (vs)
RNO	ON=O		*cis*	1625—1610 cm^{-1} (s)
			trans	1681—1653 cm^{-1} (s)
	NN=O			1460—1430 cm^{-1} (s)
	SN=O			Near 1534 cm^{-1} (s)
	CN=O		Monomer	1600—1500 cm^{-1} (s)
			Dimers *trans*	1290—1190 cm^{-1}
			cis	1425—1370 cm^{-1}
			Amine Oxides	1300—1200 cm^{-1}
NO$_3$$^-$				1410—1340 and 860—800 cm^{-1}

17.2. Nitro-compounds

17.2(a). The group C—NO$_2$. Barnes *et al.* [1] noted in 1944 that the NO$_2$ grouping gives rise to absorptions in the 1550 cm^{-1} and 1340 cm^{-1} regions, and this correlation was confirmed by Raman data [2] and by studies on the simpler nitro-paraffins [3, 4]. The correlation was extended by Colthup [5], who indicated that the C—NO$_2$ compounds absorbed at different frequencies from those containing the O—NO$_2$ group. In recent years this correlation has been placed on a much sounder footing by detailed studies on the influence of environment on the vibrations of the CNO$_2$ group, and the effects of conjugation, α-halogen substitution and other changes of environment are now known.

Early work on simple alkyl nitro-compounds [3, 4] indicated that whilst nitromethane absorbed at 1580 cm^{-1} and 1375 cm^{-1}, higher homologues showed the asymmetric absorption at somewhat lower frequencies. This has been fully substantiated by later studies. Haszeldine [24] repeated the earlier work of Smith [4] *et al.* and added some further compounds. The overall ranges for simple alkyl derivatives were 1567—1550 cm^{-1} (*as.*) and 1379—1368 cm^{-1} (*s.*). These values probably relate to the liquid state, but changes on solution are relatively small in this series. Brown [25], and Kornblum Ungnade and Smiley [26] have each examined about thirty-five alkyl nitro-compounds with essentially similar results. More recently, Lunn [59] has studied the effects of substitution at the α-carbon atom, and Geiseler *et al.* [60, 61] have reviewed the infra-red and Raman data. Additional Raman data have been provided by Popov *et al.* [62] and Slovetskii *et al.* [63].

In most cases there are sufficient differences between the frequencies of primary, secondary and tertiary nitro compounds to enable them to be differentiated. Slovetskii *et al.* [63] suggest the following ranges, based on Raman data.

Primary	ν_{as}	1554 ∓ 6	ν_s	1382 ∓ 6 cm^{-1}
Secondary		1550 ∓ 3		1370 ∓ 4
Tertiary		1539 ∓ 5		1349 ∓ 5
gem dinitro		1575 ∓ 12		1332 ∓ 5
1 : 1 : 1-trinitro		1600 ∓ 3		1302 ∓ 5

This is generally in accord with the results of other authors but some caution is needed in the use of the lower frequency symmetric mode as this is much more prone to coupling than is the antisymmetric frequency and shifts out of the listed range may be found. This is well shown by the fact that the antisymmetric frequencies show good regularities with Taft σ values [59, 64] whereas the symmetric mode does not. Popov *et al.* [62] have also noted that ν_{as} is more stable in frequency and in intensity than is ν_s. For this same reason there is no overall relationship between the frequencies of these two bands, although these do exist for separate limited series. In primary and secondary alkyl nitro compounds for example $\nu_s = 3261 - 1.213\,\nu_{as}$ and $\nu_{as} = 21.37\,\sigma^* + 1550.7$ but this does not apply to tertiary compounds where different relationships hold [59]. A further complication is the splitting which sometimes occurs in ν_s. Eckstein *et al.* [65] find that a methyl group at the α-carbon atom causes ν_s to double with the total intensity equally divided between the two bands.

Conjugation of the nitro-group by attachment to an ethylenic double bond leads to a fall in both frequencies similar to that shown by the carbonyl absorption. Shechter and Sheppard [27], for example, find that 2-methyl-1-nitropropene absorbs at 1515 cm^{-1} and 1350 cm^{-1}, whereas 2-methyl-3-nitropropene absorbs at 1555 cm^{-1} and 1366 cm^{-1}. This has been confirmed also by Brown [25], who finds the ranges 1524 ± 4 cm^{-1} and 1353 ± 6 cm^{-1} for monoalkyl nitroethylenes and slightly lower values of 1515 ± 4 cm^{-1} and 1346 ± 9 cm^{-1} for di- and tri-alkylnitroethylenes.

α-Halogen substitution has a marked effect upon nitro-group frequencies, which are similar in this respect to carbonyl compounds. In this case, however, whilst the asymmetric frequency is raised considerably the symmetric frequency is lowered by about the same amount. In extreme cases this leads to the appearance of these bands in the region normally associated with $-O-NO_2$ groupings. The effect is, of course, most marked in the fully halogenated nitromethanes. Chlorpicrin has been studied by several workers [8, 24, 28] and absorbs at 1625 cm^{-1} and at 1311 cm^{-1}. In the trifluoro-derivative it is interesting to note that the shifts are slightly less [28, 29] (1620 cm^{-1}, 1315 cm^{-1}), whereas a further rise might have been expected. In the tribromo-derivative [28] the shift is again smaller (1606 cm^{-1} and 1311 cm^{-1}). The effects of mono- and di-halogen substitution are parallel. Both Brown [25] and Haszeldine [24] have examined a number of compounds of this type, and it would seem that a single α-chlorine atom alters the frequencies to 1575 ± 5 cm^{-1} and 1348 ± 6 cm^{-1}, while two α-chlorine atoms lead to ranges of 1587 cm^{-1} ± 10 cm^{-1}, and

$1332 \text{ cm}^{-1} \pm 5 \text{ cm}^{-1}$. Although fewer data are available on nitro-compounds with other α-electronegative groups, it would seem likely that similar effects will occur.

Aromatic nitro-compounds have been studied by many workers, and although there has been a good deal of overlapping in the compounds studied, there has been some disagreement over the mean frequencies to be expected. Thus Francel [6] places these bands near 1530 cm^{-1} and 1360 cm^{-1}, and this is supported by the data of Lothrop et al. [7]. Brown [25] lists the frequencies of twelve compounds as falling in the ranges $1527 \pm 16 \text{ cm}^{-1}$ and $1348 \pm 11 \text{ cm}^{-1}$, Randle and Whiffen [30] quote average values (twenty-three compounds) of 1518 cm^{-1} and 1349 cm^{-1} and Kross and Fassel [31] (thirty-four compounds) 1523 cm^{-1} and 1344 cm^{-1}. The latter workers have also shown that the frequency shifts which accompany a change of state from the solid to solution in a non-polar solvent are very small. These differences between average values are not real and merely reflect the sensitivity of the nitro-group frequencies to the other ring substituents, so that the mean values obtained depend to a large extent on the types of compounds studied. In p-dinitrobenzene, for example, the as. frequency is 1560 cm^{-1}, whereas in sodium p-nitrophenoxide [31] it is 1501 cm^{-1}. Similarly, replacement of the methyl group of nitrotoluene by the electron-donating $N(CH_3)_2$ group alters the NO_2 group frequencies [32] from 1527 cm^{-1} and 1350 cm^{-1} to 1506 cm^{-1} and 1332 cm^{-1}. Franck, Hormann and Scheibe [33] and Borek [66], have recently related the observed asymmetric stretching frequencies to broad classifications of substituents. They assign normal and coplanar aromatic nitro-group frequencies to the range $1548-1520 \text{ cm}^{-1}$. This is the expected range, being somewhat higher than the corresponding compounds with ethylenic conjugation. The presence of a strong electronegative group in the para-position, or of a large group in the ortho-position which tends to throw the nitro-group out of the plane of the ring, leads to higher frequencies in the range $1565-1540 \text{ cm}^{-1}$ whereas electron donor groups in the ortho- or para-positions result in lower frequencies $1525-1490 \text{ cm}^{-1}$. The symmetric absorption usually falls in the $1350-1340 \text{ cm}^{-1}$ range for planar nitro groups [66] but is again subject to coupling, so that its position is not directly linked to that of the antisymmetric mode. Thus only the latter gives a good linear relation with the Hammett σ values of the substituents [64, 68]. Rao et al. [68] give this as $\nu_{as} = 1523 + 30.5 \, \sigma$.

Nitro groups in heteroaromatics such as the pyridines behave in exactly the same way as other aromatics [67]. In compounds containing two or more nitro-groups which, because of their positions, are influenced to different extents by the substituents, multiple frequencies are found. In 1 : 3-*bis*methylamino-2 : 4 : 6-trinitrobenzene [33], for example, absorptions occur at 1554 cm^{-1} and 1538 cm^{-1}, and weak peaks also occur at 1508 cm^{-1} and 1493 cm^{-1}. Multiple peaks also occur when steric effects cause a nitro group to turn out of the plane of the ring so that it shows a higher frequency [69]. Yamaguchi [70] has studied this effect and related the changes in ν_{as} to the alterations in the angle of twist.

Salts of alkyl nitro compounds exist as ions in which there is extensive delocalisation and in which the C—N bond has considerable double bond character [71, 72]. The N—O bonds give antisymmetric and symmetric frequencies in the ranges 1315—1205 and 1175—1040 cm^{-1}. The bands are usually doubled due to crystal splitting [72]. The salts of aromatic nitro compounds have been less studied but Ezumi *et al.* [73] have given some data for nitrobenzene and nitropyridine which suggest that these absorb at rather higher frequencies.

One other possible correlation for alkyl nitro-compounds relates to the CH_2 and CH_3 frequencies of adjacent methyl or methylene groups. Kornblum *et al.* [26] have noted a band at 1379 cm^{-1} in all the primary nitro-compounds they have studied, even when a methyl group is absent. This band is additional to the NO_2 frequency itself, and almost certainly originates in the CH_2 deformation frequency of the perturbed methylene group. This is similar to the characteristic frequency shown by the group CH_2COOH in acids. In secondary alkyl nitro-compounds several bands occur in this region in addition to the nitro-group frequency itself. Brown [25] has also made some observations in this region. He finds that in simple nitro alkanes the symmetrical NO_2 stretching frequency and the asymmetrical methyl deformation frequency are superimposed at 1379 cm^{-1}. However, when the methyl group is attached to the same carbon atom as the nitro-group two frequencies appear at 1395 cm^{-1} and 1370 cm^{-1}.

Similarly, the structure $(CH_3)_2\overset{|}{C}NO_2$ is characterised by strong bands at 1397 cm^{-1}, 1374 cm^{-1} and at 1351 cm^{-1}. This is no doubt the explanation of the doubling of the nitro group band at 1350 cm^{-1} reported by Eckstein *et al.* for α-methyl nitro compounds [65].

17.2(b). Covalent nitrates —O—NO$_2$. Raman evidence [3] indicates that in covalent nitrates the asymmetric and symmetric NO$_2$ frequencies are split farther apart and occur near 1640 cm^{-1} and 1260 cm^{-1}. Only a limited amount of information is available, but some results have been published on nitrocellulose [9] and on nitroglycerine [10, 23] which are in line with this and with the early studies of Lecomte and Mathieu [35] on simple alkyl nitrates. Kumler [36] has quoted values of 1656 cm^{-1} for the asymmetric stretching frequency in some nitrates of enol esters. Brown [25] has examined twenty-one nitrate esters and finds the characteristic stretching frequencies in the ranges 1639 ± 13 cm^{-1} and 1279 ± 7 cm^{-1}. These results have been fully confirmed by Carrington [74] and by Guthrie and Spedding [75], who find essentially the same ranges. Carrington has also studied the intensities of both bands, and points out that primary nitrates can be distinguished from secondary ones by the fact that the latter always show ν_s as a doublet of bands of equal intensity. This band is also split in cyclic nitrates in most cases.

17.2(c). Covalent nitramines —N—NO$_2$. The nitramines show the asymmetric NO$_2$ frequency at much the same position as the C—NO$_2$ compounds, whereas the symmetric absorption is displaced towards lower frequencies. In this case, however, the nature of the adjoining structure appears to have a more marked influence on the frequency range, perhaps due to the greater possibilities of changes in the electronic structure of the nitrogen atoms to which the nitro-group is attached. Lieber *et al.* [11] have examined seventeen *N*-nitro-compounds of various types. In fourteen of these the symmetric NO$_2$ vibration is reasonably constant within the range 1315–1260 cm^{-1}, whilst the exceptions (which absorb at higher frequencies) are either acid or salt forms in which the structure $N\overset{+}{=}N\overset{-}{=}O_2$, might be expected to have a marked influence. The asymmetric frequencies, on the other hand, fall in a wider range, and appears to be more influenced by the nature of the substituents. Nitroguanidine and related compounds with alkyl substituents absorb between 1634 cm^{-1} and 1605 cm^{-1}, whilst arylguanidines and nitrourea absorb in the range 1587–1575 cm^{-1}. Polynitramines and salts of 5-nitroaminotetrazole absorb between 1563 cm^{-1} and 1547 cm^{-1}. Kumler [37] has also found values between 1655 cm^{-1} and 1620 cm^{-1} for the asymmetric frequency in nitroguanidines. Salyamon and Yaroslaviskii [38] have shown that nitramines exist preferentially in the form $RNHNO_2$ rather than in the alternative $RN=NOOH$ structure, from studies in the overtone region, and the same laboratories [39] have made Raman studies on nitramine salts which show the symmetric stretching frequency rises to near 1400 cm^{-1} in these compounds.

Our own experience is confined to polynitramines of various types (ten in all) which follow the above classification and show the NO_2 absorptions in the ranges $1587-1530 \text{ cm}^{-1}$ and $1292-1260 \text{ cm}^{-1}$. This is presumably the region of absorption for nitramines generally, except for the cases of nitroguanidines, and ureas in which alternative prototropic forms are possible which might absorb at higher frequencies.

17.2(d). The intensities of the $R-NO_2$ stretching absorptions.

In all the above cases the two valence vibration absorptions of the NO_2 group are extremely strong bands. However, there are considerable variations in the absolute intensity from compound to compound, and the increase in the intensities on passing from mono-nitro-aromatics to di- and poly-nitro-materials is not linear with the molar NO_2-group concentration. This applies even to closely related compounds, so that, for example, the extinction coefficient of each nitro-group is about 80 per cent higher in diethylene glycol dinitrate than in nitroglycerine [10]. The intensities of these absorptions cannot therefore be used for determining the proportion of NO_2 groups in unknown compounds. In general, the asymmetric absorption is appreciable more intense than the symmetric band [74]. This is similar to the case of the ionised carboxyl group with which the nitro-group is isoelectronic. However, this is not an invariable rule, and in some instances both bands appear to be of comparable intensity. Marked alterations in the band shapes of both nitro-group absorptions occur in situations in which the nitro-group is subject to steric hindrances. In *ortho*-nitrotoluene, for example, the half-band width of the symmetric absorption is a little more than twice that of the corresponding band in *p*-nitrotoluene [40]. This may have useful diagnostic applications, although the absolute intensities of the two bands as determined from area measurements show very little change.

17.2(e). Low-frequency absorptions of the RNO_2 group.

Absorption bands in the low-frequency region are to be expected from skeletal vibrations and from deformation modes involving the RNO_2 group.

In alkyl nitro-compounds the C–N absorption has been tentatively identified by Haszeldine [24] with a strong to medium absorption in the range 920–830 cm^{-1}. Randle and Whiffen [30] have also suggested that the C–N mode in aromatic nitro-compounds may be responsible for a commonly found band near 850 cm^{-1}. This suggestion has been discussed further by Brown [24] and by Kross and Fassel [31]. More recently Green [76] has assigned a strong infra-red band in the range 874–837 cm^{-1} to the C–N stretch in alkyl nitro compounds, other than nitromethane where it falls at 818 cm^{-1}. This band appears at 852 cm^{-1} in nitrobenzene as suggested by earlier workers.

The most detailed study is that of Geiseler and Kessler [60] using both infra-red and Raman spectra. They quote a wider overall range 910–840 cm^{-1}, but this is due to the fact that the band is doubled due to the presence of *trans* and *gauche* conformers. The *trans* form absorbs between 914–868 cm^{-1} and the gauche between 890–845 cm^{-1}.

The NO_2 group also has three bending modes, which occur below 700 cm^{-1}. The out of plane bending frequency falls between 670 and 610 cm^{-1}, and the symmetric deformation frequency between 630–610 cm^{-1}. The latter is very clearly seen in the Raman spectra where it is a good group frequency for the CNO_2 group. These bands have been little studied in nitrates or nitramines.

17.3. R–N=O Stretching vibrations

In compounds such as nitrosyl chloride and bromide [14] the N=O frequency falls near 1800 cm^{-1}, but this is clearly higher than the normal value, due to the shortening of the N=O distance under the influence of the halogen atom. In nitrites the N=O valence vibration occurs near 1660 cm^{-1}. Nitroso-compounds show rather more variable N=O frequencies, but when monomeric usually absorb in the 1600–1500 cm^{-1} range. Nitrosamines absorb at still lower frequencies, usually below 1500 cm^{-1}.

17.3(a). Nitrites —O—N=O. Tarte [12, 13, 22] has shown that in fifteen nitrites which he has examined the N=O frequency appears as a double band in the ranges 1681–1653 cm^{-1} and 1625–1613 cm^{-1}. These are attributed respectively to the *trans*- and *cis*-forms of the nitrite structure. The frequencies show a remarkably steady stepwise fall as the size of the attached group is increased, the highest being with methyl nitrite at 1681 cm^{-1} and 1625 cm^{-1} and the lowest with amyl nitrite at 1653 cm^{-1} and 1613 cm^{-1}. These ranges have been confirmed by similar studies by Haszeldine *et al.* [41, 42] on a small number of alkyl nitrites. In the case of 2 : 2 : 2-trifluoroethyl nitrite, the absorption is shifted to 1736 cm^{-1} and 1695 cm^{-1}, which is a remarkable shift in view of the distance along the chain of the fluorine atoms from the N=O link. The effect is probably due to the steric arrangement, and somewhat related phenomena have been observed in certain β-fluorinated esters.

The overall relative intensities of these bands afford a clear indication of the type of substitution involved, since the proportions of *cis*- and *trans*-forms vary between the different classes of nitrite. Thus, on determining the ratio of the extinction coefficients of the *trans*- and *cis*-forms, the original value of 1 : 1 in methyl nitrite rises to 2 : 3 in ethyl nitrite and reaches a stable value of 3 : 3.5 in the higher primary nitrites. In the secondary nitrites, however, this ratio is increased to 6 : 10 and in tertiary nitrites, in which the proportion of the *cis*-form is necessarily much reduced, it increases again to 40 : 1.

The intensities in this region are considerable, and the nitrite bands are described as having an intensity rarely met with in the infra-red.

17.3(b). Nitroso-compounds >C—NO. The assignment of the —N=O stretching frequency in nitroso-compounds has been a matter of some difficulty, which arises mainly from the ease with which primary and secondary nitroso-compounds pass over into the oximes, and from the fact that the tertiary compounds dimerise readily. Thompson, Nicholson and Short [15] originally suggested that this frequency fell in the 1400—1300 cm^{-1} range, whilst Brownlie [16] assigned it as being near 1650 cm^{-1} on the basis of some nitrosopyrimidenes which he examined. More recently this problem has been studied extensively by Lüttke [43—45] and by Tarte [46], and the situation is becoming clearer.

Lüttke has studied the changes in the spectrum which occur with time when primary and secondary nitroso-compounds are volatilised. In the cases of nitrosomethane and nitroso*cyclo*hexane, for example, absorptions appear at 1564 cm^{-1} and 1558 cm^{-1} which can be associated with the —N=O frequency. These bands rapidly diminish in intensity on standing and new bands appear, so that in a relatively short time the spectra are those of formaldoxime and of *cyclo*hexanone oxime respectively. In solution these compounds exist preferentially as dimers, and nitrosomethane, for example, absorbs near 1290 cm^{-1} in this form.

In tertiary nitroso-compounds $R_1 R_2 R_3 CN{=}O$ the possibilities of oxime formation are eliminated. In these compounds also dimerisation is common in the solid state or in solution, but it is usually possible to identify the monomeric frequencies in dilute solution and in the vapour state. However, many of the compounds of this type studied by Tarte [46] and by Lüttke [32] contain one or more halogen atoms as the R groups, and this leads to a further complication in that field effects similar to those found in α-halogenated ketones are to be expected, which depend upon the molecular configuration. Doubling of the N=O absorption is therefore found in mono- and di-α-halogenated compounds, and the identification of this with the rotational isomerism has been confirmed. This effect also leads to an elevation of the frequency as in the corresponding carbonyl compounds. Nitroso*cyclo*hexane, for example, absorbs at 1558 cm^{-1}, whilst 1 : 4-dichloro-1 : dinitroso*cyclo*hexane absorbs [43] at 1570 cm^{-1}. In general, the N=O absorption appears to be particularly sensitive to the nature of its environment, falling as low as 1495 cm^{-1} in 2 : 4 : 6-trimethylnitrosobenzene and rising to 1620 cm^{-1} in $CCl_3 NO$. In this latter compound it is interesting to note that the frequency is higher than that of the corresponding trifluoro-derivative [29, 47] which suggests that induction and field effects are not the only factors operating in this case and that mesomerism involving double-bonded halogen links also plays an important part. In this connection it is perhaps significant that, unlike the carbonyl group in similar circumstances, the N=O group frequency rises steadily as more and more halogen atoms are attached at the α-position.

In general, aromatic nitroso-compounds absorb near 1500 cm^{-1}, tertiary aliphatic materials at higher frequencies near 1550 cm^{-1} and α-halogenated compounds at still higher frequencies up to 1620 cm^{-1}. In the latter cases the bands are frequently doubled due to rotational isomerism. The $\overset{\diagdown}{\underset{\diagup}{C}}{-}N{=}O$ characteristic frequency would seem therefore to be reasonably well established by these studies. An alternative suggestion for this group frequency involving the 1380–1340 cm^{-1} region has been put forward by Nakamoto and Rundle [48], but this almost certainly relates to the dimer rather than the monomer form. Gowenlock *et al.* [49] have studied dimeric tertiary nitroso compounds and assigned the N=O stretch to $1426-1370$ cm^{-1} when the two oxygen atoms are *cis* with respect to each other and $1290-1190$ cm^{-1} when they are *trans*.

17.3(c). Nitrosamines —N—N=O.

Early studies by Earl *et al.* [17] suggested that the N=O absorption in *N*-nitroso-compounds occurred near 1400 cm^{-1}. This was supported by Raman data [18], but alternative assignments in the $1600-1500$ cm^{-1} region were suggested for nitrosoguanidines and similar compounds studied by Lieber *et al.* [11]. Although the total number of compounds of this type which have been studied remains relatively small, the positions of the characteristic N=O absorption is now clearer.

Haszeldine *et al.* [41, 42, 50] and Tarte [46, 50] have each examined small numbers of these compounds in various states, and although the earlier studies [41] gave rise to some doubt as to whether dimerisation occurred in the liquid state, this has now been satisfactorily resolved [50]. There is general agreement that in the monomeric state the N=O frequency of dialkylnitrosamines occurs near 1490 cm^{-1} in the vapour. In solution in carbon tetrachloride this falls to near 1450 cm^{-1}, and a small further fall occurs in the liquid state, but dimerisation does not occur. No data are available on α-fluorinated compounds, but it is interesting to note that *N*-nitroso-*bis*-2-2-2-trifluoroethylamine shows its N=O frequency in the vapour state [42] at 1550 cm^{-1}. This is a considerable shift for a situation in which the fluorine atoms are so far removed from the N=O link, but it may be that the geometric arrangement of the molecule leads to their being relatively close in space despite the number of intervening linkages. Elevated $\diagdown$N—N=O frequencies have also been reported in the case of amides which are nitrosated on the nitrogen atom. White [51] has studied a number of compounds of this type and assigned the N=O absorptions to the $1527-1515$ cm^{-1} region. These materials also show elevated carbonyl frequencies corresponding to esters rather than to amides, so that it is quite possible that considerable dipolar interactions are taking place in this series also. The higher frequencies for nitrosoguanidines suggested by Lieber *et al.* [11] may also arise by a similar mechanism.

17.3(d). Other characteristic absorptions of R—N=O groups.

In the lower-frequency regions a few bands have been reported which arise primarily from skeletal frequencies but which may be useful in some cases for confirmatory evidence. Tarte [12, 13, 22] for example, has made some useful suggestions on alkyl nitrites.

All the fifteen nitrites he has examined exhibit absorption in the 814–751 cm^{-1} range which he ascribes to the fundamental ν_{N-O} vibration, and there is also a pair of bands in the 691–617 cm^{-1} and 625–565 cm^{-1} ranges which is assigned to the $-O-N=O$ deformation of *cis-* and *trans-*forms. Despite the rather wide frequency ranges typical of skeletal vibrations, these assignments may well be valuable in confirming the presence of nitrite groups, and in both cases the intensities are reasonably high. In addition, bands corresponding to combinations of these fundamentals have been characterised at 2300–2250 cm^{-1} ($\delta_{ONO} + \nu_{N=O}$), near 2500 cm^{-1} ($\nu_{N-O} + \nu_{N=O}$) and the first harmonic of $\nu_{N=O}$ at 3300–3200 cm^{-1}. In the case of nitrosoamines he has identified [46] related absorptions corresponding to the N–N stretching and N–N=O deformations. These occur near 1050 cm^{-1} and 660 cm^{-1} respectively. The second is therefore not much altered from the corresponding O–N=O mode. The presence of the former absorption has also been confirmed by Haszeldine in some compounds [41, 42].

No other useful absorptions have been identified in the cases of $-C-N=O$ compounds, although the C–N stretching mode has been associated with a strong diffuse band near 1100 cm^{-1}.

17.3(e). Amine oxides. Amine oxides also contain a partially multiple bonded N–O system, and can conveniently be considered under this heading. Costa [52–54] and his co-workers have studied this absorption in pyridine *N*-oxide and a variety of substituted derivatives, so have Shindo [77] and Katritzky *et al.* [78–80]. The results have been reviewed by Katritzky [81]. In pyridine oxide itself the N–O stretch appears at 1250 cm^{-1} in non-polar solvents, but this appears to be very sensitive to the electrical character of substituent groups. Strong electron donors lower this frequency appreciably and *p*-methoxy-*N*-pyridine oxide [52] absorbs at 1238 cm^{-1}. *p*-Hydroxy- and amino-groups are even more effective [53] and the band in the latter case appears to be near 1200 cm^{-1}. With strong electron-accepting substituents the effect is reversed, and in the *p*-nitro-derivative [52] the N → O absorption occurs at 1304 cm^{-1}. A small number of tertiary amine oxides have also been studied by Mathis-Noel., Wolf and Gallais [5]. In this series the N → O band is assigned to the 970–950 cm^{-1} region. This is a very considerable shift from the corresponding assignment in the pyridine series, but the absorption would be expected to move to lower frequencies when the multiple bond character is minimised.

The *N*-oxides of pyrimidines have also been examined and compared with the pyridine *N*-oxides by Wiley and Slaymaker [58] and by Katritzky [81]. The $N \rightarrow O$ absorptions occur in the 1300–1255 cm^{-1} range and show the same sensitivity to substituent effects as the pyridine oxides. Pyrazine oxide absorbs at 1318 cm^{-1}.

Compounds related to amine oxides occur in the azoxy series

$$\overset{\text{O}}{\overset{\uparrow}{}}$$

which contain the linkage $-\text{N}=\text{N}-$. Witkop and Kissman [56] have examined a small number of such compounds and noted a common absorption band in the 1310–1250 cm^{-1} region. They associate this with the $N \rightarrow O$ link, but are careful to point out that the aromatic C–N stretching absorption is to be expected in the same region. Jander and Haszeldine [57] have studied hexafluoroazoxymethane, which absorbs at 1282 cm^{-1} and 1256 cm^{-1}. In view of the close parallel with the amine oxides, this would therefore appear to be a likely assignment for the $N \rightarrow O$ absorption is azoxy-compounds.

17.4. Ionic nitrates

The nitrate ion has four fundamental modes of vibration, all of which are observable in the infra-red in suitable cases, but only two are normally sufficiently intense to be used for identification purposes. These are the ν_3 and the ν_2 vibrations, which absorb near 1390 cm^{-1} and 800 cm^{-1}.

Strong bands near these frequencies are shown by inorganic nitrates such as ammonium, thallium, barium and lead nitrates [19, 20], whilst the 1390 cm^{-1} absorption has also been observed in many Raman spectra [21]. With ionic organic nitrates the bands occur [5] in the overall ranges 1410–1340 cm^{-1} and 860–800 cm^{-1}, and this is supported by the spectra of a number of compounds, such as guanidine nitrate, examined by Lieber *et al.* [11]. The high-frequency band is always considerably stronger than the other, which is rarely more than of medium intensity.

In Raman work on inorganic solutions it has been shown [21] that the ν_3 vibration splits into two distinct bands at high concentrations, giving frequencies up to 1500 cm^{-1} on one side (Th) and down to 1315 cm^{-1} on the other (Al). This is the basis for a suggestion that there is a change of form of the ionic nitrate group at such concentrations. However, the frequencies do not fall outside the $1410-1345$ cm^{-1} range until extremely high concentrations are reached, and it is extremely doubtful whether any serious departure from this correlation will arise in organic nitrates from any similar cause.

17.5. Bibliography

1. Barnes, Gore, Liddel and Williams, *Infra-red Spectroscopy* (Reinhold, 1944).
2. Hibben, *The Raman Effect and its Chemical Applications* (Reinhold, 1939).
3. Nielsen and Smith, *Ind. Eng. Chem. (Anal.),* 1943, **15**, 609.
4. Smith, Pan, and Nielsen, *J. Chem. Phys.,* 1950, **18**, 706.
5. Colthup, *J. Opt. Soc. Amer.,* 1950, **40**, 397.
6. Francel, *J. Amer. Chem. Soc.,* 1952, **74**, 1265.
7. Lothrop, Handrick and Hainer, *ibid.,* 1951, **73**, 3581.
8. Randall, Fowler, Fuson and Dangl, *Infra-red Determination of Organic Structures* (Van Nostrand, 1949).
9. Nikitin, *Zh. Fiz. Khim.,* 1949, **23**, 786.
10. Pinchas, *Analyt. Chem.,* 1951, **23**, 201.
11. Lieber, Levering and Patterson, *Analyt. Chem.,* 1951, **23**, 1594.
12. Tarte, *Bull. Soc. Chim. (Belg.),* 1951, **60**, 227.
13. *Idem, ibid.,* p. 240.
14. Burns and Bernstein, *J. Chem. Phys.,* 1950, **18**, 1669.
15. Thompson, Nicholson and Short, *Discuss. Faraday Soc.,* 1950, **9**, 222.
16. Brownlie, *J. Chem. Soc.,* 1951, 3062.
17. Earl, Le Févre, Pulford and Walsh, *ibid.,* 1951, 2207.
18. Barredo and Goubeau, *Z. anorg. Chem.,* 1943, **251**, 2.
19. Newman and Halford, *J. Chem. Phys.,* 1950, **18**, 1276.
20. *Idem, ibid.,* p. 1291.
21. Mathieu and Lounsbury, *Discuss. Faraday Soc.,* 1950, **9**, 196.
22. Tarte, *J. Chem. Phys.,* 1952, **20**, 1570.
23. Pristera, *Analyt. Chem.,* 1953, **25**, 844.
24. Haszeldine, *J. Chem. Soc.,* 1953, 2525.
25. Brown, *J. Amer. Chem. Soc.,* 1955, **77**, 6341.
26. Kornblum, Ungnade and Smiley, *J. Org. Chem.,* 1956, **21**, 377.
27. Shechter and Sheppard, *J. Amer. Chem. Soc.,* 1954, **76**, 3617.
28. Mason and Dunderdale, *J. Chem. Soc.,* 1956, 754.
29. Jander and Haszeldine, *ibid.,* 1954, 912.
30. Randle and Whiffen, *ibid.,* 1952, 4153.
31. Kross and Fassel, *J. Amer. Chem. Soc.,* 1956, **78**, 4225.
32. Lippert and Vogel, *Z. Phys. Chem.,* 1956, **9**, 133.
33. Franck, Hormann and Scheibe, *Chem. Ber.,* 1957, **90**, 330.
34. Palm and Werbin, *Can. J. Chem.,* 1953, **31**. 1004.
35. Lecomte and Mathieu, *J. Chim. Phys.,* 1942, **39**, 57.
36. Kumler, *J. Amer. Chem. Soc.,* 1953, **75**, 4346.
37. *Idem, ibid.,* 1954, **76**, 814.
38. Salyamon and Yaroslaviskii, *Sbornik Statei Obschiy. Khim.,* 1953, **2**, 1325.
39. Salyamon and Bobovich, *ibid.,* 1332.
40. Conduit, *Private Communication.*
41. Haszeldine and Jander, *J. Chem. Soc.,* 1954, 691.

42. Haszeldine and Mattinson, *ibid.*, 1955, 4172.
43. Lüttke, *Zeit. Electrochem.*, 1957, **61**, 302.
44. *Idem, J. Phys. Radium,* 1954, **15**, 633.
45. Schindler, Lüttke and Holleck, *Chem. Ber.*, 1957, **90**, 157.
46. Tarte, *Bull. Soc. Chim. Belges*, 1954, **63**, 525.
47. Mason and Dunderdale, *J. Chem. Soc.*, 1956, 759.
48. Nakamoto and Rundle, *J. Amer. Chem. Soc.*, 1956, **78**, 1113.
49. Gowenlock, Spedding, Trotman and Whiffen, *J. Chem. Soc.*, 1957, 3927.
50. Haszeldine and Jander; Tarte, *J. Chem. Phys.*, 1955, **23**, 979.
51. White, *J. Amer. Chem. Soc.*, 1955, **77**, 6008.
52. Costa and Blasina, *Z. Phys. Chem.*, 1955, **4**, 24.
53. Costa, Blasina and Sartori, *ibid.*, 1956, **7**, 123.
54. Sartori, Costa and Blasina, *Gazz. Chim.*, 1955, **85**, 1085.
55. Mathis-Noel, Wolf and Gallais, *Compt. Rend. Acad. Sci. (Paris)*, 1956, **242**, 1873.
56. Witkop and Kissman, *J. Amer. Chem. Soc.*, 1953, **75**, 1975.
57. Jander and Haszeldine, *J. Chem. Soc.*, 1954, 919.
58. Wiley and Slaymaker, *J. Amer. Chem. Soc.*, 1957, **79**, 2233.
59. Lunn, *Spectrochim. Acta*, 1960, **16**, 1088.
60. Geiseler and Kessler, *Ber. Bunsenges. Phys. Chem.*, 1964, **68**, 571.
61. Geiseler and Kessler, *Ber. Bunsenges. Phys. Chem.*, 1966, **70**, 918.
62. Popov and Shylapochnikov, *Optics and Spectroscopy*, 1963, **15**, 174.
63. Slovetskii, Shylapochnikov, Shevelev, Fainzil'berg and Novikov, *Izvest. Akad. Nauk. SSSR Otd. Khim. Nauk.*, 1961, 330.
64. Uhlich and Kresze, *Zeit. Analyt. Chem.*, 1961, **182**, 81.
65. Eckstein, Glucinski, Sobotka and Urbanski, *J. Chem. Soc.*, 1961, 1370.
66. Borek, *Naturweiss.*, 1963, **50**, 471.
67. Katritzky and Simmons, *Rec. Trav. Chim. Pays Bas.*, 1960, **79**, 361.
68. Rao and Venkataraghaven, *Can. J. Chem.*, 1961, **39**, 1757.
69. Conduit, *J. Chem. Soc.*, 1959, 3273.
70. Yamaguchi, *J. Chem. Soc. Japan*, 1959, **80**, 155.
71. Brookes and Jonathan, *Spectrochim. Acta*, 1969, **25A**, 187.
72. Feuer, Savides and Rao, *Spectrochim. Acta*, 1962, **19**, 431.
73. Ezumi, Miyazaki and Kutota, *J. Phys. Chem.*, 1970, **74**, 2397.
74. Carrington, *Spectrochim. Acta*, 1960, **16**, 1279.
75. Guthrie and Spedding, *J. Chem. Soc.*, 1956, 953.
76. Green, *Spectrochim. Acta*, 1961, **17**, 486.
77. Shindo, *Chem and Pharm. Bull. Tokyo,* 1956, **4**, 460; 1958, **6**, 117.
78. Katritzky, Beard and Coats, *J. Chem. Soc.*, 1959, 3680.
79. Katritzky and Garner, *J. Chem. Soc.*, 1958, 2198.
80. Katritzky and Hands, *J. Chem. Soc.*, 1958, 2195.
81. Katritzky and Ambler, Katritzky and Taylor, in *Physical Methods in Heterocyclic Chemistry.* Vols. 2 and 4. (Academic Press, New York, 1963 and 1971).

18

Organo-Phosphorus Compounds

18.1. Introduction and table

The widespread use of these compounds in commerce as insecticides, oil additives, plasticisers, etc., coupled with the possibilities of studying phosphate metabolism in nucleic acids, lecithins and similar products, has stimulated a good deal of work in this field, and a number of correlations are now available. These have been summarised in correlation charts for: (*a*) inorganic phosphorus links [22], and (*b*) organo phosphorus compounds [23] by Corbridge, and by Thomas and Chittenden [36]. Useful reviews of P=O frequencies have been given by Hudson [37], and by Thomas [59].

A full list of the correlations which have been proposed is given in Tably 18. It should be stated at once, however, that not all of these are of equal value, and that, whilst many of them are based upon the examination of large numbers of compounds, others are only very tentative correlations based on a few compounds or on compounds of a limited type.

The basis of each of these correlations is discussed below, and some indication is given of the degree of reliability which the correlation may be expected to show.

18.2. P=O Stretching vibrations

Meyrick and Thompson [1] were the first to examine the infra-red and Raman spectra of a number of organo-phosphorus compounds, and as a result of studies on five phosphonates they suggested that

Table 18

Phosphorous–Carbon Links

(a) P=O (free)	$1350-1250$ cm^{-1} (exceptionally lower, down to 1175 cm^{-1}) (s.)
P=O (hydrogen bonded)	$1250-1110$ cm^{-1} (v.s.)
(b) P–O–C (aromatic)	$1240-1190$ cm^{-1} (s.) and $995-850$ cm^{-1} (s.)
(c) P–O–C (aliphatic)	$1050-990$ cm^{-1} (v.s.) (exceptionally lower) and near 1150 cm^{-1}
(d) P–O–Ethyl	$1170-1150$ cm^{-1} (w.) and $1092-1008$ cm^{-1}
(e) P–O–Methyl	1190 ± 10 cm^{-1} (w.) and $1060-1015$ cm^{-1}
(f) P–O–P	$970-900$ cm^{-1}
(g) P–OH	$2700-2560$ cm^{-1} (broad and shallow)

Phosphorus–Hydrogen Links

P–H	$2457-2270$ cm^{-1} (m.)

Phosphorus–Carbon Links

(a) P–Phenyl	$1450-1425$ cm^{-1} (m.)
(b) P–Alkyl	No useful correlations
P–CH$_3$	$1320-1280$ cm^{-1}

Phosphorus–Sulphur Links

P=S	$802-660$ cm^{-1} and $730-550$ cm^{-1} (w.)

Phosphorus–Nitrogen Links

(a) P–NH$_2$ and P–NH–	Normal for NH$_2$ and NH structures
(b) P–N–	$1655-870$ cm^{-1}

Phosphorus–Halogen Links

(a) P–Cl	$580-440$ cm^{-1} (s.)
(b) P–F	$940-740$ cm^{-1} (s.)

Ionic Groups

Phosphates PO$_2$$^-$	$1323-1092$ cm^{-1} and $1156-990$ cm^{-1}
Phosphates PO$_3$$^{2-}$	$1140-1055$ cm^{-1} and $1010-970$ cm^{-1} (s.)
Phosphonates PO$_3$$^{2-}$	$1124-970$ cm^{-1} and $1000-962$ cm^{-1} (s.)

the moderately intense band shown in all cases in the $1260-1250$ cm^{-1} region originated in the P=O stretching vibration. This assignment was supported by Raman data on POCl$_3$[2] (in which the P=O vibration occurs at 1295 cm^{-1}) and on other phosphites and phosphonates [3].

This work was extended and amplified by Gore [4] and by Daasch and Smith [5], who independently studied considerable numbers of phosphonates and phosphates. Gore [4] was able to show that the suggestion of the 1250 cm^{-1} region for this vibration was reasonably in accord with theory, as the application of Gordy's rule would indicate a frequency of 1300–1100 cm^{-1}, depending on the P–O bond length used. Furthermore, he examined a considerable number of phosphates and compared them with the corresponding thiophosphates. In each case the most obvious point of difference was the presence of a band in the 1300–1250 cm^{-1} region in the phosphates. This finding was fully confirmed by Daasch and Smith [5], who also obtained evidence that the frequency of this vibration was largely determined by the number of electronegative substituents on the phosphorus atom. In compounds with three electronegative substituents, such as phosphates or fluorphosphonates, the P=O frequency was in the approximate range 1310–1275 cm^{-1}, falling to 1275–1250 cm^{-1} in compounds with only two electronegative substituents, such as hydrogen and alkyl phosphonates. With phosphinates and similar materials, absorptions near 1240 cm^{-1} were found, whilst in the extreme cases of triphenyl- and trimethyl-phosphine oxides the frequencies fell to 1190 cm^{-1} and 1176 cm^{-1}, respectively. The latter two compounds are also of interest in illustrating the apparent lack of influence of the conjugation of the aromatic rings on the P=O frequencies. In general, the correlation of this band with the P=O group was established by comparison with the corresponding phosphites and similar trivalent phosphorus compounds in which this group is absent.

This correlation was further confirmed by Bellamy and Beecher [6, 7, 8] working mainly on phosphates and phosphonates, and has also been supported by the work of Holmstedt and Larsson [9], Bergmann, Littauer and Pinchas [10], Harvey and Mayhood [24], Maarsen [25], Emeleus *et al.* [26, 27] and Thomas [28]. The results on over 900 compounds containing the P=O link have been reviewed and systematised by Thomas and Chittenden [18]. The overall frequency range, even in the absence of hydrogen bonding is very wide and the absorption is usually only of moderate intensity and is sometimes split into a doublet. The positive identification of this group from the P=O band alone is therefore unreliable. On the other hand, once the presence of the group is known, the strict dependence of the frequency upon the sum of the inductive effects of the substituents can reveal much about the nature of the attached groups.

The variations of the frequency with substitution is well illustrated by the values for phosphoryl compounds with three identical substituents. These range from 1414 cm^{-1} for trifluoro, 1290 cm^{-1} for trichloro, $1290-1314$ cm^{-1} for tri O-aryl, $1258-1286$ cm^{-1} for tri O-alkyl, to $1151-1183$ cm^{-1} for trialkyl. As will be seen later, compounds with mixed substituents give intermediate values determined by their individual contributions. Cyclic phosphoryl compounds have been studied by Jones and Katritzky [19] who find no changes in the P=O frequencies with alterations in the ring size. This is in contrast to the behaviour of the carbonyl compounds but is to be expected in view of the very small amplitude of the phosphorus atom as compared with the oxygen atom.

The electronegativity dependence of this absorption has been put on a quasi-quantitative basis by Bell *et al.* [30]. They have shown that for the halogen-substituted phosphine oxides there is a linear relation between the sum of the halogen electronegativities and the phosphoryl frequency. This has been extended to the derivation of effective group electronegativities for other substituents which appear to be of fairly general application. Goubeau and Leutz [20] have also studied electronegativity relations and correlated $\nu_{P=O}$ with the Pauling electronegativity and with the position of the atom concerned in the periodic table. Others have sought for correlations with Taft σ values as measures of the inductive effects. Griffin [21] relates the P=O frequencies of phosphonic acid derivatives with the sum of the Taft σ values of the substituents, and Goldwhite *et al.* [38] have derived a similar linear relationship between the square root of the absolute intensity and the sum of the Taft σ values. However the most directly useful of these studies is that of Thomas and Chittenden [18, 36]. Starting from the linearity of the frequencies of the trihalides with electronegativity, they go on to derive average values for other substituents which give the best additive agreement with all the known data. These they call π values, and by the use of the equation $\nu_{P=O} = 930 + 40\Sigma\pi$ one can derive predicted frequencies for any compound which correlate closely with experiment. Although these π values are empirical numbers and not electronegativities they are nevertheless a true relative measure of the inductive properties of the substituents in phosphoryl compounds. The fact that aryl groups have higher π values than alkyl and that the triaryl phosphine oxides have higher frequencies than the trialkyl derivatives is consistent with this as aryl groups have a greater inductive effect than alkyl. In this case the geometry of the phosphoryl compounds is such that the P=O bond is not coplanar with the rings and conjugation or resonance cannot occur. It is for the same reason that the nitrogen substitution raises $\nu_{P=O}$ whereas in amides $\nu_{C=O}$ is lowered. These π values are not applicable in situations in which the phosphoryl group is strongly hydrogen-bonded.

The P=O frequency is subject to shifts of 50–80 cm^{-1} when the molecule contains OH or NH groups with which the P=O group is able to form hydrogen bonds [5, 7, 8, 18, 26, 27]. The bonding is exceptionally strong in the case of OH compounds, as shown by the accompanying large frequency shift of the OH stretching vibration (see later). In addition to undergoing a low-frequency shift, the intensity of the P=O vibration is considerably increased. This is not commonly the case with hydrogen-bonded carbonyl groups, except for compounds of the type —CO—C=C(OH)—, in which the resonant structure is responsible for the intensity increase. It is probable that a similar resonance structure of the group $P\overset{\displaystyle O}{\underset{\displaystyle OH}{\diagup\!\!\!\diagdown}}$ is responsible for the observed intensity increase in this case also. Thomas [18] gives the range of 1115–1250 cm^{-1} as probable for phosphoryl groups in hydroxy compounds.

Hydrogen bonding of this type appears to be limited to hydroxy- and amino-compounds, and the P=O frequency of hydrogen phosphonates, for example, is normal. There is, indeed, evidence from heats of mixing etc. that many other types of organo-phosphorus compounds show association effects [11], but the P=O group may not be involved, as its infra-red frequency is relatively constant. On steric grounds it would appear to be difficult for the hydroxy-group in these compounds to form an intramolecular hydrogen bond, and association in these cases is through dimer formation, as with carboxylic acids. However, the association is extremely strong, and in the few cases which have been studied in dilute solutions in non-polar solvents, the shifted P=O frequency persists, but is accompanied also, in many cases, by a second absorption corresponding to the unbonded P=O vibration [7]. With the amino compounds the bonds are also intermolecular, and the NH and P=O frequencies revert to normal in dilute solution. One interesting application of this correlation is in the demonstration that dialkyl hydrogen phosphites exist wholly in the phosphonate form:

Thus Meyrick and Thompson [1] reported strong P=O and P–H absorptions in a number of compounds of this type, and the same effects have been noted by Bellamy and Beecher [6] and by Nyquist [39]. The absence of P–OH absorptions and the fact that the P–O absorption frequency corresponds to an unbonded phosphoryl group confirm that not more than a very small proportion of the whole can exist as hydrogen phosphite. The basicity of the P=O link has been measured by hydrogen-bonding methods in which the frequency shift of some standard XH donor such as $CDCl_3$ is measured. Gramstad [40, 41] has shown that $\delta\nu$ can then be related to the Taft σ values of the phosphorus substituents. These therefore determine both the frequency and the polarity. The intensity of the P=O band is also related to the $\delta\nu$ values of proton donors and is therefore similarly determined by the polarity [41].

The intensity of the P=O absorption is reasonably strong, but varies somewhat with the nature of the attached groupings. The comparison of absolute intensities for different classes of materials is, however, complicated by the fact that this band very frequently occurs as a doublet. In some instances this has been clearly shown to be due to rotational isomerism, but the effect persists in some compounds in which this is unlikely and coupling effects with CH modes have been suggested as another cause [25, 28].

18.3. P–O–C Stretching vibrations

18.3(a). P–O–C (Aromatic) vibrations. Just as there are marked differences between the absorption frequencies of C–O–C (alkyl) and C–O–C (aryl) linkages, so are there marked differences between the two types of P–O–C linkage.

The differentiation between the two classes of P–O–C linkage has been clearly demonstrated by studies on related series of compounds such as triphenyl, diphenyl-ethyl, phenyl-diethyl and triethyl phosphates. Examination of these materials at the same concentration in the same cell reveals a gradual weakening and ultimate disappearance of a strong band near 1200 cm^{-1}, whilst, at the same time, a band at 1030 cm^{-1} that is absent in triphenylphosphate shows a corresponding increase in strength throughout the series [6].

On the basis of this and other related series [6, 7], and especially of the work of Nyquist [42] who has correlated the data on over 100 compounds, the P–O–C (aromatic) absorption has been assigned to the 1200 cm^{-1} region, and it has been shown that a band in this region is present in all the aromatic phosphates, phosphonates and phosphites so far examined, as well as in some aryl thiophosphates and similar materials, whilst no band is present in wholly aliphatic compounds. The band falls between 1240 and 1190 cm^{-1} in most cases [42, 43], but Thomas [43] points out that the overall range can be as wide as 1242–1110 cm^{-1} in unusual cases.

This band is always accompanied by a second absorption which has been attributed to either the symmetric stretch of the P–O–C system (the antisymmetric mode being that at 1200 cm^{-1}), or to a separate P–O stretch which is not so coupled. Thomas [43] strongly favours the latter explanation which is supported by the persistence of this band in both P–O–P and P–OH compounds, and by the fact that in the latter the frequency is a linear function of the π values of the substituents. Nyquist [42] places this band between 994 and 914 cm^{-1} and notes that the band is at a lower frequency in trivalent phosphorus compounds. Thomas [43] widens this range to 905–996 cm^{-1} for pentavalent compounds and suggests 860–850 cm^{-1} for trivalent. He points out that this band is often complex and the values quoted therefore refer to the strongest band.

In unsaturated esters of the type P–O–C=C or P–O–$\overset{\overset{\text{O}}{\|}}{\text{C}}$ the same correlation applies.

18.3(b). P–O–C (Alkyl) vibrations. The studies described above also serve to establish the position of the P–O–C (alkyl) absorption as being near 1030 cm^{-1}, whilst the same correlation had also been put forward earlier by Daasch and Smith [5]. Altogether, a very large number of alkylphosphates and similar materials have now had their spectra determined [1, 4–10, 42, 43], and in all cases an extremely strong absorption is found in the range 1050–950 cm^{-1}, with the great majority absorbing near 1030 cm^{-1}. Bergmann *et al.* [10] described this band as being absent in tri*iso*propylphosphate, and assign the 995 cm^{-1} band found to other causes. However, the spectra of other branched-chain materials described by Meyrick and Thompson [1] and by Bellamy and Beecher [6, 7] are in line with the general correlation, so that there is no reason to suppose that branched-chain materials are in any way exceptional. Recently Thomas [28, 43] has noted that in some 1000 organo-phosphorus compounds examined this absorption falls in the range 1050–1000 cm^{-1} for methyl and ethyl esters. With longer-chain alkyl groups, however, the frequency is somewhat lower, falling to 950 cm^{-1} in extreme cases unless a fluorine or other electronegative substituent is also present. This assignment is, of course, applicable only to the P—O—C vibration in phosphates, and the change in the valency state to trivalent phosphorus leads to a shift to lower frequencies.

As with the P—O-Aromatic group, a second band, this time near 1150 cm^{-1} is associated with the P—O—C system. As before there has been much controversy on whether these bands should be assigned as antisymmetric and symmetric vibrations as suggested by Bergmann et al. [30] or whether they can be regarded as isolated P—O and C—O stretching modes. Among those who prefer this second interpretation, there is also controversy as to which band is due to which link. Thus Thomas [28] and McIvor et al. [31] assign the 1030 cm^{-1} band to the P—O—(C) vibration and the higher-frequency band to the C—O—(P) mode. Maarsen [25], on the other hand, identifies the C—O—(P) mode with the 1030 cm^{-1} region and assigns the other at lower frequencies. Nyquist [42] follows Marsden and assigns the C—O stretch in the range 1050—950 cm^{-1} with the P—O between 850 and 780 cm^{-1}. Thomas [43] has reviewed the arguments and on the basis of over 1000 spectra containing this group, prefers to retain his original assignment of the P—O stretch as an individual band near 1030 cm^{-1}. His choice is based on the arguments already given for the parallel assignment for the P—O-Aromatic link.

With the two frequencies so close together it would be surprising if there were to be no coupling between them and their sensitivity to small changes of structure suggests that this occurs. For the purposes of structural diagnosis however, the precise origins of the bands are of less importance than their positions, and here at least there is a great deal of unequivocal data. It is generally agreed that in addition to the strong band between 1060—950 cm^{-1} there is a second medium intensity band in the range 1200—1087 cm^{-1}. This range can be subdivided according to the degree of substitution at the carbon atom attached to the oxygen. Methyl groups have this band between 1200—1168 cm^{-1}, CH_2 groups have it between 1170—1105 cm^{-1} and carbons with only a single hydrogen between 1190—1087 cm^{-1} [43].

18.3(c). P—O-Ethyl vibrations etc. The broad assignments discussed above can be refined considerably for a single group such as

P—O-Ethyl. The band is at 1029 cm^{-1} in triethyl phosphate, and this only falls to 1026 cm^{-1} in the corresponding thiophosphate. It is at this same point in the trivalent ethoxy compound. The overall range for this band in many compounds, is given by Thomas [43] as 1042—1008 cm^{-1} but this includes many with highly polar substituents and the norm is close to 1050 cm^{-1}.

The higher frequency band appears between 1170—1152 cm^{-1} as compared with values down to 1105 cm^{-1} for longer chains. This band is single in ethyl esters but multiple for most longer chain systems.

18.3(d). P—O-Methyl vibrations. The P—O—CH$_3$ group [43] has its higher frequency band close to 1190 cm^{-1} although the overall range is 1200—1168 cm^{-1}. The lower frequency band lies between 1015 and 1060 cm^{-1} with the trivalent compounds in the 1015—1034 cm^{-1} region.

18.4. Pyrophosphates

A good deal of work has been carried out by different groups in an attempt to identify absorptions characteristic of the pyrophosphate P—O—P linkage. Bergmann *et al.* [10] have examined three pyrophosphates and compared their spectra with those of the original phosphates. They find that in each case the former contain one more band than the latter and that this is in the range 970—930 cm^{-1}. They point out that the data on *n*-butyl pyrophosphate given by Daasch and Smith [5] also indicate a band at 950 cm^{-1}. They have therefore assigned the asymmetric P—O—P vibration as being the origin of this absorption.

Holmstedt and Larsson [9], on the other hand, have tentatively suggested that the P—O vibration of the pyrophosphate grouping occurs in the 710 cm^{-1} region. In organic compounds this absorption has been studied by Corbridge and Lowe [32]. In this class of compound they obtained evidence for the presence of P—O—P absorptions near 900 cm^{-1} and also near 700 cm^{-1}, thus providing support for both assignments. However, much of their data related to polypyrophosphates, and it has been suggested that in organophosphorus pyrophosphates the 900 cm^{-1} band occurs at higher frequencies. Harvey and Mayhood [24] suggest the 950—910 cm^{-1} region for this absorption as a result of studies on fourteen compounds of this type, and McIvor *et al.* [31] confirm this on a further range of compounds with an extension of the upper frequency limit to 970 cm^{-1}. Thomas [28] and Simon *et al.* [33] have quoted similar data. More recently Thomas [43] has reviewed the spectra of some 60 pyrophosphates. He finds no regular absorptions in the 710 cm^{-1} region which might be attributable to this group, whereas a strong band is present in all cases in the range 980–900 cm^{-1} with more than half absorbing between 945–925 cm^{-1}. He assigns this band to the P–O stretch, and believes that it is possible to identify the pyrophosphate system by a combination of the bands due to this and to the P–O–C groups. As with the P–O–C system, the band appears at lower frequencies in trivalent phosphorus systems, and in thiopyrophosphates [31, 43].

18.5. P–OH Stretching vibration

The OH valency vibration is, of course, relatively independent of the nature of the attached group. However, in acid compounds of phosphorus containing the $\overset{O}{\underset{}{P}}$—OH group, hydrogen bonding effects are even greater than those of carboxylic acids, so that the OH vibrations occupy a very characteristic position. In all the acids examined by Daasch and Smith [5] and by Bellamy and Beecher [7, 8] no OH bands could be found in the normal region, but, instead, a broad shallow absorption appeared in the 2700–2560 cm^{-1} range. In solutions in non-polar solvents this band persists. The band is clearly to be associated with the OH group, as it disappears on salt formation, whilst the P=O frequency also provides support for the suggestion that very strong hydrogen bonds are involved. Similar results have also been reported for synthetic kephalins [8, 12] and for naphthyl acid phosphates [13]. Inorganic phosphoric acids and related compounds behave similarly [22], and Thomas [28] has also confirmed that thiophosphorus acids absorb in this region, although he notes that in structures involving the $\overset{S}{\underset{}{P}}$—OH group the frequency is raised, reflecting the weaker association forces. On the other hand, in acids such as monosodium trifluoromethyl phosphonate the band occurs at lower frequencies [26, 27] near 2350 cm^{-1}, reflecting the enhanced hydrogen bonding strength resulting from the fluoromethyl substitution.

This, and the other P–OH bands have been studied in detail by Thomas *et al.* [43–45]. They have shown that at very high dilutions the hydrogen bonds can be broken showing that they are indeed intermolecular. Compounds with a P=O structure show two other
$$\overset{\mid}{O}H$$
bands between 2350–2080 cm^{-1} and 1740–1600 cm^{-1}, but the last of these is absent from P=S compounds. This band pattern is
$$\overset{\mid}{O}H$$
typical of many other systems with very strong hydrogen bonds and there is still much doubt as to their origins. However, the absence of the 1600 cm^{-1} band in the sulphur compounds is consistent with the other evidence that these contain rather weaker hydrogen bonds.

The P–O stretching band of these systems has been assigned by Thomas [43] in the 1000 cm^{-1} region, on the basis of a large number of compounds. This is supported by the finding that the frequency varies with the inductive properties of the substituents in exactly the same way as the P=O frequency. It gives a linear relationship with Thomas's π values and this is represented by the equation $\nu_{P-O(H)} = 650 + 40\Sigma\pi$. It is interesting to note that this line has the same slope as that for the P=O group and differs only in the constant term. The overall frequency range is considerable (1040–909 cm^{-1}) but the use of this relationship enables the band to be identified and separated from other P–O stretching bands in the same region.

The OH deformation band is not well defined. Emeleus *et al.* [26, 27] assigned it as near 940 cm^{-1} in some fluorinated compounds, and others have suggested it occurs near 1030 cm^{-1} where it is likely to be obscured by other absorptions [22]. Ryskin and Stavitskaya [46] prefer an assignment near 1280 cm^{-1} and this has been supported by deuteration studies by Thomas [43]. However, it is clear that this absorption is of very much less use for identification than is the corresponding band of dimeric carboxylic acids.

18.6. P–H Stretching vibration

A number of inorganic and organic phosphorus compounds containing the P–H linkage have been examined by various workers [1, 5–7, 27, 28, 32, 36, 47–51], and in these a sharp absorption band of medium intensity occurs in the range

$2460-2270$ cm^{-1}. This is the expected region for the P–H valence vibration, and phosphine itself absorbs [14] at 2327 cm^{-1} and 2421 cm^{-1}, so that the assignment can be regarded as being fully substantiated. In almost all cases the band is found to be sharp, and the fact that the P=O absorption occurs in very much the expected region for the appropriate dialkyl phosphonate suggests that the P–H link is not capable of entering into hydrogen bonding to any appreciable extent.

The overall range for 109 compounds is given by Thomas [43] as $2457-2270$ cm^{-1}. This is a much wider spread than is found for most XH stretching frequencies, except silicon, and it must reflect the considerable variations in charge density at the phosphorus atom which occur with structural changes. There is no simple relationship with the inductive effects of the substituents but the overall range can be subdivided into relatively narrow regions for related groups of compounds. Thus dialkyl phosphonic esters absorb in the narrow range $2445-2431$ cm^{-1} [39] and the range for all published compounds of this group is only $2445-2380$ cm^{-1} [49]. Monoalkyl phosphonic acids absorb at slightly higher frequencies [49] in the range $2457-2427$ cm^{-1}. Esters of phosphinic acid absorb between 2355 and 2310 cm^{-1}. The derivatives of phosphine show a somewhat larger spread, but dialkyl phosphines absorb at $2300-2270$ cm^{-1} and the corresponding phosphine oxides between $2380-2294$ cm^{-1}. It is therefore possible to obtain much useful structural information from the precise position of the P–H stretching band.

The P–H deformation modes appear to be less useful. Nyquist [39] has identified a characteristic band in dialkyl phosphonates in the $968-979$ cm^{-1} region which is strong and is deuterium sensitive, and he assigns this to the δPH vibration. However in a wider range of compounds this band appears to extend over the range $1015-897$ cm^{-1} and it is not of much value for identification purposes.

18.7. Phosphorus–carbon links

18.7(a). Aromatic. Daasch and Smith [5] have examined a considerable series of aryl phosphorus compounds containing the P-aryl group, and have found in all cases bands in the regions $1450-1435$ cm^{-1} and $1005-995$ cm^{-1} which are sharp and of moderate intensity. They associate these with the P-phenyl link, but point out also that they might equally well arise from ring vibrations which have become activated in some way. Later studies [29] have confirmed that this is so and that both bands originate in the ring modes rather than in the P—C bond. Insofar as the 1000 cm^{-1} band is concerned similar and indistinguishable bands appear in compounds with P—O—C (aromatic) and P—N—C (aromatic) even when these compounds have no P—Aromatic link. However, it does seem that there is sufficient difference between the frequencies of the 1440 cm^{-1} band to enable one to differentiate between the different possibilities. Thomas and Chittenden [29] have reviewed the data for a very large number of compounds and summarised the frequencies as follows: P—O—Ph 1445—1458 cm^{-1} >P—Ph 1425—1450 cm^{-1} >P—N—Ph 1425—1405 cm^{-1}. There is of course some overlap, but as the P—O—Ph system can be identified by other bands, this difficulty can be overcome.

18.7(b). Aliphatic. No useful correlations for the P—C linkage have been formulated, and as the P—C stretching bands of compounds such as trimethyl phosphine oxide [15] appear to lie in the 750—650 cm^{-1} range in which many other types of organo-phosphorus compounds also absorb, it is doubtful whether any such correlations could be developed.

In the case of P—CH$_3$ compounds a characteristic absorption should arise due to the symmetric CH$_3$ deformation mode. This has been identified as falling [22, 24, 28, 29, 31] in the range 1330—1270 cm^{-1}, where it occurs as a strong-to-medium intensity band in most cases. In pentavalent phosphorus compounds this band usually occurs in the upper half of this range, whereas it falls nearer 1285 cm^{-1} in the trivalent series [28, 29]. This band is not, however, entirely specific for this group, despite its narrow frequency range, as N(CH$_3$)$_2$ groups which are commonly found in organo-phosphorus insecticides also absorb near this point. It has been suggested that differentiation between the two possibilities may be possible from measurements of the half-band widths, which are appreciably smaller in the case of the P—CH$_3$ bands [24, 28]. An alternative and more discriminating method is the identification of the P—CH$_3$ rocking mode which occurs in the 935—874 cm^{-1} range for pentavalent phosphorus compounds [29, 52] except when a hydrogen atom is attached to the same phosphorus, when the frequency falls to 842—850 cm^{-1}. The band doubles when more than one methyl group are attached to the same phosphorus atom. With trivalent phosphorus the range is lower; 872—862 cm^{-1}.

No useful correlations exist for other P—alkyl groups. The

P—O—Ethyl group has one or two weak bands in the 1250 cm^{-1} region, but they are not sufficiently strong or sufficiently characteristic to be of value.

18.8. Phosphorus—sulphur links

Infra-red spectral correlations covering the P=S and P—S— bands would be of considerable value in the organo-phosphorus insecticide field as, combined with studies on the corresponding phosphorus—oxygen links, the method might permit the study of the complex isomerisation occurring on heating thiophosphates. However, although a good deal of work has been done in this field, the characteristic absorptions appear to be very weak and are, therefore, of very limited value.

18.8(a). P=S Stretching vibrations. The compounds $PSCl_3$ and $PSBr_3$ absorb at 753 cm^{-1} and 718 cm^{-1}, and these bands are absent from the corresponding oxychlorides. This suggests that the P=S absorption should occur in the 750 cm^{-1} region, and this has been supported by Gore's calculated value of 752 cm^{-1} based on Gordy's rule. Since then this absorption has been extensively studied and the literature has been reviewed by Thomas [50] and by Bellamy [53]. Two separate bands in the overall ranges 802—658 and 730—550 cm^{-1} have been identified as being in some way connected with the P=S absorption, as both vanish when phosphorothionates are converted into phosphorothiolates. There has been much discussion as to the origins of these two bands and some authors assign the higher frequency to P=S whilst others prefer the lower. A low frequency band of this kind is likely to be very extensively coupled so that it is probable that both have some P=S content, and that neither is an unmixed vibration. In the Raman spectra the P=S band is expected to be exceptionally strong, and as the lower frequency band is found to be the more intense it is likely that this has the greater P=S content. Neither of the two bands shows any systematic shifts which reflect changes in the inductive properties of the substituents, and this is not unexpected if they do indeed arise from mixed modes. Clittenden and Thomas [50] have tabulated the positions of both bands in a wide variety of different phosphorus compounds and these can be useful if the likely form of the compound under study is known. Otherwise, the wide frequency variations and the fact that the lower frequency band often appears with low intensity in the infra-red render this correlation of very limited value.

18.9. Phosphorus-nitrogen links

18.9(a). P—NH$_2$ and P—NH— vibrations. A number of phosphoramidates and similar materials have been examined [7, 9, 32, 54, 55]. The NH stretching and deformation vibrations are very little affected by the presence of the phosphorus atom and occur at very much the same positions as in normal amines. Thus P—NH$_2$ compounds show two, and sometimes three NH stretching bands in the 3330—3150 cm^{-1} region when examined as solids. Dilute solutions in which the hydrogen bonds are broken absorb near 3400 cm^{-1}. Similarly, P—NH— compounds show normal behaviour in this region, but Nyquist [54] finds that when the NH group is attached to the P=O link, there are two free NH stretching bands in dilute solutions. He associates these with the *cis* and *trans* forms. This is of course parallel with the behaviour of the secondary amides. P—NH$_2$ compounds, also, exhibit their deformation absorption in the region 1570—1550 cm^{-1}, except in isolated instances when, as in benzyl hydrogen phosphoramidate, a zwitterion type structure is involved [7].

18.9(b). P—N Stretching vibrations. Several workers have attempted to identify a characteristic P—N stretching frequency. Early workers [9, 23–25, 34] attempted to identify a characteristic P—N absorption in the 750 cm^{-1} range. Although many compounds with this link absorb near here, others do not and it must be accepted that this bond will be so coupled with those next to it that no reliable group frequency can be expected. Thomas [55] has reviewed the data, and concludes that this bond absorbs over a very wide range between 1055 cm^{-1} and 870 cm^{-1} and that it is best identified where possible by indirect correlations such as the PNH$_2$ group *etc.* The 750 cm^{-1} assignment has been supported by Harvey and Mayhood [24] and by Corbridge [23], to the extent that they are agreed in finding a common absorption when this group is present, in the 750—680 cm^{-1} range. Laarsen [33] has also suggested a

modification to include the P—N—C structure, with a C attached above the N, which he believes to absorb in the 820–780 cm^{-1} region. Any correlation for the P–N link must be expected to be mass sensitive, and therefore liable to considerable frequency shifts with minor alterations in structure. The correlation is not therefore a particularly useful one, and in a number of cases it has not proved possible to identify the band in question. Maarsen [25] confirms this and attributes it to obscuration by other absorptions, such as the C–P stretching band, but Thomas [28] has also noted a number of instances in which the P–N– link shows no absorption in the range quoted, and this is in accord with our own experience.

In cyclic structures in which the P=N link arises, absorptions are found at higher frequencies, which may be associated with this group. Daasch [35] has assigned absorptions in phosphonitrilic chlorides to the 1300–1200 cm^{-1} region, and salts of phosphonitrilic acids [22] also show strong bands in the range 1300–1100 cm^{-1}. In compounds of this kind in which there is much aromatic character these frequencies are best described as ring modes rather than P=N bands. Thomas [55] lists all the available data which show that the frequencies are sensitive to ring size and to the nature of the substituents.

In compounds in which resonance of this kind is not possible a P=N band can be found at higher frequencies in the 1400–1325 cm^{-1} range [55, 56].

18.10. Phosphorus–halogen links

18.10(a). P–Cl vibrations. Phosphorus trichloride [16] has two absorptions at 511 cm^{-1} and at 488 cm^{-1} which correspond to P–Cl stretching modes, and $POCl_3$ and $PSCl_3$ exhibit two bands in this spectral region. Daasch and Smith [5] have examined seven other compounds containing the P–Cl band, all but one of which contained more than one chlorine atom. The observed frequency ranges were 572–500 cm^{-1} and 500–433 cm^{-1}. These findings have been confirmed by later studies by Corbridge [23], Thomas [28, 49] and McIvor *et al.* [31], and many others [36, 42, 47, 57]. Thomas [49] has reviewed the data for over 90 compounds and gives the overall range 587–420 cm^{-1}. This refers to the main strong band. In many cases this is accompanied by a second weaker band which in some classes of compound is at higher frequency and in others at lower. Within this wide overall range very much narrower limits have been defined for specific structural units. Dialkyl phosphorochloridates for example show a reasonably strong band between $565-557$ cm^{-1}. Thomas [49] lists a number of such classes although the numbers studied are often small. Dichloro compounds show two bands in the same range.

18.10(b). P—F vibrations. In PF$_3$ the first fundamental [16] is at 890 cm^{-1}, and this suggests a likely upper limit for this absorption. Compounds containing this link have been studied by a number of authors [6, 23, 28, 36, 49]. As is to be expected the range is a wide one, which Thomas puts at $940-740$ cm^{-1}, although this can be subdivided into pentavalent phosphorus and trivalent phosphorus compounds. The former absorb between $940-794$ and the latter between $800-740$ cm^{-1}.

18.10(c). P—Br and P—I linkages. Phosphorus links with bromine or iodine have been little studied. The stretching frequencies of the P—Br links in PBr$_3$ [16] are at 400 and 380 cm^{-1}.

18.11. The ionic phosphate group

In inorganic phosphates of the type $M_3{}^+PO_4{}^{---}$ a strong band is found [17] in the region $1050-1000$ cm^{-1} and there is also evidence of a second absorption in the 980 cm^{-1} region [14, 10] in some cases. With the hydrogen phosphates a distinct but progressive shift is observed towards higher frequencies, and many compounds of the type $HPO_4{}^=$ absorb in the $1070-1050$ cm^{-1} range and $H_2PO_4{}^-$ in the $1090-1030$ cm^{-1} range [17]. In both cases weaker absorptions are also shown in the $1000-900$ cm^{-1} region (see Chapter 21). These preliminary correlations have been confirmed and extended by detailed studies by Corbridge and Lowe [22, 23, 32] on inorganic phosphate salts. Correlations based in general on at least eight salts of any one class are now available for *ortho*-phosphates (mono-, di- and tri-basic), phosphofluoridates, methyl phosphonates, monobasic phosphoramidates, hypophosphates, pyrophosphates, etc., etc. These have been very conveniently summarised in the form of a correlation chart [23] limited to salts of phosphorus oxy-acids, and this has useful applications in analytical work.

With organophosphoric acids the position is further complicated by the fact that the electronegativity of the organic substituents may have a direct influence on the bond lengths of the resonance hybrid arising from the $P{\overset{\displaystyle O}{\diagdown}}{-}O^-$ structure. The $PO_2{}^-$ anion for example has its *as.* mode between $1323-1092$ cm^{-1} and the *sym.* mode between $1156-990$ cm^{-1}. In addition to substituent effects these frequencies can change by as much as 50 cm^{-1} with changes in the metal cation [58]. However these ranges become much smaller for any single related series, and there is also a general trend for the frequencies to follow the π values. In dialkyl phosphate ions for example, the bands lie in the $1284-1124$ and $1121-1064$ cm^{-1} range. The lower frequency band is of course more difficult to identify because of the occurrence of the P—O—C stretch in the same range.

The PO_3^{2-} ion has also been studied by several groups and the results have been summarised by Thomas [58]. This group has strong bands at $1140-1055$ and $1010-970$ cm^{-1} for phosphate ions and between $1124-970$ and $1000-962$ cm^{-1} for phosphonate ions. As is usual in the infra-red, the intensity of the higher frequency antisymmetric band is considerably greater than that of the symmetric mode.

18.12. Bibliography

1. Meyrick and Thompson, *J. Chem. Soc.*, 1950, 225.
2. Yost and Anderson, *J. Chem. Phys.*, 1934, 2, 624.
3. Arbusov, Batuev and Vinogradova, *Compt. rend. Acad. Sci. U.S.S.R.*, 1946, 54, 599.
4. Gore, *Discuss. Faraday Soc.*, 1950, No. 9, 138.
5. Daasch and Smith, *Analyt. Chem.*, 1951, 23, 853 and *N.R.L. Report*, 3657.
6. Bellamy and Beecher, *J. Chem. Soc.*, 1952, 475.
7. *Idem, ibid.*, p. 1701.
8. *Idem, ibid.*, 1953, 728.
9. Holmstedt and Larsson, *Acta Chem. Scand.*, 1951, 5, 1179.
10. Bergmann, Littauer and Pinchas, *J. Chem. Soc.*, 1952, 847.
11. Kosolapoff and McCullough, *J. Amer. Chem. Soc.*, 1951, 73, 5392.
12. Baer, Maurukas and Russell, *ibid.*, 1952, 74, 152.
13. Friedman and Seligman, *ibid.*, 1951, 73, 5292.
14. Herzberg, *Infra-red and Raman Spectra of Polyatomic Molecules* (Van Nostrand, 1945), p. 302.
15. Daasch and Smith, *J. Chem. Phys.*, 1951, 19, 22.
16. Herzberg, *Infra-red and Raman Spectra of Polyatomic Molecules* (Van Nostrand, 1945), p. 164.
17. Miller and Wilkins, *Analyt. Chem.*, 1952, 24, 1253.
18. Thomas and Chittenden, *Spectrochim. Acta*, 1964, 20, 467.
19. Jones and Katritzky, *J. Chem. Soc.*, 1960, 4376.
20. Goubeau and Lentz, *Spectrochim. Acta*, 1971, 27A, 1703.
21. Griffin, *Chem. and Ind.*, 1960, 1058.
22. Corbridge and Lowe, *J. Chem. Soc.*, 1954, 4555.
23. Corbridge, *J. App. Chem.*, 1956, 6, 456.
24. Harvey and Mayhood, *Can. J. Chem.*, 1955, 33, 1552.

25. Maarsen, *Thesis, University of Amsterdam*, 1956.
26. Emeleus, Haszeldine and Paul, *J. Chem. Soc.*, 1955, 563.
27. Bennett, Emeleus and Haszeldine, *ibid.*, 1954, 3598.
28. Thomas, *Chem. and Ind.*, 1957, 198, and private communication.
29. Thomas and Chittenden, *Spectrochim. Acta*, 1965, **21**, 1905.
30. Bell, Heisler, Tannebaum and Goldenson, *J. Amer. Chem. Soc.*, 1954, **76**, 5185.
31. McIvor, Grant and Hubley, *Can. J. Chem.*, 1956, **39**, 1611.
32. Corbridge and Lowe, *J. Chem. Soc.*, 1954, 493.
33. Simon and Stolzer, *Chem. Ber.*, 1956, **89**, 2253.
34. Laarsen, *Acta Chem. Scand.*, 1952, **6**, 1470.
35. Daasch, *J. Amer. Chem. Soc.*, 1954, **76**, 3403.
36. Thomas and Chittenden, *Chem. and Ind.*, 1961, 1913.
37. Hudson, in *Structure and Mechanisms in Organo Phosphorus Compounds*, Chapter 3 (Academic Press, New York, 1965).
38. Goldwhite and Previdi, *Spectrochim. Acta*, 1970, **26A**, 1403.
39. Nyquist, *Spectrochim. Acta*, 1969, **25A**, 47.
40. Gramstad and Mundheim, *Spectrochim. Acta*, 1972, **28A**, 1405.
41. Gramstad, *Spectrochim. Acta*, 1970, **26A**, 426.
42. Nyquist, *Applied Spectroscopy*, 1957, **II**, 161.
43. Thomas and Chittenden, *Spectrochim. Acta*, 1964, **20**, 489.
44. Thomas, Chittenden and Hartley, *Nature*, 1961, **192**, 1283.
45. Thomas and Chittenden, *J. Opt. Soc. Amer.*, 1962, **52**, 829.
46. Ryskin and Stavitskaya, *Optics and Spectroscopy*, 1959, **7**, 418.
47. Larsson, *Svensk. Kemisk. Tidskrift*, 1958, **70**, 405.
48. Schindlbauer and Steininger, *Monats. Chem.*, 1961, **92**, 868.
49. Chittenden and Thomas, *Spectrochim. Acta*, 1965, **21**, 861.
50. Chittenden and Thomas, *Spectrochim. Acta*, 1964, **20**, 1679.
51. Wolf, Miquel and Mathis, *Bull. Soc. Chim. France*, 1963, 825.
52. Mallion, Mann, Tong and Wystrach, *J. Chem. Soc.*, 1963, 148.
53. Bellamy, *Advances in Infrared Group Frequencies* (Methuen, London, 1968).
54. Nyquist, *Spectrochim. Acta*, 1963, **19**, 713.
55. Chittenden and Thomas, *Spectrochim. Acta*, 1966, **22**, 1449.
56. Kabachnik, Gilyarov and Popov, *J. Gen. Chem. USSR*, 1962, **32**, 1581.
57. Quinchon, Maryvunme, LeSech and Gryszkiewcz-Trochimowski, *Bull. Soc. Chim. France*, 1961, 735.
58. Thomas and Chittenden, *Spectrochim. Acta*, 1970, **26A**, 781.

19

Halogen Compounds

19.1. Introduction and table

The only characteristic absorptions shown by halogen compounds in the rock-salt region are those arising from the C—X stretching mode. This is a single-bond skeletal mode, and it is consequently subject to considerable frequency alterations as a result of interactions with neighbouring groups. This is especially the case with the lighter halogens, fluorine and chlorine, in which the masses involved are not very different from the rest of the carbon skeleton. The interaction effects are reduced in organic bromides in which the C—Br stretching absorption is therefore rather more stable in position. Alkyl halides studied in solution or in the liquid state show multiple C—X stretching frequencies. These arise from the co-existence of rotational isomers with different fundamental frequencies. In simple alkyl halides such forms are limited to *trans*- and *gauche*-configurations only, but in compounds such as 1 : 2-dichlorethane or in long-chain halides the number of possible isomers is increased. A substantial volume of work has been carried out on individual molecules of this type, and a review of the data by Mizushima [12], and detailed studies by Brown and Sheppard [13, 14] provided the first systematic characterisation of these frequencies. They established that in the solid states in which only one rotational form is allowed the *trans*-form is generally the more stable and has the higher C—X stretching frequency. Parallel results are found also in the axial and equatorial C—X stretching modes of chlorinated *cyclo*hexanes [15] and sterols [16] which have useful diagnostic applications.

Since that time there have been many Raman studies, particularly of chlorinated hydrocarbons, in which the C—Cl bonds are very easily identified by their great intensity, and these, coupled with further infra-red measurements have led to a series of useful correlations which enable one to identify not only the halogen present in any given type of substitution, but also to identify the substitution at the carbon carrying the halogen. The characteristic frequency ranges of the various rotational conformations have also been clarified so that it is possible to decide which of several possible conformers is present in the haloalkanes. These correlations are, however, of little or no use in either fluoro-alkanes or in chlorinated benzenes.

With fluorine substitution, interaction effects are at their greatest, and C—F stretching absorptions occur anywhere within the range $1400-1000$ cm^{-1}, depending on the nature and degree of fluorination. In heavily fluorinated compounds such as the fluorocarbons this results in a series of very intense absorptions appearing over the whole of this range. The intensity of the C—F absorption is exceptionally high, and fluorocarbons can be recognised as such from this property alone. Fluorine substitution results in very considerable high-frequency shifts in adjacent CH, C=C and C=O stretching vibrations, and further details of this will be found in the appropriate chapters dealing with these groupings. It is clear, therefore, that the only correlations likely to be reasonably precise and widely applicable are those which arise from some particular vibration occurring in a constant environment. Some possible correlations along these lines are suggested below for the CF—CF$_3$ grouping.

Chlorine substitution also gives rise to interaction effects, but they are relatively small, and the influence of chlorine on neighbouring C=C bonds, for example, is slight. The C—Cl stretching absorption in compounds containing only a single such link is therefore usually confined to a narrow frequency range, for any given conformation. However, the differences between conformers can be considerable, and amount to about 70 cm^{-1} in the cases of the two forms allowed in straight chain 1-chloro-alkanes. With secondary or tertiary chloroalkanes a number of rotational forms can occur giving in the extreme cases as many as six different C—Cl stretching bands. The positions of each and the probable frequency ranges are now established so that these bands are useful in the identification of the actual rotamers allowed.

A parallel situation holds for bromo- and iodo-alkanes. The Raman data have been well summarised by Dollish *et al.* [20] and the infra-red by Bentley *et al.* [21] Altona [22] has also reviewed the correlations available for alkyl and *cyclo*alkyl halogenides. The notation for the characterisation of the individual rotational forms was first suggested by Mizushima and later revised and extended by Krimm *et al.* [23, 24]. The substitution is first classified into primary (P), secondary (S) or tertiary (T), and then subdivided by the use of subscripts which identify the atom or atoms which are *trans* to the halogen. A straight chain primary compound would therefore have two bands in solution which would be classifed as P_C and P_H. If, however, the next nearest neighbour carbon atom is twisted away from the planar zig-zag position, as in 1-chloro-2-methyl propane, a slightly different C—Cl frequency is found and this form is classified as $P_H{}'$. It will be evident that in secondary and tertiary alkanes a multiplicity of forms can occur, which in the case of the secondary series amount to the six different ways in which any pair of C, H or H' can be combined.

The correlations dicussed are shown in Table 19 below.

Table 19

Carbon—Halogen Stretching Frequencies: Haloalkanes (cm^{-1})

Conformation	Cl	Br	I
Primary			
P_H	660−648	565−560	510−500
$P_H{}'$	690−679	625−610	590−578
P_C	730−720	650−640	602−584
Secondary			
S_{HH}	615−605	540−535	495−483
$S_{HH'}$	637−627	590−575	560−545
$S_{H'H'}$	690−680	670−650	−
S_{CH}	675−655	620−605	585−575
$S_{CH'}$	670	−	−
S_{CC}	760−740	700−680	−
Tertiary			
T_{HHH}	580−560	520−510	495−485
$T_{HHH'}$	540	−	−
T_{CHH}	630−590	590−580	580−570
$T_{CHH'}$	590	−	−
Other Halogen Correlations			
C—F stretch	$1400−100 \text{ cm}^{-1}$ very intense.		
CF$_3$ Phenyl	$1321 \mp 9, 1179 \mp 7, 1140 \mp 9 \text{ cm}^{-1}$		
CF—CF$_3$	$745−730, 1365−1325 \text{ cm}^{-1}$		

19.2. C—F Linkages

In simple molecules the presence of a single fluorine atom usually results in the appearance of a moderately intense absorption in the $1100-1000$ cm^{-1} region. A considerable number of simple fluorinated compounds have been fully studied [1, 2, 4], and it is found that in general this frequency rises with further fluorine substitution and is split into two peaks arising from symmetric and asymmetric vibrations. With larger molecules containing a considerable proportion of fluorine very intense absorption occurs over the $1400-1000$ cm^{-1} range. The study of individual compounds of this type has been a popular subject, and a considerable number of papers have been published in this field without, however, any general correlations being formulated.

A typical series which exhibit this frequency shift is the fluoromethane group described by Thompson and Temple [3]. Their assignment of the CF stretching mode is as follows: CCl_2FH, 1072 cm^{-1}; CCl_3F, 1102 cm^{-1}; CCl_2F_2, 1155 cm^{-1} and 1095 cm^{-1}, and $CClF_3$, 1210 cm^{-1} and 1102 cm^{-1}. With even a small increase in complexity up to 1 : 1 : 1-trifluoroethane, multiple peaks appear due to interaction, and this compound shows bands at 1290 cm^{-1}, 1278 cm^{-1}, 1266 cm^{-1}, 1230 cm^{-1} and 1135 cm^{-1} which cannot be fully assigned.

In the fully fluorinated hydrocarbons the spectra are very complex, and a whole series of very intense bands appear in the $1350-1100$ cm^{-1} regions, most of which are probably associated in some way with C—F stretching vibrations. The spectra of a number of such materials have been given by Thompson and Temple [5]. They are immediately recognisable both by the extreme intensities shown and by the lack of any appreciable absorption at frequencies higher than 1350 cm^{-1}, but no detailed correlation rules can be formulated, and it is clear from the evidence of interactions that it is unlikely that any simple correlations will be developed in the future. This may, however, be possible if they relate to larger molecular units in which the environment around a specific linkage is reasonably constant. Barnes *et al.* [6] have suggested that the $=CF_2$ group may have characteristic absorptions at 1340 cm^{-1} and 1200 cm^{-1}, and this is supported by many of the spectra of simple fluorinated compounds in which this group is present, but its value is much reduced by the appearance of bands at these points in many saturated fluorocarbons.

19.3. C—Cl Linkages

The spectra of chlorinated compounds are generally more regular than those of fluorinated materials, and the C—Cl stretching band appears in a narrower, although still broad, range. There is also less interaction and interference with the characteristic absorptions of neighbouring groups such as C=O or C=C. As before, a very large number of simple chlorine-containing molecules have been examined [1, 7, 12—14, 20—26]. In primary straight chain compounds, only the P_C and P_H forms can occur. These absorb in the narrow frequency ranges 730—720 cm^{-1} (P_C) and 660—650 cm^{-1} (P_H). When the chain is branched at the 2-position the P_H form is replaced by the P_H' as the methyl group is rotated away from the planar zig-zag position. This results in a band in the 690—680 cm^{-1} range. With 2 methyl groups at this point, as in 1-chloro-2 : 2-dimethyl propane, only the P_C form can occur and this compound has a single band at 722 cm^{-1}.

With secondary haloalkanes, six different configurations are theoretically possible but in many cases the actual numbers are reduced by steric factors. Nevertheless the band positions for each characteristic form have been successfully identified. The individual frequencies are listed in the table. 2-Chloropropane exists only in the HH form and has only a single band at 614 cm^{-1}. 3-Chloro-2 : 2 : 4 : 4-tetramethyl pentane, in contrast has only the CC form and absorbs at 758 cm^{-1}. Longer straight chain secondary compounds, such as 3-chloro pentane show multiple bands, in this case due to four forms. The band from the HH form is usually prominent in these cases.

In tertiary compounds in which rotation can occur about two or three axes, up to ten rotamers are theoretically possible but usually steric factors limit the numbers to two or three. The frequency of the THHH form is at 560 cm^{-1} and that of the TCHH at 620 cm^{-1}. In cyclic structures, such as *cyclo*hexanes, chlorine substitution can be either axial or equatorial, and the two configurations show different frequencies as before. The connection between these cases and the simple alkyl halides has been discussed by Sheppard [17, 28]. In *cyclo*hexane the C—Cl absorptions are at 742 cm^{-1} (equatorial) and 688 cm^{-1} (axial) [15], and similar results are shown by halogenated steroids. In the latter case, however, there are also some differences which are characteristic of the position of substitution [16]. Equatorial C–Cl stretching frequencies of 2-, 3- and 7-chlorosteroids, for example, are 755 cm^{-1}, 782–750 cm^{-1} and 749 cm^{-1} respectively; the corresponding axial frequencies being 693 cm^{-1}, 730–617 cm^{-1} and 588 cm^{-1}. The characteristic frequencies are also a function of ring size. Ekejiuba and Hallam [27] have presented a review of the data. In five-membered rings the equatorial band is near 624 cm^{-1} and the axial near 588 cm^{-1}. In four-membered rings these fall to 618 (equatorial) and 528 (axial).

With multiple chlorine substitution at the same carbon the frequencies are of course split into symmetric and antisymmetric modes. In the simple case of 1 : 1-dichloro ethane this gives rise to two bands at 689 and 641 cm^{-1}, but in more complex cases multiple bands occur. In general there is an upwards frequency shift as the number of halogens increases. In carbon tetrachloride for example the intense infra-red band is at 797 cm^{-1}.

The vicinal dihalogen alkanes exist in the liquid or solution state as mixtures of the *trans* and *gauche* forms. The *trans* form usually predominates and becomes the only form in the solid state. A review of the stretching frequencies of compounds of this series is given by Altona and Hageman [8].

19.4. C–Br Linkages

Work on simple molecules indicates that the C–Br absorption occurs in the region of 600–500 cm^{-1} but, as before, two or more bands are shown by alkyl bromides in solution, and in cyclic structures the frequency also varies with the orientation. Early studies by Mortimer *et al.* [9] and by Brown and Sheppard [13, 14] identified the *trans* band of alkyl bromides near 650 cm^{-1} and the *gauche* near 500 cm^{-1}. Detailed infra-red [29], and Raman studies [20] have since elaborated these correlations to a point where the individual frequencies of the various rotational isomers have been identified, just as they have been for the chlorine series. The individual values for the rotamers are summarised in the table. The overall spread is very wide (720–480 cm^{-1}), but the ranges for individual species are quite narrow. It is therefore possible to identify the substitution pattern and the specific forms allowed within that pattern.

In bromo-*cyclo*alkanes, the frequency is doubled corresponding to equatorial and axial forms. In *cyclo*hexane rings the equatorial band is near 690 cm^{-1} and the axial near 660 cm^{-1}. As the ring size diminishes these frequencies fall. In *cyclo*butanes, for example, the corresponding values are 534 and 418 cm^{-1}. In the sterol series [16, 30] a limited number of compounds have been studied, and the results indicate that equatorial substitution leads to bands in the $750-700 \text{ cm}^{-1}$ region, whereas axial substitution results in absorption in the $590-690 \text{ cm}^{-1}$ range. In some dibromosteroids axial substituents show an even wider range and absorb near 550 cm^{-1}. The reasons for this wide range and for the abnormally high frequencies of the equatorial form compared with alkyl bromides are not known.

19.5. C—I Linkages

The iodo-alkanes have been studied in detail by Bentley *et al.* [29]. The overall range is $600-465 \text{ cm}^{-1}$, but the ranges for the individual rotamers are again much smaller, nd the specific forms can therefore be identified. The values for the separate forms are tabulated above.

19.6. Other correlations

It has been suggested above that in view of the considerable coupling which occurs between C—C and C—F links, the only correlations which are likely to be of any value for such bonds will be those relating to larger structural units in which these effects are stabilised. Several correlations of this type have in fact been described.

Thus in the grouping CF_3Ph the asymmetric and symmetric deformation frequencies of the CF_3 group appear at stable positions and are not much influenced by the nature or positions of other substituents in the aromatic ring. They occur at $1321 \pm 9 \text{ cm}^{-1}$ (symmetric); $1179 \pm 7 \text{ cm}^{-1}$ and $1140 \pm 9 \text{ cm}^{-1}$ (asymmetric) [11], the asymmetric frequency being split due to the influence of the aromatic ring. These values are therefore useful for the recognition of the CF_3-aryl group.

A second limited correlation of this type relates to the C—F stretching frequencies of the CF_3 group when attached to a fluoromethylene link. The structure CF_3CF_2-X has a characteristic frequency in the range $1365-1325 \text{ cm}^{-1}$ which is reasonably strong when X is a halogen or carbonyl group [18, 19]. The CF_3CF_2- group has also a characteristic frequency [10], which is probably a C—F deformation frequency in the range $745-730 \text{ cm}^{-1}$.

19.7. Bibliography

1. Herzberg, *Infra-red and Raman Spectra of Polyatomic Molecules* (Van Nostrand, 1945).
2. Torkington and Thompson, *Trans. Faraday Soc.*, 1945, **41**, 236.
3. Thompson and Temple, *J. Chem. Soc.*, 1948, 1422.
4. *Idem, ibid.*, p. 1428.
5. *Idem, ibid.*, p. 1432.
6. Barnes, Gore, Stafford and Williams, *Analyt. Chem.*, 1948, **20**, 402.
7. Cf. Randall, Fowler, Fuson and Dangl, *Infra-red Determination of Organic Structures* (Van Nostrand, 1949).
8. Altona and Hageman, *Rec. Trav. Chim. Pays Bas*, 1969, **88**, 33.
9. Mortimer, Blodgett and Daniels, *J. Amer. Chem. Soc.*, 1947, **69**, 822.
10. Bellamy and Branch, *Nature*, 1954, **173**, 633.
11. Randle and Whiffen, *J. Chem. Soc.*, 1955, 1311.
12. Mizushima, *Structure of Molecules and Internal Rotation* (Academic Press, New York, 1954).
13. Brown and Sheppard, *Trans. Faraday Soc.*, 1954, **50**, 1164.
14. *Idem, Proc. Roy. Soc.*, 1955, **A231**, 555.
15. Larnaudie, *Compt. Rend. Acad. Sci. (Paris)*, 1952, **235**, 154; 1953, **236**, 909.
16. Barton, Page and Shoppee, *J. Chem. Soc.*, 1956, 331.
17. Sheppard, *Chem. and Ind.*, 1957, 192.
18. Hauptschein, Stokes and Nodiff, *J. Amer. Chem. Soc.*, 1952, **74**, 4005.
19. Tiers, Brown and Reid, *ibid.*, 1953, **75**, 5978.
20. Dollish, Fateley and Bentley, *Characteristic Raman Frequencies of Organic Compounds*, (Wiley, Interscience, New York, 1974).
21. Bentley, Smithson and Rozek, *Infrared Spectra and Characteristic Frequencies 700–300 cm^{-1}* (Interscience, New York, 1968).
22. Altona, *Tetrahedron Letters*, 1968, **19**, 2325.
23. Shipman, Folt and Krimm, *Spectrochim. Acta*, 1962, **18**, 1603.
24. Shipman, Folt and Krimm, *Spectrochim. Acta*, 1968, **24A**, 437.
25. Klaboe, *Spectrochim. Acta*, 1970, **26A**, 87.
26. Gates, Mooney and Willis, *Spectrochim. Acta*, 1967, **23A**, 2043.
27. Ekejiuba and Hallam, *J. Mol. Structure*, 1970, **6**, 341.
28. Sheppard, *Advances in Spectroscopy* Vol. I. (Interscience, 1959).
29. Bentley, McDevitt and Rozek, *Spectrochim. Acta*, 1964, **20**, 105.
30. Page, *Chem. And Ind.*, 1958, 58.

20

Organo-Silicon Compounds

20.1. Introduction and table

Infra-red absorption bands arising from linkages involving silicon atoms are about five times more intense than the bands from the corresponding carbon linkages. The reasons for this have been discussed by Wright and Hunter [1]. This fact is of considerable help in analytical work, and infra-red absorption spectra have been very extensively employed in the development and characterisation of organo-silicon polymers and related materials. Many of the basic correlations in this area were identified as a result of the early studies of Wright and Hunter [1], Young, Servais, Currie and Hunter [2], Richards and Thompson [3], and Clark, Gordon, Young and Hunter [4] and their results have been fully confirmed and extended by later workers. The frequencies of bonds involving silicon appear to be largely unaffected by the physical state, except where hydrogen bonding occurs. Simon and McMahon [18] have confirmed this by a comparison of the spectra of some alkyl silanes and siloxanes in the gaseous, liquid and solid phases. The correlations listed have been supported by a considerable number of fundamental studies on single molecules, or related groups, such as silane [6, 7], tetramethyl silane [8, 9], and methyl [19, 23, 24] and halogenated silanes [10, 11, 20, 21, 23]. Frequencies which have been suggested as being characteristic for specific groupings including silicon, are listed in Table 20, and the evidence on which they are based is discussed in the text.

Table 20

Frequencies of Silicon-Containing Groups (cm^{-1})

$Si(CH_3)_3$	1260, 1250.	near 840 and 755, 715−680, 660−435.
$Si(CH_3)_2$	1258 ∓ 5,	near 850, and 800.
$Si(CH_3)$	1258 ∓ 5,	near 765.
$Si(CH_2)$	1250−1200,	760−670.
SiC_6H_5	1125−1100.	
Si−O−Si	Cyclic	Trimers 1020−1010.
		Tetramers 1090−1080.
		Larger Rings 1080−1050.
	Open Chain	
		1093−1076, 1055−1020.
Si−O−C		1110−1080.
SiH	Overall	2280−2050.

			Alkyl or aryl
SiH$_3$	Alkyl 2153−2142.		947−930, 930−910.
	Aryl 2157−2152.		
SiH$_2$	Dialkyl 2138−2117.		940−925.
	Diaryl 2147−2130.		
SiH	Trialkyl 2105−2095.		845−800.
	Triaryl 2132−2112.		

SiF	1000−830.
Si−Cl	Usually one band between 600−550, overall 650−370.

20.2. Si−C Linkages

Vibrations involving the stretching of the Si−C link occur in the $900-500$ cm^{-1} region of the spectrum, and are considerably influenced by the nature of the substituent groupings. Thus the C−Si stretching vibration in tetramethyl silane has been assigned to 860 cm^{-1}, whereas that of tetraethyl silane is given at 733 cm^{-1}. The frequency ranges for more limited systems are of course much smaller and these are useful. The *as* and *s* stretching frequencies of the Si−C bonds in the $Si(CH_3)_3$ group, for example, fall in a narrow range, so also do the Si−C bands of a single methyl group attached to an SiH$_2$ group. Yet another frequency range is shown for O−Si−CH$_3$. The SiCH$_3$ group is best identified by a near invariant band near 1260 cm^{-1} but the numbers of such groups attached to the silicon can then be determined from the positions of the Si−C bands and from the CH$_3$ rocking modes.

20.2(a). SiCH$_3$ stretching vibrations. *−Si(CH$_3$)$_3$*. In the series of

open-chain compounds examined by Wright and Hunter [1] (the range from hexamethyl disiloxane to octadecamethyl octasiloxane) a strong band is shown at 841 cm^{-1} which decreases in intensity as the chain length is increased, and which is uniformly absent from the corresponding cyclic compounds (hexamethyl*cyclo*trisiloxane to hexadecamethyl*cyclo*octasiloxane). A second band whose intensity varies with chain length occurs in the same compounds at 756–754 cm^{-1}. These are clearly associated with vibrations involving the $Si(CH_3)_3$ end-groups of the open-chain materials. These findings were confirmed by Richards and Thompson [3] and by Clark *et al.* [4]. These bands have both been identified with the CH_3 rocking mode [5, 19] and as will be seen the number and positions of the bands is a characteristic of the number of methyl groups attached to the silicon atom [5].

The Si–C stretching vibrations in this system have been identified by Kriegsmann *et al.* [12, 13]. Two bands occur in the ranges 715–680 and 660–485 cm^{-1} corresponding to the anti-symmetric and symmetric Si–C stretching modes.

$-Si(CH_3)_2-$. All the compounds examined by Wright and Hunter [1] show absorption in the 800 cm^{-1} region when $-Si(CH_3)_2-$ groups are present. This is a methyl rocking mode, and a second band is always associated [5] with it near 855 cm^{-1}. The 800 cm^{-1} band appears in the range 814–802 cm^{-1} in the cyclic polymers, and is practically invariant at 800 cm^{-1} in the open-chain materials and methyl silanes. Richards and Thompson [3] have shown that the band persists at this point also, in the dimethyl siloxanes. The position of the Si–C stretching bands in this series is less well defined but in dimethyl silane [17] these occur at 730 and 650 cm^{-1}.

$-Si(CH_3)$. Kuivilia *et al.* [5] identify the methyl rocking frequency of this system as being close to 765 cm^{-1}. However, the Si–C stretching frequency can also fall in this region. Ebsworth *et al.* [17] give values of 760–735 cm^{-1} for the monohalosilanes, but the band position is apparently sensitive to the nature of adjacent groups. In the group $OSiCH_3$ the Si–C stretch rises to 850–840 cm^{-1} when this is an end group. If it is a side chain alkoxy group the frequency falls [31] to 800–770 cm^{-1}. In cyclic siloxanes the frequency depends to some extent on ring size but the band is in the overall range 815–790 cm^{-1}. The band is strong but the very great intensity of the Si–O stretching bands can give the illusion that it is not.

20.2(b). Si—CH₃ deformation vibrations. All the compounds examined by Wright and Hunter exhibit a strong band at 1258 cm^{-1}. A similar band is exhibited by tetramethylsilane [8, 9] at 1250 cm^{-1}. This band has been studied by a number of workers [5, 25, 32] and there is now no doubt that it is due to the symmetric methyl deformation. The band is remarkably constant in position, falling within one or two wave numbers of 1258 cm^{-1} in silanes with methyl substituents. It is reasonably strong and is therefore easily recognised. It does rise in frequency up to 1275 cm^{-1} when the silicon atom is linked to a metal [5] or to some other very electropositive group. The band is close to its normal position in siloxane derivatives. Unlike the corresponding symmetric methyl deformation in alkanes the band does not double when two methyl groups are attached to the same atom but does so with three. However a second band is found in this same region in many compounds including cyclic materials with only single methyl groups on any one silicon [2, 3, 23, 32].

This lower frequency band is of medium intensity and is less constant in position than the one at 1258 cm^{-1}. It is due to the deformation mode of the CH_2 group adjacent to the silicon, and the frequency falls progressively from 1250 cm^{-1} towards 1200 cm^{-1} as the chain length is increased [32]. This same group is also responsible for a band in the 760–670 cm^{-1} range due to the methylene rocking vibration. The SiC_2H_5 group itself absorbs in the 1250–1230 cm^{-1}, 1020–1000 cm^{-1} and 970–945 cm^{-1} regions [32]. The antisymmetric deformation mode of the methyl group is of less interest. It occurs as a relatively weak band near 1410 cm^{-1} and as such it is of very limited use for identification purposes.

20.2(c). Si—Phenyl vibrations. There have been numerous attempts to identify frequencies which would characterise the silicon phenyl bond [2, 3, 25, 33, 34]. Most of these have concentrated their interest on bands near 1430 cm^{-1} and 1100 cm^{-1}. However the first of these is a typical normal band of monosubstituted aromatics. Indeed the ring vibrations are all essentially normal [34] and the only one which is at all sensitive to the silicon substitution is that near 1100 cm^{-1}. This is an in-plane deformation of the ring which has some contribution from the Si—C stretch. It is therefore stronger than usual and is easy to identify from its characteristic aromatic sharpness. It occurs in the 1125–1100 cm^{-1} range so that in siloxanes it often appears as a shoulder on the side of the very much stronger Si—O bands.

Lee Smith *et al.* [34] have commented on the fact that in silicon-phenyl compounds the 1600 cm^{-1} band is more intense than that at 1470 cm^{-1}. This is not usual but does occur in other aromatics, (see Chapter 5).

20.3. Si—O Stretching vibrations

The Si—O—Si bond gives rise to at least one very intense band in the 1100—1000 cm^{-1} region due to the antisymmetric stretching mode. However the corresponding mode of the Si—O—C group also appears in this region so that the two cannot always be differentiated. In the cases of the polymeric siloxanes, only the first structure is present unless the silicon atom has an alkoxy substituent. There are clear differences between cyclic and open-chain polysiloxanes which reflect the differing bond angles.

20.3(a). Cyclic siloxanes. Wright and Hunter [1] have given the spectra of the six cyclic materials from hexamethyl*cyclo*trisiloxane to hexadecamethyl*cyclo*-octasiloxane. In the trimer the Si—O band appears at 1018 cm^{-1}, whilst in the remainder it falls in the range 1076—1056 cm^{-1}, the frequency showing a small but steady fall throughout the series with increasing ring size. Richards and Thompson [3] confirmed these findings for the first five members of the series, and also examined two cyclic aromatic siloxanes $(SiPh_2O)_3$ and $(SPh_2O)_4$. The first of these materials gave a strong band at 1015 cm^{-1} close to that of the trimer of the methyl series, whilst the tetramer gave its strongest band in this region nearer to 1100 cm^{-1}.

Young *et al.* [2] studied the trimers and tetramers only, but covered a wider range of substituents on the silicon atom. Thus, in cyclic materials $(SiR_1R_2O)_n$ they examined the trimers and tetramers of the dimethyl, diethyl diphenyl, methyl-phenyl and ethyl-phenyl series together with a mixed diphenyl tetraethyl trimer. They found that there was remarkably little difference in the position of the Si—O—Si absorption bands with variation in the nature of the substituents, although there was a considerable shift as between the trimers and tetramers. Thus they found all the trimers to exhibit the Si—O—Si absorption band in the narrow range 1020—1010 cm^{-1}, whereas all the tetramers absorbed in the range

$1093-1081$ cm^{-1}. Kriegsmann [14] has also studied cyclic dimethyl siloxanes of various ring sizes with similar results to Wright and Hunter [1]. The ability to distinguish between trimers and tetramers in this way is valuable as these are the ones most commonly produced by the hydrolysis of the chlorides. The occurrence of only one strong band in this region is also diagnostic, as open chain polymers have two, except in some cases where the chain is branched.

20.3(b). Open-chain polymeric and monomeric siloxanes. In the series of open-chain polymethyl siloxanes from hexamethyl-disiloxane, to octadecamethyloctasiloxane, Wright and Hunter [1] find the Si—O absorption band within the range $1055-1024$ cm^{-1}, and again there is a small fall in the frequency for each unit increase of the chain length. However this band is accompanied by a second strong band between $1093-1076$ cm^{-1}, the frequency now rising with increasing degrees of polymerisation. These results have been confirmed by Richards and Thompson [3] and by Okawara [31] using alkoxy polysiloxanes. These contain not only the strong pair of Si—O—Si bands but an additional band at 1100 cm^{-1} which can be attributed to the Si—O—C bond.

The Si—O—C bond is best studied in alkoxy silanes and similar molecules in which confusion with the Si—O—Si frequencies can be avoided [27, 28, 35—38]. In simple alkoxy silanes containing only the SiOCH$_3$ group the band is at $1107-1100$ cm^{-1}. This is the antisymmetric mode in the view of some workers and the symmetric mode is assigned in the range $947-1100$ cm^{-1}. However, as this last is close to the Si—O stretch in the silanols it could equally well be that the 1100 cm^{-1} band is primarily a C—O stretch, similar to that in normal ethers. This is supported to some extent by its sensitivity to changes at the carbon atom. Even the SiOC$_2$H$_5$ group shows some difference and has two bands in the range $1100-1075$ cm^{-1}. With more complex compounds the number of bands increases still further. Liu [36] lists a number of compounds which all show one band in the $1100-1080$ cm^{-1} range but have a second in the $1060-1040$ cm^{-1} range. Verrijn Stuart *et al.* [38] find two bands in the same region in *iso*proxy and normal propoxy trichlorsilanes but only one at 1064 cm^{-1} in compounds in which the carbon of the Si—O—C bond is part of a ring. Frequencies higher than 1100 cm^{-1} are found in other cases such as those listed by Shostakovski *et al.* [37] in which a triple bond is attached to the carbon atom. The

Si—O—C bands are assigned in these compounds at 1150 and 1028—1044 cm^{-1}

Silanols have been reported [26] as 900—885 cm^{-1} with lower values near 820 cm^{-1} for aromatic silanols. These values are in line with those reported by Kessler *et al.* [35] for Si—O—C systems, and the aromatic values are comparable with the range 810—780 cm^{-1} reported for simple silyl esters with the grouping CO—O—SiH$_3$ [39].

20.4. Si—H Stretching and bending vibrations

In silane itself the SiH stretching frequency [6, 7] is at 2187 cm^{-1}, and this fixes the approximate region in which this group can be expected. In fact the stretching frequency is very sensitive to the electronegativity of the groups attached to the silicon, so that trichlorosilane [15] absorbs at 2274 cm^{-1} whereas trimethyl silane absorbs at 2118 cm^{-1} [40]. There is a relationship between ν_{SiH} and the effective electronegativities of the substituents, as Smith and Angelotti [40] have shown, and Thompson [41] has converted this into a relationship with Taft σ^* values. This takes the form $\nu_{SiH} = 2106 + 17.5\Sigma\sigma^*$ and can be valuable in giving reasonable approximations of the expected stretching frequencies for any given structure.

The antisymmetric and symmetric stretching frequencies of the SiH$_3$ and SiH$_2$ groups are so close together as to be indistinguishable except in a few special cases. However, like the methyl group, SiH$_3$ often shows two deformation frequencies whereas the SiH$_2$ group has only one.

The SiH$_3$ group. A useful collection of data on monoalkyl and aryl silanes is that of Kniseley *et al.* [33]. Alkyl silanes absorb between 2153—2142 cm^{-1} and the aryl counterparts at slightly higher frequencies 2157—2152 cm^{-1}. Conjugation with triple bonds such as C≡N or C≡C raises the frequency higher and splits the degeneracy in some cases so that the two separate stretching frequencies can be seen. Thus [43] the compound SiH$_3$ C≡CSiH$_3$ absorbs at 2190 and 2170 cm^{-1}. Vinyl silanes have been studied by Frankiss [42].

The SiH$_3$ deformation frequencies occur at 947—930 and 930—910 cm^{-1} in both alkyl and aryl silanes [32, 33]. Duncan [44] quotes a band between 947 and 912 cm^{-1} in compounds with a triple bond attached to the silicon, but does not report a second band in this region except for vinyl silane and disilylether. In silyl esters also, the two bands are not well resolved [39]. They occur near 950 cm^{-1} but are separated by only 3 cm^{-1} in the trifluoroacetate. The rocking frequencies of this group occur near 700 cm^{-1} and appear to absorb in a reasonably narrow range between 720 and 680 cm^{-1}.

The SiH$_2$ group. There is a small further fall in the SiH stretching frequency when a second alkyl group is introduced. Dialkyl silanes absorb between 2138–2117 cm^{-1}, and diaryl silanes between 2147–2130 cm^{-1} [17, 25, 29, 32, 33]. The frequency rises sharply if halogens are substituted for the alkyl groups. Difluorosilane absorbs at 2246 cm^{-1} and methyl chlorosilane [17] at 2200 cm^{-1}. The deformation frequency is less affected by the substituents, and in alkyl or aryl silanes it falls in the 940–928 cm^{-1} range close to the SiH$_3$ deformations. Replacement of one methyl by fluorine raises this to 975 cm^{-1}. Methyl chlorosilane absorbs at 960 cm^{-1}.

The SiH group. Numbers of tri-substituted silanes with three identical substituents have been studied by Smith and Angelotti [40] in connection with their electronegativity relationships, and many additional data are available from studies on siloxanes and similar materials. Triaryl silanes absorb between 2132–2112 cm^{-1} and trialkyl compounds between 2100–2095 cm^{-1}. Mixed alkyl/aryl materials have corresponding intermediate values. The effects of the electronegativity of the substituents is well shown by the values of trifluro (2282 cm^{-1}), tribromo (2236 cm^{-1}) and trimethoxy (2203 cm^{-1}) silanes. Compounds with mixed groups have the appropriate intermediate values.

The SiH group does not absorb in the 950–910 cm^{-1} region [33] but has a bending frequency in the 846–800 cm^{-1} range [45, 46].

20.5. Si–Halogen vibrations

The general order of the Si–Halogen frequencies can be seen from the values for this bond in the silyl halides [44]. These are Si–F 872 cm^{-1}, Si–Cl 551 cm^{-1}, Si–Br 430 cm^{-1} and Si–I 360 cm^{-1}. Data on a small number of Si–F compounds suggest [17, 22] that the frequency range is considerable, and that this link can absorb anywhere between 1000 and 830 cm^{-1}. Many Si–Cl bonds absorb between 600 and 550 cm^{-1}, but again the spread is wide, perhaps as great as $650-370 \text{ cm}^{-1}$, and there are two bands in many cases even when only one Si—Cl bond is present. Data on such compounds have been given by a number of workers [16, 32, 35, 47], but the relationships between the frequency ranges and the various rotational forms have not been clarified as they have been for the C—Cl bond. In a small number of chloro alkoxy silanes the Si—Cl bands appear [35] in the narrow ranges $575-570 \text{ cm}^{-1}$ and $525-510 \text{ cm}^{-1}$, but in trimethyl chloro silane [47] there is only a single band at 487 cm^{-1}. The dichloro and trichloro compounds show two bands in the $550-570$ and $475-445 \text{ cm}^{-1}$ regions, due to the antisymmetric and symmetric modes.

20.6. Si—OH vibrations

The free OH stretching frequency of silanol is significantly higher than that of methanol. This is consistent with the usual rise in OH frequencies when the oxygen atom is attached to an element of low electronegativity. The free OH frequencies of alkyl and aryl silanols have been examined by West and Baney [48, 49] and by Nillius and Kriegsmann [50]. The aryl compounds absorb at slightly lower frequencies than the alkyl, but although this difference is only about 11 cm^{-1} there are large differences in the acidities and basicities of the two [48, 49]. In cyclohexane solution, trimethyl silanol absorbs at 3697 cm^{-1} and the triaryl compounds at 3686 cm^{-1}. Mixed derivatives show intermediate values. Both compounds are sensitive to solvent effects, and even solutions in chloroform show reductions in ν_{OH} of about 25 cm^{-1}.

In the liquid state these compounds associate in the usual way and the bonded OH frequencies fall to about 3250 cm^{-1}. The values in other situations depend of course upon the hydrogen bond strengths, but the band invariably broadens and intensifies, as in normal alcohols.

The frequencies of the Si—O stretching bonds have already been discussed in Section 20.3 above.

20.7. The influence of Si—C links on CH vibrations

The possible influence of the Si—C link on CH vibrations on the same carbon atom has been considered by Wright and Hunter [1] by Young et al. [2] and by Smith [32]. Both Wright [1] and Smith comment that the CH stretching band in all his methyl-type polymers occurs at precisely the same point as in the methyl group of hydrocarbons, but that the intensity of this band is of the order of only one-third to one-fourth of that in corresponding $C-CH_3$ compounds. The CH bending vibration occurs at a very slightly lower frequency at 1412 cm^{-1} in all the compounds studied in solution. Young *et al.* [2] confirm these findings on both methyl and ethyl polymers, and comment on the invariability in both position and in relative intensity shown by both of these bands when the materials are studied in solution.

With the aromatic compounds, as has been noted above, all the bands appear in the normal positions of aromatics [34]. The only band with any identifiable Si−Phenyl content is that between 1125 and 1100 cm^{-1} as noted earlier.

20.8. Bibliography

1. Wright and Hunter, *J. Amer. Chem. Soc.*, 1947, **69**, 803.
2. Young, Servais, Currie and Hunter, *ibid.*, 1948, **70**, 3758.
3. Richards and Thompson, *J. Chem. Soc.*, 1949, 124.
4. Clark, Gordon, Young and Hunter, *J. Amer. Chem. Soc.*, 1951, **73**, 3798.
5. Kuivilia and Maxfield, *J. Organomet. Chem.*, 1967, **10**, 71.
6. Steward and Nielsen, *J. Chem. Phys.*, 1934, **2**, 712; *Phys. Rev.*, 1935, **47**, 828.
7. Tindal, Straley and Nielsen, *Proc. Nat. Acad. Sci. U.S.*, 1941, **27**, 208.
8. Young, Koehler and McKinney, *J. Amer. Chem. Soc.*, 1947, **69**, 1410.
9. Rank, Saksena and Shull, *Discuss. Faraday Soc.*, 1950, **9**, 187.
10. Bailey, Hale and Thompson, *Proc. Roy. Soc.*, 1938, **A167**, 555.
11. Scott and Frisch, *J. Amer. Chem. Soc.*, 1951, **73**, 2599.
12. Kriegsmann, *Z. Electrochem.*, 1957, **61**, 1088.
13. Kriegsmann, *Z. Anorg. Chem.*, 1958, **294**, 113.
14. Kriegsmann, *Z. Anorg. Chem.*, 1959, **298**, 232.
15. Gibian and McKinney, *J. Amer. Chem. Soc.*, 1951, **73**, p. 1431.
16. Herzberg, *Infra-red and Raman Spectra of Polyatomic Molecules* (Van Nostrand, 1945), p. 167.
17. Ebsworth, Onyszchuk and Sheppard, *J. Chem. Soc.*, 1958, 1453.
18. Simon and McMahon, *J. Chem. Phys.*, 1952, **20**, 905.
19. Smith, *J. Chem. Phys.*, 1953, **21**, 1997.
20. Newman, O'Loane, Polo and Wilson, *ibid.*, 1956, **25**, 855.
21. Andersen and Bak, *Acta Chem. Scand.*, 1954, **8**, 738.
22. Kriegsmann, *Z. Electrochem.*, 1958, **62**, 1033.
23. Kaye and Tannenbaum, *J. Org. Chem.*, 1953, **18**, 1750.
24. Cerato, Lauer and Beachell, *J. Chem. Phys.*, 1954, **22**, 1.
25. Harvey, Nebergall and Peake, *J. Amer. Chem. Soc.*, 1954, **76**, 4555.
26. Andrianor and Ismaylor, *J. Organomet. Chem.*, 1967, **8**, A35.
27. Kreshov, Mikhailenkov and Yakimovich, *Zhur. Anal. Khim.*, 1954, **9**, 208.
28. *Idem, Zhur. Fiz. Khim.*, 1954, **28**, 538.
29. West and Rochester, *J. Org. Chem.*, 1953, **18**, 303.
30. Kakudo, Kasai and Watase, *J. Chem. Phys.*, 1953, **21**, 1894.

31. Okawara, *Bull. Chem. Soc. Japan*, 1958, **31**, 154.
32. Lee Smith, *Spectrochim. Acta*, 1960, **16**, 87.
33. Kniseley, Fassel and Conrad, *Spectrochim. Acta*, 1959, **15**, 651.
34. Lee Smith and McHard, *Analyt. Chem.*, 1959, **31**, 1174.
35. Kessler and Kriegsmann, *Z. Anorg. Chem.*, 1961, **342**, 1710.
36. Liu, *J. Chinese Chem. Soc.*, 1962, **9**, 273.
37. Shostakovskii, Shergiva, Komarov and Maroshin, *Izvest. Akad. Nauk. SSSR. Ser. Khim.*, 1964, 1606.
38. Verrijn Stuart, LaLau and Breederveld, *Rec. Trav. Chim. Pays Bas*, 1955. **74**, 747.
39. Robiette and Thompson, *Spectrochim. Acta*, 1965, **21**, 2023.
40. Smith and Angelotti, *Spectrochim. Acta*, 1959, **15**, 412.
41. Thompson, *Spectrochim. Acta*, 1960, **16**, 238.
42. Frankiss, *Spectrochim. Acta*, 1966, **22**, 295.
43. Gokhale and Jolly, *Inorg. Chem.*, 1964, **3**, 946.
44. Duncan, *Spectrochim. Acta*, 1964, **20**, 1807.
45. Kaplan, *J. Amer. Chem. Soc.*, 1954, **76**, 5880.
46. Kessler and Kriegsmann, *Z. Anorg. Chem.*, 1966, **342**, 63.
47. Rudakova and Pentin, *Optics and Spectroscopy*, 1966, **21**, 240.
48. West and Baney, *J. Phys. Chem.*, 1960, **64**, 822.
49. West and Baney, *J. Amer. Chem. Soc.*, 1959, **81**, 6145.
50. Nillius and Kriegsmann, *Spectrochim. Acta*, 1970, **26A**, 121.

21

Inorganic Ions

21.1. Introduction and table

Most early work on the vibrational spectra of inorganic ions was carried out using Raman spectra, because of the problems of sample handling in the infra-red. However, a small amount of work was done using reflection spectra [1], and the development of the nujol mull technique stimulated many studies by Lecomte and his co-workers [2–5, 10]. The introduction of the pressed potassium bromide disc technique with the associated equipment for fine grinding has led to a stimulation of interest in this field.

From the point of view of correlations for analytical and identification work, the most important and detailed paper is that of Miller and Wilkins [6], who examined and reproduced the spectra of one hundred and fifty-nine inorganic salts, and compiled a chart showing correlations for a large number of different ionic groups. This was later supplemented by another detailed paper which covered some of the lower lying frequencies [45]. In the last fifteen years there have been many detailed infra-red and Raman studies on single inorganic crystals and on groups of compounds, and an extensive literature has been built up. Nakamoto [46] has published a book solely devoted to inorganics, and Lawson [47] has provided a bibliography of this field. The book by Adams [48] and the extensive review by Cotton [49] offer a good coverage of the literature on metal ions in coordination compounds.

The infra-red method will not of course permit the complete analysis of complex mixtures except possibly in the case of certain minerals in which matching of the whole spectrum is possible but, as

Miller and Wilkins have pointed out, it affords a valuable adjunct to inorganic analysis, especially when coupled with emission analysis and X-ray diffraction data. A number of analyses of rocks and minerals have also been carried out in this way [7, 12], and detailed studies on silicate rocks have revealed sufficient differences between various types to enable infra-red methods to be used in their identification.

The correlations discussed are listed in Table 21 below.

Table 21

Carbonates (CO_3^{--})	$1490-1410$ cm^{-1} (v.s.), $880-860$ cm^{-1} (m.)
Sulphates (SO_4^{--})	$1130-1080$ cm^{-1} (v.s.), $680-610$ cm^{-1} (m.-w.)
Nitrates (NO_3^-)	$1380-1350$ cm^{-1} (v.s.), $840-815$ cm^{-1} (m.)
Nitrites (NO_2^-)	$1250-1230$ cm^{-1} and $1360-1340$ cm^{-1} (v.s.), $840-800$ cm^{-1} (w.)
Ammonium (NH_4^+)	$3300-3030$ cm^{-1} (v.s.), $1430-1390$ cm^{-1} (s.)
Cyanide, thiocyanates, cyanates and complex ions	$2200-2000$ cm^{-1} (s.)
Phosphate (PO_4^{---}, HPO_4^{--}, $H_2PO_4^-$)	$1100-1000$ cm^{-1} (s.)
Silicates (all types)	$1100-900$ cm^{-1} (s.)

21.2. Carbonates and bicarbonates

Hunt et al. [7], and Miller and Wilkins [6] have examined the infra-red spectra of twelve different carbonates in addition to a number of basic carbonates, bicarbonates and similar materials, and Louisfert and Pobeguin [11] have examined a number of different crystal forms of calcium carbonate.

The normal carbonates are characterised by a strong band in all cases in the range $1490-1410$ cm^{-1} and by a medium intensity band in the range $880-860$ cm^{-1}. In the last case, however, lead carbonate appears to be an exception, giving only an extremely weak absorption at 840 cm^{-1}. Hunt et al. [7] have related the position of this second absorption to the atomic weight of the cation, and have found a progressive logarithmic displacement to lower frequencies with increase in atomic weight. The carbonates of Miller and Wilkins (except lithium) also follow this progression. In calcium carbonate these bands occur at 876 cm^{-1} and 1430 cm^{-1}, and the former has been used for the estimation of carbonate in phosphates [13, 14]. Vibrational assignments for the individual carbonate frequencies have been given by Louisfert [15] and by Decius [16]. Studies have also been made on the nature of bone carbonates using infra-red methods [17]. The 1450 cm^{-1} band arises from the antisymmetric CO_3- stretching frequency. The corresponding symmetric band is found near 1090 cm^{-1} in the Raman spectra but is forbidden in the infra-red if the site symmetry is D_3. It is absent from the spectrum of calcite but appears in that of aragonite where the site symmetry is changed to C_s. Changes in site symmetry can also lead to doubling of bands in one compound whilst they remain single in others.

In addition, many carbonates show absorption in the 750–700 cm^{-1} region. Not all of them absorb in this region, but when they do so the observed frequencies are often useful in the identification of minerals. For calcite and dolomite, for example, the bands are 17 cm^{-1} apart [7].

Considerable spectral changes occur with basic carbonates and with bicarbonates. In the former case the spectra are more complex and the 1430 cm^{-1} absorption is almost always doubled. In addition, the number of bands between 1110 cm^{-1} and 700 cm^{-1} increases in most cases from two to five, whilst the bonded OH groupings are evident at 3300 cm^{-1}. Here also the various basic carbonates show sufficient differences amongst themselves to enable individuals to be identified in many cases.

Bicarbonates do not absorb in the 1431 cm^{-1} range. Instead two widely separated bands appear on either side. However, these are very variable in position and the most reliable means of identification is the pair of bands in the 665–655 cm^{-1} and 710–690 cm^{-1} regions [45]. The first of these is of medium intensity but the second is strong in all cases.

21.3. Sulphates

Miller and Wilkins [6] have examined ten different inorganic sulphates, and two others have been reported by Hunt *et al.* [7]. In all cases a very strong band is shown in the range 1130–1080 cm^{-1} accompanied by a considerably weaker band, in most cases, in the region 680–610 cm^{-1}. These findings were confirmed by Miller *et al.* [45]. Sulphato complexes absorb in the same general regions [50]. With calcium sulphate dihydrate the strong band was centred at 1140 cm^{-1}, and after heating at $170°C$ overnight a group of three bands was found in this region. This illustrates the influences which both hydration and crystal symmetry may have on such spectra and possible variations from such causes must be taken into account in assessing the spectra.

Unlike the carbonates, there did not appear to be any direct connection between the observed absorption frequencies and any property of the positive ion, despite the fact that the variations from one sulphate to the next were usually sufficiently large to enable any particular one to be identified. Lecomte *et al.* [18] have also studied the sulphate ion and report essentially similar conclusions. They note in particular [19] that the band near 610 cm^{-1} is single with Mg, K or Mn, but is doubled with Ca, Cu, Be or Ba.

Bisulphates have not been fully studied, but from the samples examined [6] it is clear that the strong 1110 cm^{-1} absorption is not shown, and there are again indications that it is replaced by a pair of widely separated medium intensity bands on either side. Sulphites do not absorb in this region.

21.4. Nitrates and nitrites

Ten inorganic nitrates examined by Miller and Wilkins [6] all show characteristic absorptions in the ranges $1380-1350 \text{ cm}^{-1}$ (v.s.) and $840-815 \text{ cm}^{-1}$ (m.). Again, no correlation could be found between the position of these bands and the nature of the positive ion. Vratny [51], Buijs and Schutte [52], and Ferraro [53] have confirmed this, but point out that there are differences between the spectra of monovalent and transition metal nitrates which arise from the differences in symmetry between the two. Monovalent metal nitrates show only the 1370 and 840 cm^{-1} bands but the other modes, absorbing at $1050-1000$ and $750-700 \text{ cm}^{-1}$ are allowed in some other cases. There is also a general trend of the 1370 cm^{-1} band to shift in relation to the degree of covalency of the metal nitrate bond [51]. There is a good deal of support for this correlation from data on organic nitrates, such as guanidine nitrate, in which the NO_3 group exists in the ionic form. These also absorb in the given regions, although the spectral ranges in these cases are sometimes rather wider (Chapter 17). Nitrites also show a weak to medium intensity band in the $840-815 \text{ cm}^{-1}$ region, but they can be readily differentiated from nitrates by the presence of a strong absorption [5, 6, 54] in the region $1250-1230 \text{ cm}^{-1}$ and between

$1360-1340$ cm^{-1}. These are the symmetric and antisymmetric frequencies. A vibrational assignment for sodium nitrite has been given by Tarte [20], and Hafele [41], and Vratny *et al.* [54] have also discussed the spectra of some crystalline nitrates.

21.5. The ammonium ion

Eighteen different ammonium salts were examined by Miller and Wilkins [6]. Two characteristic absorptions are shown, in all cases. One of these corresponds to the NH stretching vibration and occurs in the 3200 cm^{-1} region, whilst there is a second absorption, also strong, in the range $1430-1390$ cm^{-1}.

The number and position of bands in the NH stretching region reflects the possibilities of hydrogen bonding within the crystal, and in cases where multiple peaks are shown it may also indicate the formation of several hydrogen bonds of different strengths. In ammonium perchlorate the NH stretching frequency is as high as 3300 cm^{-1}, and this indication of little or no hydrogen bonding may be correlated with the fact that this compound does not absorb water to form a hydrate. Most ammonium salts show some degree of bonding and the NH frequencies of quite a high proportion fall between 3200 cm^{-1} and 3100 cm^{-1}. Ammonium nitrate is an extreme case in which strong bonds are formed, and this exhibits a triplet absorption [6] at 3160 cm^{-1}, 3060 cm^{-1} and 3030 cm^{-1}. The spectrum of ammonium azide has been studied by Pimentel *et al.* [21], and Waldron and Hornig [22] have examined some crystalline ammonium hydrates.

The lower frequency absorption near 1410 cm^{-1} is reasonably strong and the band is sharp and clear-cut so that, coupled with the stretching absorption, the identification of this ion is usually relatively simple.

21.6. Cyanides

Inorganic cyanides have been studied both as simple salts and as coordination complexes. Rao *et al.* [55] have listed data on the former, as have El Sayed and Sheline [56], and Nakamoto has given a table of a very large number of frequencies for coordination compounds [46]. The frequencies depend upon the electronegativity, oxidation number and coordination number of the atom to which the C≡N group is attached [56]. In salts such as sodium cyanide where the bonding is almost wholly ionic, the frequency is near 2080 cm^{-1}, but as the covalency of the metal C≡N bond is increased the frequency rises towards the values of organic nitriles. Thus the silver salt absorbs at 2151 cm^{-1} and the mercury salt at 2180 cm^{-1}. In most coordination compounds [6, 24–27, 46] the values lie between these two extremes. Solution studies in water have also been made on complexes of silver, copper and gold [28–30]. Some remarkable changes can accompany changes of state. Sodium and potassium cyanide show a strong band between 2080 cm^{-1} and 2070 cm^{-1}. In the fused state this moves to 2250 cm^{-1}, and there has been some discussion [23] as to whether this represents isomerism to the *iso*cyanide. Cyanates also absorb close to this region, and Rao *et al.* [53] quote a range of 2220–2130 cm^{-1} for a small number of these. Thiocyanates have been studied in small numbers [6, 55] and these absorb between 2105–2060 cm^{-1}. The range 2200–2000 cm^{-1} therefore can be regarded as covering all the various forms in which the —C≡N grouping occurs in inorganic compounds.

21.7. Phosphates

The ionic phosphate absorption is reported on the basis of Raman work to absorb at 1080 cm^{-1} and 980 cm^{-1} [8]. However, the second of these is forbidden in the infra-red and only appears in exceptional cases where the symmetry is disturbed. Only a single band is shown by most phosphates. This is, however, very strong and broad, and in a series of nine orthophosphates studied by Miller and Wilkins [6] was found between 1030 cm^{-1} and 1000 cm^{-1}. The only other fundamental allowed in the infra-red appears near 500 cm^{-1}.

More recently, the fundamental frequencies of all types of phosphate ions, such as metaphosphates, pyrophosphates, hypophosphites, etc., have been studied by several groups of workers, notably by Corbridge [31, 32], by Lecomte [33, 34] and by Tsuboi [42]. Their findings have been condensed in the form of a correlation chart for phosphorus oxyacids by Corbridge [32], and sufficient data are now available to enable most of the individual types to be differentiated. These absorptions have already been discussed in Chapter 18, to which further reference should be made. The spectra of some phosphate high polymers in molten, glaceous and crystalline states have been reported by Bues and Gehrke [35].

21.8. Silicates

Fifteen different silicates have been examined by Hunt *et al.* [7]. All show very intense absorption bands in the $1100-900 \ cm^{-1}$ range. Both the numbers and positions of these bands vary considerably from one mineral to the next, but compounds in which the silicate ions are similar show related spectra. Tremolite $((OH)_2 Ca_2 Mg_5 (Si_4 O_{11})_{12})$ and actinolite $((OH)_2 Ca_2 (MgFe)_5 (Si_4 O_{11})_{12})$, for example, show very closely related spectra with several strong bands in this region, whereas hornblende, $Ca(MgFe)_3 Si_4 O_{12}$; $CaMg_2(AlFe)_2 Si_3 O_{12}$ has only one band at $1010 \ cm^{-1}$. It has been shown that these differences are sufficient to give very useful leads in mineral ore analysis, and the various classes of silicates can be identified by their characteristic band patterns [9]. Studies have also been made on the spectra of materials such as vitreous porcelain [36] and on individual crystalline silicates such as afwillite [37].

21.9. Other correlations

Miller and Wilkins [7] have examined also numbers of sulphites, chromates, borates, silicates, bromates, iodates, vanadates, manganates, etc., and have made tentative assignments of characteristic frequencies in each case. These have been largely confirmed by later studies, especially those concerned with the complete vibrational assignments of single compounds. Nakamoto [46] summarises much of this data. Thus for planar structures such as the sulphite ion, chlorates, bromates and iodates, the v_1 and v_3 frequencies are very close together in the infra-red so that only one strong band is usually seen. Sulphites show this between 960 and $910 \ cm^{-1}$ although ammonium sulphite is an exception and absorbs at $1105 \ cm^{-1}$. The deformation frequencies occur near 633 and $496 \ cm^{-1}$. With chlorates the main stretching band is also near $960 \ cm^{-1}$, with lower values for bromates (near $830 \ cm^{-1}$) and iodates (near $800 \ cm^{-1}$).

Tetrahedral ions such as perchlorates, vanadates and chromates show only a single broad band in the normal infra-red region. Perchlorates absorb near $1110 \ cm^{-1}$, chromates near $885 \ cm^{-1}$ and vanadates near $825 \ cm^{-1}$.

Several studies have been made of boric esters [38, 39, 43] and of boric acid [40]. In the latter the B—O stretching frequency occurs near $1400 \ cm^{-1}$ and the deformation mode near $810 \ cm^{-1}$, but both are liable to alterations with structure. In simple borates the B–O link appears [43] at $1340 \pm 10 \text{ cm}^{-1}$, which suggests some double bond character.

Deformation frequencies of ions XO_4 have also been studied, and differences arising from alterations in the element X have been noted [19, 46]. Data are also available on some inorganic peroxides and superoxides [44].

21.10. Bibliography

1. Schaefer and Matossi, *Das Ultrarot Spektrum* (Springer, Berlin, 1930).
2. Lecomte, *Anal. Chim. Acta*, 1948, **2**, 727.
3. *Idem, Cahiers Phys.*, 1943, **17**, 1.
4. Duval, Lecomte and Morandat, *Bull. Soc. chim. (France)*, 1951, **18**, 745.
5. Duval and Lecomte, *Compt. rend. Acad. Sci. (Paris)*, 1951, **232**, 2306.
6. Miller and Wilkins, *Analyt. Chem.*, 1952, **24**, 1253.
7. Hunt, Wisherd and Bonham, *ibid.*, 1950, **22**, 1478.
8. Herzberg, *Infra-red and Raman Spectra of Polyatomic Molecules* (Van Nostrand, 1945), p. 162.
9. Launer, *Amer. Mineralogist*, 1952, **37**, 764.
10. Duval and Lecomte, *Compt. Rend. Acad. Sci. (Paris)*, 1952, **234**, 2445.
11. Louisfert and Pobeguin, *ibid.*, **235**, 287.
12. Hunt and Turner, *Analyt. Chem.*, 1953, **25**, 1169.
13. Pobequin and Lecomte, *Compt. Rend. Acad. Sci. (Paris)*, 1953, **236**, 1544.
14. Pobequin, *J. Phys. Radium*, 1954, **15**, 410.
15. Louisfert, *Compt. Rend. Acad. Sci. (Paris)*, 1955, **241**, 940.
16. Decius, *J. Chem. Phys.*, 1954, **22**, 1956.
17. Underwood, Toribara and Neuman, *J. Amer. Chem. Soc.*, 1955, **77**, 317.
18. Pascal, Duval, Lecomte and Pacault, *Compt. Rend. Acad. Sci. (Paris)*, 1951, **233**, 118.
19. Duval and Lecomte, *ibid.*, 1954, **239**, 249.
20. Tarte, *Ann. Soc. Sci. Bruxelles*, 1956, **70**, 244.
21. Dows, Whittle and Pimentel, *J. Chem. Phys.*, 1955, **23**, 1475.
22. Waldron and Hornig, *J. Amer. Chem. Soc.*, 1953, **75**, 6079.
23. Brugel, Daumiller and Rommel, *Angew. Chem.*, 1956, **68**, 440.
24. Bonino and Fabbri, *Atti Accad. Naz. Lincei. Rend. Classe. Sci. Fis. mat. Nat.*, 1956, **21**, 246.
25. *Idem, ibid.*, 1955, **19**, 386; 1956, **20**, 414.
26. Bonino and Salvetti, *ibid.*, 1956, **20**, 150.
27. Emschwiller, *Compt. Rend. Acad. Sci. (Paris)*, 1954, **238**, 1414.
28. Jones and Penneman, *J. Chem. Phys.*, 1954, **22**, 965.
29. Jones, *ibid.*, 1954, **22**, 1135.
30. Penneman and Jones, *ibid.*, 1956, **24**, 293; 1958, **28**, 169.
31. Corbridge and Lowe, *J. Chem. Soc.*, 1954, 493, 4555.
32. Corbridge, *J. App. Chem.*, 1956, **6**, 456.
33. Duval and Lecomte, *Compt. Rend. Acad. Sci. (Paris)*, 1955, **240**, 66.
34. *Idem, Mikrochim. Acta*, 1956, 454.
35. Bues and Gehrke, *Z. Anorg. Chem.*, 1956, **288**, 307.
36. Avgustinik, Setkina and Fedorova, *Zhur. Fiz. Khim.*, 1954, **28**, 637.
37. Petch, Sheppard and Megaw, *Acta cryst.*, 1956, **9**, 29.

38. Steele and Decius, *J. Chem. Phys.*, 1956, **25**, 1184.
39. Duval and Lecomte, *J. Opt. Soc. Amer.*, 1954, **44**, 261.
40. Bethell and Sheppard, *Trans. Faraday Soc.*, 1955, **51**, 9.
41. Hafele, *Z. Physik.*, 1957, **148**, 262.
42. Tsuboi, *J. Amer. Chem. Soc.*, 1957, **79**, 1351.
43. Werner and O'Brien, *Austral. J. Chem.*, 1955, **8**, 355.
44. Brame, Cohen, Margrave and Meloche, *J. Inorg. Nuclear, Chem.*, 1957, **4**, 90.
45. Miller, Carlson, Bentley and Jones, *Spectrochim. Acta*, 1960, **16**, 135.
46. Nakamoto, *Infrared Spectra of Inorganic and Coordination Compounds*, (Wiley, New York, 1963).
47. Lawson, *Infrared Absorption of Inorganic Substances*, (Reinhold, New York, 1961).
48. Adams, *Metal Ligand and Related Vibrations*, (Arnold, London, 1966).
49. Cotton, in *Modern Coordination Chemistry, Chapter 5*, (Wiley, New York, 1960).
50. Nakamoto, Fugita, Tanaka and Kobayashi, *J. Amer. Chem. Soc.*, 1957, **79**, 4904.
51. Vratny, *Applied Spectroscopy*, 1959, **13**, 59.
52. Buijs and Schutte, *Spectrochim. Acta*, 1962, **18**, 307.
53. Ferraro, *J. Mol. Spectroscopy*, 1960, **4**, 99.
54. Vratny, Tsai and Gugliota, *Nature*, 1960, **188**, 484.
55. Rao, Ramachandran and Shankar, *J. Sci. Ind. Research (India)*, 1959, **18B**, 169.
56. ElSayed and Sheline, *J. Inorg. and Nuclear Chem.*, 1958, **6**, 187.
57. Rocchiccioli, *Compt. Rend. Acad. Sci. Paris*, 1957, **244**, 2704.

22

Organo-Sulphur Compounds

22.1. Introduction and table

A considerable number of correlations are available for absorption bands arising from sulphur linkages, but only a limited number of them are likely to find any general application in structural analysis. Linkages including sulphur and hydrogen or sulphur and oxygen have not been as fully studied as the corresponding OH and CO groups, but the absorptions arising from them are well-defined and can usually be recognised without undue difficulty. The sulphur–hydrogen bands occur in a region in which few other materials absorb, whilst the sulphur–oxygen absorptions can usually be recognised from their intensity and characteristic complex appearance. In the case of the sulphur–oxygen stretching modes of the group XSO_2Y it has been shown [37] that the asymmetric and symmetric frequencies are related by a simple linear function, and a cross check of an identification of this group is possible in this way.

On the other hand, the C–S, S–S and similar linkages are not readily identified because either the absorption is extremely weak or the position of the band is too variable. Usually both factors apply. However, a certain amount of useful data concerning these links may be obtained in certain cases and details of the correlations proposed are therefore included. Both these and any other vibrations involving bonds to sulphur atoms are invariably strong in the Raman spectra unless forbidden by symmetry. This therefore provides the most certain method of identification in doubtful cases.

The correlations discussed are listed in Table 22.

Table 22

SH			$2950-2550$ cm^{-1} (usually w.)
S–S	Disulphides	Alkyl	$525-510$ cm^{-1} (w.) Branched chains up to 545 cm^{-1}
		Aryl	$540-520$ cm^{-1}
C–S			$715-620$ cm^{-1}
C=S	Thioketones	Alkyl	$1270-1245$ cm^{-1}
		Aryl	$1225-1205$ cm^{-1}
	Dithioesters		$1225-1190$ cm^{-1}
	Thioamides		$800-690$ cm^{-1}
C–SO–C			$1070-1035$ cm^{-1} strong
O–SO–O			near 1225 cm^{-1} (s.)
N–SO–N			near 1120 cm^{-1} (s.)
S–SO–S			near 1105 cm^{-1} (s.)
C–SO$_2$–C		Alkyl	$1330-1307$ cm^{-1} Straight chain 1330–1317 cm^{-1} $1152-1136$ cm^{-1}
		Aryl	$1376-1358$ cm^{-1} All bands strong $1169-1159$ cm^{-1}
	SO$_2$ deformation		$610-545$ cm^{-1}
C–SO$_2$–Cl			$1410-1360$ cm^{-1} (s.) $1195-1168$ cm^{-1}
OSO$_2$–Cl			$1452-1408$ cm^{-1} (s.) $1260-1225$ cm^{-1}
CSO$_2$–NR$_2$			$1358-1336$ cm^{-1} (s.) $1169-1152$ cm^{-1}
CSO$_2$–OR			$1380-1347$ cm^{-1} $1193-1182$ cm^{-1}
ROSO$_2$–OR			$1415-1380$ cm^{-1} (s.) $1200-1185$ cm^{-1}
RSO$_3$Na			$1192-1175$ cm^{-1} (s.) $1065-1050$ cm^{-1}

22.2. SH Stretching vibrations

Several early workers studied the position of the SH stretching absorption band. Bell [1, 2] was able to show that a number of simple mercaptans such as propyl, butyl and *iso*amyl mercaptans gave a well-defined but rather weak absorption at $2650-2550$ cm^{-1}. He was also able to show that the bands near 2550 cm^{-1} in thiophenol and thio-*p*-cresol were absent from the corresponding sulphides.

Confirmation of this assignment of the SH mode was forthcoming from further work by Ellis, who found the first overtone in the 5000 cm^{-1} region [3], and from Williams [4], who extended the series of compounds examined by Bell and showed that other mercaptans absorbed near 2630 cm^{-1}.

These early assignments have been fully supported by later

workers [5, 6, 7]. Trotter and Thompson [5], and Sheppard [6] have each examined numbers of simple mercaptans and compared them with the corresponding sulphides and in all cases one of the major points of difference has been the disappearance of the SH absorption, which the former workers assign as being near 2575 cm^{-1}, whilst Sheppard finds a range of $2590-2560$ cm^{-1} for the compounds he has examined.

Data on numerous thiols have been given by David and Hallam [38] and by Miller and Krishnamurphy [39] and the frequencies of many individual molecules have been listed by Bellamy for the infra-red [40], and by Dollish *et al.* [70] for the Raman. These fully support the narrower range of $2590-2550$ cm^{-1} for all thiols except of course hydrogen sulphide itself. This range also includes the thioacids which self-associate only very weakly and usually appear as monomers in solution. Thioacetic acid [8] for example absorbs at 2550 cm^{-1}. The self-association of these compounds has been studied by Ginzburg and Loginova [41]. In general the SH link is not capable of the extensive degree of hydrogen bonding which occurs with OH and NH groupings. There is very little change in frequency of SH absorptions on passing from the liquid state to dilute solutions, so that any intermolecular bonding effects must be very small. Small shifts suggestive of hydrogen bonding have been observed in solution in certain bases and other compounds, indicating the formation of weak S-H ... N. bonds. Gordy and Stanford [9] studied the SH stretching frequencies of thiophenol, butyl and benzyl mercaptans and of thioacetic acid in a variety of different solvents. In ether and similar solvents the frequency was unaffected, but a slight fall was observed in aniline and in *iso*propyl ether, whilst the largest shifts (up to 80 cm^{-1}) were observed with pyridine, α-picoline and dibenzylamine. Josien *et al.* [42] have also described studies indicating hydrogen bonding by thiophenol, and similar results have been obtained with dithiophosphoric acids [43]. Plant *et al.* [69] have not found any evidence of hydrogen bonding between SH links in the liquid state.

In the Raman the SH stretch is invariably strong but the intensity in the infra-red is variable and usually weak. A moderately strong band is shown by thiophenol, but in alkyl mercaptans the band can be so weak as to be virtually undetectable. It is also obscured in compounds containing COOH groups which. exhibit general absorption in this region. However, if allowance is made for these factors the presence or absence of a band in this region can afford decisive evidence for the occurrence of a mercapto-group. Studies in this region have been used, or example, to determine whether certain thiols of quinoline exist in the mercaptan or thio-keto form [10], whilst the absence of any SH absorption from the spectrum of mercapto-benz-thiazole is one of the strongest pieces of evidence for the existence of this substance as a thio-ketone under normal conditions.

22.3. S—S Stretching vibrations

The —S—S— linkage absorption is unlikely to be of any value in infra-red analysis except in a very few special instances. It occurs as a very weak band in the 500—400 cm^{-1} region, and is identifiable only because the corresponding Raman line is very strong. Venkateswaren [11] has attributed the 470 cm^{-1} Raman line of liquid sulphur to the —S—S— stretching mode, and such infra-red and Raman studies as have been made on organic disulphides indicate that the S—S link absorbs in the same region in these compounds.

Trotter and Thompson [5] assigned weak bands at 517 cm^{-1} to the S—S stretching mode in dimethyl and diethyl disulphides respectively, but in di-*n*-propyl disulphide only a very feeble absorption near 500 cm^{-1} could be detected. Similarly Sheppard [6] assigned the 510 cm^{-1} band of di*cyclo*hexyl disulphide to this vibration, but was unable to detect corresponding bands in di-*n*-butyl or di-*tert.*-butyl disulphides. More recently Allum *et al.* [71] have studied the S—S stretching frequencies of numbers of alkyl and aryl disulphides. They find that in straight chain compounds the S—S group has two bands in the narrow range 525—510 cm^{-1}. The range for diaryl compounds is a little higher and they suggest 540—520 cm^{-1}. Chain branching also raises the frequency and di-*tert.*butyl disulphide absorbs at 543 cm^{-1}. With trisulphides and longer sulphur chains, the frequency falls below 500 cm^{-1} and tends towards the value for liquid sulphur.

These correlations must be used only with the greatest caution in the infra-red as the bands are weak and fall in regions where other absorptions can occur. In the Raman spectra however the great intensity allows a positive identification.

22.4. C—S Stretching vibrations

The initial study of a group of compounds containing the C—S linkage was made by Trotter and Thompson [5], who determined the infra-red spectra of the seven simple mercaptans from methyl to *tert.*-butyl mercaptan, and correlated their findings with the published Raman data [11, 13–15]. Samples were examined in the vapour or liquid state, and they noted that considerable shifts in the C–S frequency resulted from relatively small changes in structure. The C–S frequency for methyl mercaptan, for example, was assigned at 705 cm^{-1} whilst that of ethyl mercaptan occurred at 660 cm^{-1}. In the case of some of the higher homologues a number of bands were found in this region which made the assignment difficult. Three sulphides and three disulphides were also examined, and the C–S frequencies were found to lie in the same overall range 705–587 cm^{-1}. In a fourth disulphide in which the C–S link was conjugated with an aromatic ring the C–S frequency was found to be lowered to 570 cm^{-1}. On the basis of this work and of the earlier work of Thompson and Duprè on ethylene disulphide [16], Thompson [17] later proposed 700–600 cm^{-1} as an approximate correlation for the C–S–C linkage.

This work was later extended by Sheppard [6], who compared the infra-red and Raman spectra of a wider range of mercaptans, sulphides and disulphides. He noted a progressive lowering of the frequency in primary, secondary and tertiary C–S compounds, and proposed a subdivision of the 700–600 cm^{-1} range as follows:

	$CH_3 S-$	705–685 cm^{-1}
Primary	$RCH_2 -S-$	660–630 cm^{-1}
Secondary	$R'R''CH-S$	630–600 cm^{-1}
Tertiary	$R'R''R'''C-S$	600–570 cm^{-1}

*cyclo*Hexyl derivatives were found to be an exception to this classification, although pentamethylene sulphide absorbs in the appropriate position for a primary sulphide (654 cm^{-1}). This sub-division of the range is therefore applicable only to saturated acyclic materials and may well be applicable only to alkyl mercaptans, sulphides and disulphides similar to those on which it is based. Sheppard also noted a progressive small frequency shift throughout the series mercaptan → sulphide → disulphide. These shifts were in all cases towards lower frequencies, but were of the order of 10 cm^{-1} and did not affect the overall classification. A limited number of compounds in which the C–S linkage was conjugated with a double bond were also examined, and in these a low-frequency shift of the order of 60 cm^{-1} was found. Carbonyl sulphide appeared to be an exception to this, but the

assignment in this case was in doubt. An increase in the intensity of
the C–S vibration was noted following conjugation, di*cyclo*hexenyl
sulphide showing a strong band at 634 cm^{-1} against a medium
intensity band in *cyclo*hexyl *cyclo*hexenyl sulphide at the same
point. These results have been reviewed by Allum *et al.* [71] and
these authors have given C–S frequencies for a considerable number
of compounds [71]. They point out that there is often more than
one band from this group but that one of these usually falls between
715–620 cm^{-1}. Dollish *et al.* [70] give a useful table of assignments
for organic sulphides.

Sheppard has also directed attention to the marked coincidence of
corresponding C–S and C–Cl frequencies in this region. The C–Cl
frequencies are known from Raman work to show a progression
through the series Methyl > Et ≏ Pr > *sec.*-Pr > *tert.*-Butyl ≏ Allyl
≏ Benzyl, and the corresponding C–S compounds show not only the
same progression, but almost the same frequencies. Assuming the
decreasing frequency effect in C–Cl compounds corresponds to a
reduction in the force constant, it would appear that the small
alteration in mass in passing from Cl to S has little effect and that the
same order of decrease holds for sulphur compounds.

The assignment of the C–S band in aromatics has been discussed
by Cymmerman and Willis [12] and by Allum *et al.* [71]. It appears
that the C–S vibration couples very extensively with ring modes and
that in consequence no useful group frequencies can be identified.

As with the S–S vibration this absorption is usually weak in the
infra-red and identifications of the group frequency should therefore
not be attempted unless supported by Raman data.

22.5. C=S Stretching vibrations

The identification of the position of the C=S absorption has been a
matter of some difficulty. In carbon disulphide the C=S stretching
modes have been assigned [19] to 1522 cm^{-1} and 650 cm^{-1}, whilst
in carbonyl sulphide [20] it is given at 859 cm^{-1}, but these are
clearly unusual cases in which the carbon is doubly unsaturated, and
they do not afford any guide to the likely position of the C=S
vibration in saturated thioesters and similar compounds.

In the infra-red the C=S band is not particularly strong, and its
identification is complicated by the incidence of coupling effects and
by the tendency of simple thioketones to dimerise. Early
studies [6, 18, 21] did not therefore lead to any useful correlations.

The first wide ranging studies were those of Mecke *et al.* [44, 45] who compared the carbonyl and thioketone spectra of over seventy different compounds of various types. Preliminary calculations indicated that the ratio ν_{co}/ν_{cs} would be about 1.5 and that the C=S frequency would be found in the $1200-1050$ cm^{-1} region. This was confirmed experimentally by the comparison of the spectra of the individual pairs. Just as with the carbonyl group, the C=S absorption was found to be sensitive to the nature of the surrounding structure, but the relative effects of various substituents were not always the same. Thus halogen atoms attached to the carbon did not raise the C=S frequency as much as oxygen atoms in thioesters, whereas the reverse is true of the carbonyl series. In consequence, the ratio between the carbonyl and thiocarbonyl frequencies varies over the range $1.6-1.14$. The general ranges found by these workers for various classes of thiocarbonyl compounds are as follows: Cl_2CS 1121 cm^{-1}, $(RS)_2CS$ $1058-1053$ cm^{-1}, $(RO)_2CS$ $1212-1234$ cm^{-1} $\underset{\underset{|}{}\underset{|}{}}{>\!C\!=\!C\!-\!C}\!=\!S$ ($1156-1143$ cm^{-1}), $(Phenyl)_2C=S$ $1226-1219$ cm^{-1} and thioureas, which show the widest variation, $1400-1130$ cm^{-1}. The numbers of compounds studied in each class naturally varied, and as will be seen some of these assignments have been amended by later workers.

Thioketones have been studied more intensively by Andrieu *et al.* [72] using Raman data to confirm their assignments. Aliphatic thioketones which are not dimerised absorb between 1270 and 1245 cm^{-1} and the band position within this range is a function of the sum of the Taft σ^* values of the substituents. With aryl thioketones the frequency falls to $1225-1205$ cm^{-1}, but there is now no relationship between the frequencies and the Hammett σ values of these substituents [76]. These findings have been fully supported by other workers [73, 75].

Dithioesters have been studied by several workers [49, 77, 78] and there is general agreement that these absorb between $1225-1190$ cm^{-1}. Trithiocarbonates absorb at lower frequencies [48] with ethylene trithiocarbonate at 1083 cm^{-1}. Xanthates have a series of bands in the ranges $1250-1200$, $1140-1110$, and $1070-1020$ cm^{-1}. All have some C=S content but none of them can be unequivocably assigned to this structure.

Thiocarbonyl halides absorb between $1368-1139$ cm^{-1} depending on the electronegativity of the halogen atoms [47, 79, 80]. The difluoro compound [80] absorbs at

1368 cm^{-1}, the fluorobromide at 1130 cm^{-1} and the chloro-bromide at 1125 cm^{-1}.

The most difficult C=S absorption to characterise has been that in thioamides and thioureas. Assignments have ranged from 1570 cm^{-1} to 690 cm^{-1} and there has been much controversy on this issue. The available data have been widely reviewed by Rao *et al.* [75] and by Jensen and Nielsen [46]. The problem has been resolved by a combination of Raman studies and of normal coordinate calculations. These show that there are a number of bands in most cases which show some C=S character, and these are widely distributed over the spectrum. None can be described as a pure C=S mode and in most cases the band with the largest contribution from this mode has not much more than 50% of this. Thioformamide is an exception in that the band at 843 cm^{-1} is almost a pure C=S vibration, and this indicates the region in which the band with most C=S character is to be expected. In thioureas this band ranges [81–83] from 700 cm^{-1} to 753 cm^{-1} depending on the substitution, and the band is at 699 cm^{-1} in *N*-methyl thioacetamide. However it must be stressed that this band is of very little use for the identification of the C=S group in the infra-red.

22.6. Thioureides —N—C=S and —NH—C=S

Randall *et al.* [7] drew attention to a strong band between 1613 and 1471 cm^{-1} in all the compounds they examined which contained the N—C=S grouping. They assigned this to a C—N stretching mode. This is in fact the case, except in secondary thioamides where the band is due to a coupled C—N and NH mode [75]. The band is usually between $1490–1470 \text{ cm}^{-1}$ in the dithiocarbamates which have been studied by Mann [22] and by Chatt, Duncanson and Venanzi [50]. The latter workers assign this absorption, which they place in the wider range $1542–1480 \text{ cm}^{-1}$ to the C—N stretching mode shortened under the influence of canonical forms such as

$$X \overset{\overset{\bar{S}}{\diagup}\ \diagdown}{\underset{\diagdown S \diagup}{}} C = \overset{+}{N} R$$

They have also studied the changes in this frequency which accompany alterations in the valency and geometry of the element X. The highest frequencies in solution are shown by divalent planar complexes of Pt, Ni, Pd or Cu, and there seems to be a tendency for divalent tetrahedral complexes to absorb at slightly lower frequencies. This trend is still more marked in octahedral complexes of trivalent elements such as cobalt.

22.7. S=O Stretching vibrations

22.7(a). Sulphoxides $\overset{\text{O}}{\underset{\text{S}}{\|}}$. Large numbers of sulphoxides have been studied and recent references up to 1968 are given by Bellamy [40]. Useful reviews of the data and discussions on the relationship between the frequencies and the electronegativities of the substituents are given by Simon and Kriegsmann [28] and by Steudel [84]. The earliest systematic studies were those of Barnard, Fabian and Koch [24], who examined seven sulphoxides, of which five were obtained in a high state of purity. In all cases they found a very strong band to occur near 1050 cm^{-1} which they ascribed to the S=O link by analogy with sulphur monoxide [25] in which the S=O stretching vibration occurs at 1124 cm^{-1}.

This band was found to be not only very intense but also remarkably constant in position in dilute solution. Thus both *cyclo*hexylmethylsulphoxide and phenylmethylsulphoxide absorb strongly at 1055 cm^{-1} in dilute solution in carbon tetrachloride, and even in unsaturated materials such as diallyl sulphoxide (1047 cm^{-1}) the frequency is not diminished. This is due to the fact that the π clouds of any attached double bonds do not lie in the same plane as those of the S=O link so that conjugation does not occur. The frequency then depends only on the inductive properties of the substituents, so that in some cases aromatic substitution leads to a small frequency rise [51].

These general findings have been fully substantiated by later workers. With alkyl and aryl sulphoxides the S=O frequency falls in the 1070—1035 cm^{-1} range, but in the latter there is usually a Fermi-resonance interaction with a ring band which results in a second strong band near 1090 cm^{-1}. The S=O frequency is not as sensitive as the carbonyl to the effects of bond angles, due to the greater mass and reduced amplitude of the sulphur atom. Inclusion in rings does not therefore influence the frequency much, unless the ring size is very small.

The sulphoxide bond is therefore readily identifiable in both the infra-red and Raman by the presence of a strong band in the 1050 cm^{-1} region, and Cymmerman and Willis [12] have used the

absence of such absorptions to demonstrate that the group
—SO—SO— does not exist and that such compounds as were thought
to have this structure actually take the form S—SO$_2$.

This linkage is highly polar and is therefore sensitive to changes of
state and to hydrogen bonding. Shifts of 10—15 cm^{-1} to lower
frequencies occur on passing to the condensed phase and larger shifts
occur in hydrogen bonding solvents. Amstutz *et al.* [29] have
suggested that chelation occurs in *ortho* hydroxy diphenylsulphoxide
as the S=O frequency is reduced to the exceptionally low value of
994 cm^{-1}.

22.7(b). Covalent sulphites —SO—O— and —O—SO—O—.

Sulphinic
acids RSOOH and their esters have been examined by Detoni and
Hadži [53, 54] and Simon, Kriegsmann and Dutz [55] have studied
a small number of dialkyl sulphites (RO)$_2$S=O. These have been
amplified by further data by Simon and Kriegsmann [28],
Steudel [84] and Dorris [26]. In the sulphinic acids a strong band is
found in solution near 1090 cm^{-1} which is very little affected by
deuteration, and in the corresponding esters it is at somewhat higher
frequencies (1126—1136 cm^{-1}). When a further oxygen atom is
attached to the sulphur, as in dialkyl sulphites, the frequency would
be expected to rise further, and in the small number of such
compounds which have been studied [28, 55] there is a strong band
near 1210 cm^{-1} which is assigned to the S=O frequency.

The latter assignment is indirectly supported by data on comparable cyclic compounds. De la Mere *et al.* [56] have examined a
number of cyclic sulphites in the solid state and find strong
absorptions in the range 1213—1215 cm^{-1} for 1 : 2-compounds
(five-ring) and 1195—1208 cm^{-1} for 1 : 3-compounds (six-ring).
These ranges are those observed for the limited number of
compounds studied, and the group frequencies will necessarily fall
over a wider range. Samant and Emerson [57] have confirmed these
findings, although there is some difference in the values quoted which
probably arises from the fact that they were working in solution.
Under these conditions they report cyclic sulphites as absorbing
between 1210 cm^{-1} and 1220 cm^{-1}, but they were unable to
detect any differences between five- and six-membered ring systems.
Dorris [26] gives a value of 1231 cm^{-1} for ethylene sulphite which
is very close to the value of 1225 cm^{-1} he finds for dimethyl
sulphite. Aromatic rings attached to the oxygen atoms raise the
effective electronegativity and diaryl sulphites absorb near
1240 cm^{-1}.

22.7(c). S=O Groups attached to other elements. The S=O frequency is determined wholly by the inductive effects of the substituents and these are related to their effective electronegativities. Probably the most useful measure of these is the π values derived by Thomas in relation to the P=O vibration which is also solely determined by these factors. If therefore one considers compounds with two identical substituents the values of others with mixed substituents can be derived with reasonable accuracy.

The thionyl halides range from 1308 cm^{-1} for thionyl fluoride, through 1233 for thionyl chloride [52] to 1121 cm^{-1} for the bromide [85]. Nitrogen substitution also raises $\nu_{S=O}$ as there is no resonance. Sulphinamides have been studied by Smith and Wu [86], by Keat *et al.* [87] and by Steudel [84]. With *N*-alkyl groups on both sides of the sulphur atom the frequency is close to 1120 cm^{-1}, but lower values are found for structures such as R—NHSO—R (1060–1037 cm^{-1}) [86]. In these the bands are usually doubled which may indicate dimerisation as in the amides. Two sulphur atoms attached to the S=O group give frequencies close to 1105 cm^{-1}.

As indicated above combinations of two different elements give values which are close to the mean of the values for the corresponding pair of symmetrically di-substituted compounds. Substitution with a nitrogen and a fluorine leads to absorption near 1215 cm^{-1} and this is similar to the value for an oxygen atom and a chlorine. The precise value of $\nu_{S=O}$ can therefore be very informative as to the substitution pattern.

22.7(d). Sulphones —SO$_2$—. The first work on the characteristic infra-red frequencies of the sulphones was carried out by Schrieber [23]. He examined the solution spectra of thirteen sulphones and compared them with the spectra of the corresponding sulphides. In every case he found two bands to be present in the sulphones which were absent from the sulphides. These were at 1160–1120 cm^{-1} and at 1350–1300 cm^{-1}. These bands are very intense, and are readily identifiable by the fact that in the solid state they usually show signs of splitting into a strong group of bands of closely related frequencies. Their assignment to the S=O stretching modes is directly confirmed by the assignments for these vibrations in sulphur dioxide [30] which occur at 1151 cm^{-1} and 1361 cm^{-1}. The actual ranges of Schrieber's compounds were 1159–1128 cm^{-1} and 1352–1313 cm^{-1}, and he therefore proposed the ranges 1160–1120 and 1350–1300 cm^{-1} for the characterisation of normal sulphones. As with the sulphoxides, conjugation does not appear to have any marked effect on this frequency and, in fact, the majority of the aromatic sulphones examined showed the lower-frequency band towards the top of the range, usually about 1150 cm^{-1}. All Schrieber's measurements were made in solution, and he did not study the effect of changes of state.

Barnard, Fabian and Koch [24] compared the spectra of seven sulphones in the solid and solution state and found that, as with the sulphoxides, the solid materials absorb at frequencies 10–20 cm^{-1} lower than the solutions in carbon tetrachloride. Otherwise their findings supplement those of Schrieber, and all their compounds absorb in the expected ranges 1160–1120 cm^{-1} and 1350–1300 cm^{-1}. The shifts observed from molecular aggregation or hydrogen bonding effects appear to be a little less in the case of the lower frequency band. Thus *cyclo*hexylmethylsulphone absorbs at 1144 and 1321 cm^{-1} in solution and at 1138 and 1309 cm^{-1} in the liquid state. These findings were very fully confirmed by many subsequent studies [29, 31, 32, 35, 51, 58, 59]. Robinson [35] in particular has covered a wide range of compounds, as have Feairheller and Katon [59]. Bellamy [40] gives many other references up to 1968. The general consensus of opinion is that alkyl sulphones absorb in the ranges 1330–1307 and 1152–1136 cm^{-1} and that the first of these can be subdivided, with straight chain alkyls absorbing between 1300–1317 cm^{-1} and branched chain alkyls at the lower end of the overall range. Aromatic sulphones absorb between 1376–1358 and 1169–1159 cm^{-1} as is to be expected from the lack of conjugation and the higher inductive effects of the aromatic rings. Alkyl/aryl sulphones absorb at intermediate values 1334–1325 and 1160–1150 cm^{-1}.

The SO_2 group has both scissoring and wagging modes which are active in the infra-red. The more useful is the scissoring mode which occurs as a strong band between 610–545 cm^{-1} [55, 59, 60, 88]. A second band in the range 525–495 cm^{-1} appears in all saturated compounds and this corresponds to the wagging mode [59].

These correlations can therefore be regarded as being well established and, in view of the very great intensity shown and the splitting effects frequently observed, the identification of these bands can usually be made with reasonable confidence. A useful cross check on the identification of an SO_2 group is available through the work of Bellamy and Williams [37], who have pointed out that the asymmetric and symmetric stretching frequencies of the SO_2 group are related linearly.

22.7(e). Sulphonyl chlorides. The substitution of a halogen atom directly on to the sulphur atom of a sulphone would be expected to result in frequency shifts of both of the characteristic bands towards higher frequencies. This expectation is realised in practice. The main studies on sulphonyl chlorides are due to Robinson [35], and to Geisler and Bindernagel [61] (alkyl derivatives) and Malewski *et al.* [89] (aromatics). Aryl sulphonyl chlorides absorb between 1410–1368 and 1195–1168 cm^{-1}, with the highest values corresponding to strongly electron-attracting substituents in the ring. The values for benzene sulphonyl chloride are almost exactly the mean values of diphenyl sulphone and SO_2Cl_2. Alkyl sulphonyl chlorides absorb towards the lower end of the range for aromatic compounds.

With oxygen and chlorine substitution, the frequencies rise further, and Robinson [35] suggests the ranges 1452–1408 and 1260–1225 cm^{-1}. Fluorine substitution will of course raise the frequencies still higher as indicated by the values for SO_2F_2 of 1502 and 1269 cm^{-1}.

22.7(f). Sulphonamides. In normal amides the presence of a tertiary nitrogen atom with two spare electrons directly attached to the carbonyl group results in a lengthening of the C=O bond and a shift to lower frequencies. This effect does not occur with sulphonamides where the orientation of the π clouds of the multiple bonds is such that resonance with the nitrogen lone pair is not possible. The frequencies therefore follow the inductive effect of the nitrogen atom. As with the phosphorus compounds where the P=O frequency is similarly affected, the effective electronegativity of nitrogen atoms carrying substituents is markedly lower than would be suggested by the Pauling value. Sulphonamides therefore absorb at slightly higher values than the sulphones. For reasons which are not yet understood the substitution of nitrogen atoms on both sides of the SO_2 group does not raise the frequency further [90].

Adams and Tjepkema [33] have published the infra-red spectra of sixteen $N : N'$-di-substituted sulphonamides, and noted that a strong band occurs in all cases in the range 1180–1160 cm^{-1} which they ascribe to the $-SO_2-$ grouping. Schrieber [23] has pointed out that, in addition, these spectra also have a strong band in the 1360–1330 cm^{-1} region corresponding to the second sulphone band.

Further studies by Adams [63] have confirmed his earlier findings, as has an extensive study of this group by Baxter, Cymerman-Craig and Willis [64], who have examined twenty-five primary, secondary and tertiary sulphonamides in dilute solutions and in the solid state. The SO_2 frequencies are the same for all three types, and occur in solution in the ranges 1370–1333 cm^{-1} and 1178–1159 cm^{-1}. Further data have been given by Robinson [35] and by Momose [91]. These suggest that the overall frequency ranges can be narrowed to 1358–1336 and 1169–1152 cm^{-1}. As with the sulphones, the symmetric frequency is unchanged on passing over to the solid state, but the asymmetric frequencies fall by 10–20 cm^{-1}. The NH_2 stretching frequencies of primary sulphonamides were very similar to those of primary amines, but in secondary compounds the NH stretch was found near 3390 cm^{-1}, which is about 40 cm^{-1} lower than in the corresponding amines.

22.8. Covalent sulphates and sulphonates

An oxygen linkage attached to the SO_2 grouping would be expected to result in a shift in the SO_2 bands towards higher frequencies, in the same way as the CO vibration of esters occurs at slightly higher frequencies than in ketones. A still further frequency shift would be expected when two oxygens are so linked in covalent sulphates. Schrieber [23] finds two *para*-toluene-sulphonates to absorb at 1375–1370 cm^{-1} and at 1185 cm^{-1} in solution, whilst he quotes dimethyl and diethyl sulphates as absorbing at 1193 and 1187 cm^{-1} respectively. A series of five esters of *para*-toluene sulphonic acid absorbs in the ranges [65] 1375–1350 cm^{-1} and 1192–1170 cm^{-1}; very similar values are given by the sultones [66], in which the $-SO_2-O-$ group is included in a six-membered ring system. Simon *et al.* [68] find absorption near 1350 cm^{-1} and 1176 cm^{-1} in a few alkyl esters of sulphonic acids. In agreement with this, the more recent studies of Freeman and Hambly [92] on a number of alkyl sulphonates suggest ranges of 1380–1347 and 1193–1182 cm^{-1}.

With covalent sulphates the presence of the additional oxygen atom raises the frequency still further. The numbers studied are relatively small but Detoni and Hadži [93] proposed ranges of 1415–1380 and 1200–1185 cm^{-1}.

22.9. Sulphonic acids and salts

Only a very limited amount of data is available on sulphonic acids and their salts. Colthup [18] has suggested the ranges 1260–

1150 cm^{-1}, 1080–1010 cm^{-1} and 700–600 cm^{-1}. Schrieber [22] reports three sulphonic acids as absorbing between 1182 cm^{-1} and 1170 cm^{-1}, but does not comment on any other bands. Haszeldine and Kidd [67] have examined twelve sulphonic acids and salts, and quote strong absorptions in the ranges 1190–1170 cm^{-1} and 1064–1040 cm^{-1}. More recent measurements suggest that rather lower frequencies are the more general rule. Detoni and Hadži [93] and others give the overall ranges 1350–1340 and 1165–1150 cm^{-1} for sulphonic acids, and 1192–1175 and 1063–1053 cm^{-1} for a series of sodium salts.

22.10. Bibliography

1. Bell, *Ber.*, 1927, **B60**, 1749.
2. *Idem., ibid.*, 1928, **B61**, 1918.
3. Ellis, *J. Amer. Chem. Soc.*, 1928, **50**, 2113.
4. Williams, *Phys. Rev.*, 1938, **54**, 504.
5. Trotter and Thompson, *J. Chem. Soc.*, 1946, 481.
6. Sheppard, *Trans. Faraday Soc.*, 1950, **46**, 429.
7. Randall, Fowler, Fuson and Dangl, *Infra-red Determination of Organic Structures* (Van Nostrand, 1949).
8. Sheppard, *Trans. Faraday Soc.*, 1949, **45**, 693.
9. Gordy and Stanford, *J. Amer. Chem. Soc.*, 1940, **62**, 497.
10. Hannan, Lieblich and Renfrew, *ibid.*, 1949, **71**, 3733.
11. Venkateswaren, *Indian J. Physics*, 1930, **5**, 219.
12. Cymerman and Willis, *J. Chem. Soc.*, 1951, 1332.
13. Kohlrausch, Dadieu and Pongratz, *Wien. Ber.*, 1932, **141**, 11*a*, 276.
14. Kopper, Seka and Kohlrausch, *ibid.*, p. 465.
15. Köppl and Kohlrausch, *ibid.*, 1933, **142**, 11*b*, 477.
16. Thompson and Duprè, *Trans. Faraday Soc.*, 1940, **36**, 805.
17. Thompson, *J. Chem. Soc.*, 1948, 328.
18. Colthup, *J. Opt. Soc. Amer.*, 1950, **40**, 397.
19. Herzberg, *Infra-red and Raman Spectra of Polyatomic Molecules* (Van Nostrand, 1945), p. 276.
20. Bailey and Cassie, *Proc. Roy. Soc.*, 1932, **A135**, 375.
21. Thompson, Nicholson and Short, *Discuss. Faraday Soc.*, 1950, **9**, 222.
22. Mann, *Trans. Inst. Rubber Ind.*, 1951, **27**, 232.
23. Schrieber, *Analyt. Chem.*, 1949, **21**, 1168.
24. Barnard, Fabian and Koch, *J. Chem. Soc.*, 1949, 2442.
25. Herzberg, *Molecular Spectra and Molecular Structure*, Vol. 1 (Van Nostrand, 1950), p. 573.
26. Dorris, *Applied Spectroscopy*, 1970, **24**, 492.
27. Bender and Wood, *J. Chem. Phys.*, 1955, **23**, 1316.
28. Simon and Kriegsmann, *Zeit. Phys. Chem.*, 1955, **204**, 369.
29. Amstutz, Hunsberger and Chessick, *J. Amer. Chem. Soc.*, 1951, **73**, 1220.
30. Herzberg, *Infra-red and Raman Spectra of Polyatomic Molecules* (Van Nostrand, 1945), p. 285.
31. Ross and Raths, *J. Amer. Chem. Soc.*, 1951, **73**, 129.
32. Cope, Morrison and Field, *ibid.*, 1950, **72**, 59.

33. Adams and Tjepkema, *ibid.*, 1948, **70**, 4204.
34. Barnes, Gore, Liddel and Williams, *Infra-red Spectroscopy* (Reinhold 1944), p. 87.
35. Robinson, *Can. J. Chem.*, 1961, **39**, 247.
36. Flett, *J. Chem. Soc.*, 1953, 347.
37. Bellamy and Williams, *J. Chem. Soc.*, 1957, 863.
38. David and Hallam, *Trans. Faraday Soc.*, 1964, **60**, 2013.
39. Miller and Krishnamurphy, *J. Org. Chem.*, 1962, **72**, 645.
40. Bellamy, *Advances in Infrared Group Frequencies* (Methuen, London, 1968).
41. Ginzburg and Loginova, *Optics and Spectroscopy*, 1966, **20**, 130.
42. Josien, Dizabo and Saumagne, *Bull. Soc. Chim. France*, 1957, 423.
43. Meneefe, Alford and Scott, *J. Chem. Phys.*, 1956, **25**, 370.
44. Mecke, Mecke and Lüttringhaus, *Z. Naturforsch.*, 1955, **105B**, 367.
45. Mecke and Mecke, *Chem. Ber.*, 1956, **89**, 343.
46. Jensen and Nielsen, *Acta Chem. Scand.*, 1966, **20**, 597.
47. Moule and Subraminiam, *Can. J. Chem.*, 1969, **47**, 1012.
48. Jones, Kynaston and Hales, *J. Chem. Soc.*, 1957, 614.
49. Marvel, Radzitzky and Brader, *J. Amer. Chem. Soc.*, 1955, **77**, 5997.
50. Chatt, Duncanson, and Venanzi, *Suomen. Kem.*, 1956, **29B**, 75.
51. Price and Gillis, *J. Amer. Chem. Soc.*, 1953, **75**, 4750.
52. Vogel-Hogler, *Acta Phys. Austriaca*, 1948, **1**, 323.
53. Detoni and Hadži, *J. Chem. Soc.*, 1955, 3163.
54. *Idem, Bull. Sci. Yougoslavie*, 1955, **2**, 44.
55. Simon, Kriegsmann and Dutz, *Chem. Ber.*, 1956, **89**, 2390.
56. De la Mere, Klyne, Millen, Pritchard and Watson, *J. Chem. Soc.*, 1956, 1813.
57. Samant and Emerson, *J. Amer. Chem. Soc.*, 1956, **78**, 454.
58. Price and Morita, *ibid.*, 1953, **75**, 4747.
59. Feairheller and Katon, *Spectrochim. Acta*, 1964, **20**, 1099.
60. Simon, Kriegsmann and Dutz, *Chem. Ber.*, 1956, **89**, 1883.
61. Geisler and Bindernagel, *Z. Electrochem.*, 1959, **63**, 1140; 1960, **64**, 421.
62. Dudley, Cady and Eggers, *J. Amer. Chem. Soc.*, 1956, **78**, 2.
63. Adams and Cosgrove, *ibid.*, 1954, **76**, 3584.
64. Baxter, Cymerman-Craig and Willis, *J. Chem. Soc.*, 1955, 669.
65. Tipson, *J. Amer. Chem. Soc.*, 1952, **74**, 1354.
66. Philbin, Stuart, Timoney and Wheeler, *J. Chem. Soc.*, 1956, 4414.
67. Haszeldine and Kidd, *ibid.*, 1954, 4228.
68. Simon, Kriegsmann and Dutz, *Chem. Ber.*, 1956, **89**, 2378.
69. Plant, Tarbell, and Whiteman, *J. Amer. Chem. Soc.*, 1955, **77**, 1572.
70. Dollish, Fateley and Bentley, *Characteristic Raman Frequencies of Organic Compounds* (Wiley, New York, 1974).
71. Allum, Creighton, Green, Minkoff and Prince, *Spectrochim. Acta*, 1968, **24A**, 927.
72. Andrieu and Mollier, *Spectrochim. Acta*, 1972, **28A**, 785.
73. Andrieu, Mollier and Lozach, *Bull. Soc. Chim. France*, 1965, 2457.
74. Lozach and Guillouzo, *Bull. Soc. Chim. France*, 1957, 1221.
75. Rao and Venkataraghaven, *Spectrochim. Acta*, 1962, **18**, 541.
76. Rao and Venkataraghaven, *Can. J. Chem.*, 1961, **39**, 1757.
77. Bellamy and Rogasch, *J. Chem. Soc.*, 1960, 2218.
78. Bak, Hansen Nygaard and Pedersen, *Acta Chem. Scand.*, 1958, **12**, 1451.
79. Brema and Moule, *Spectrochim. Acta*, 1972, **28A**, 809.
80. Dows, *Spectrochim. Acta*, 1963, **19**, 1163.

81. Ritchie, Spedding and Steele, *Spectrochim. Acta*, 1971, **27A**, 1597.
82. Aitkin, Duncan and Steele, *J. Chem. Soc.*, (A), 1971, 2695.
83. Rao and Chaturvedi, *Spectrochim. Acta*, 1971, **27A**, 520.
84. Steudel, *Z. Naturforsch.*, 1970, **25B**, 156.
85. Stammreich, Foeneris and Tavares, *J. Chem. Phys.*, 1956, **25**, 1277.
86. Smith and Wu, *Applied Spectroscopy*, 1968, **22**, 346.
87. Keat, Ross and Sharp, *Spectrochim. Acta*, 1971, **27A**, 2219.
88. Otting and Neuberger, *Chem. Ber.*, 1962, **95**, 540.
89. Malewski and Weigmann, *Spectrochim. Acta*, 1962, **18**, 725.
90. Vandi, Moeller and Audrieth, *J. Org. Chem.*, 1961, **26**, 1136.
91. Momose, Ueda and Shoji, *Chem. Pharm. Bull. Japan*, 1959, **7**, 734.
92. Freeman and Hambly, *Austral. J. Chem.*, 1957, **10**, 227.
93. Detoni and Hadži, *Spectrochim. Acta*, 1957, **11**, 601.

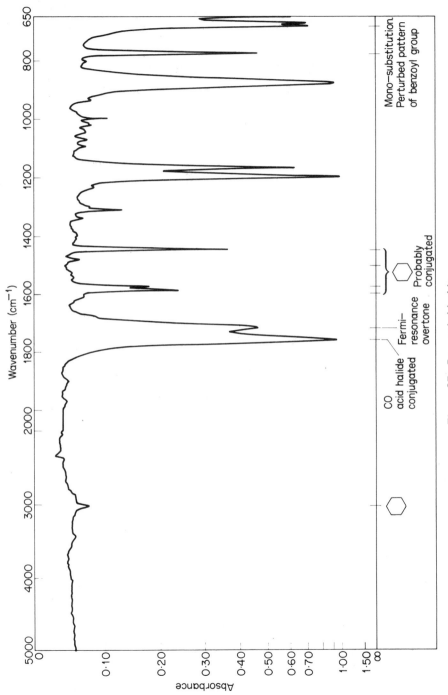

Figure 25. Benzoyl chloride

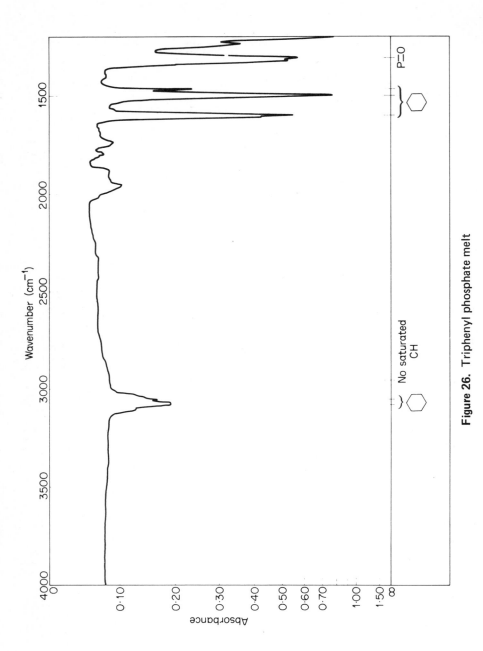

Figure 26. Triphenyl phosphate melt

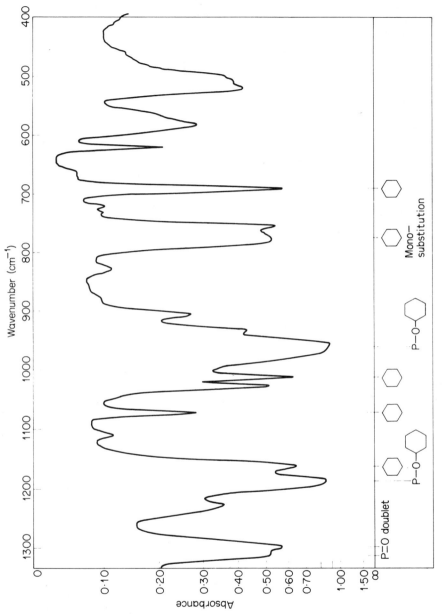

Figure 27. Triphenyl phosphate melt

413

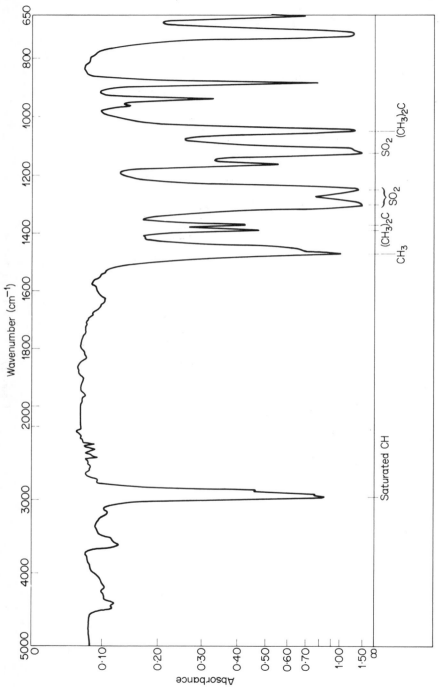

Figure 28. Di-*iso*propyl sulphone melt

414

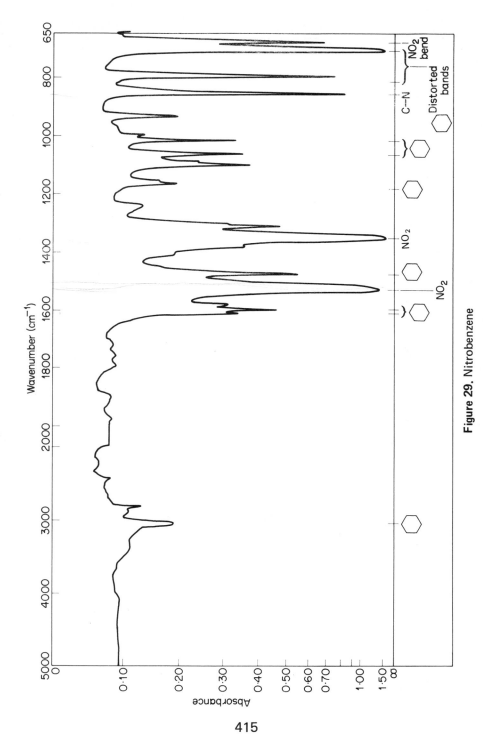

Figure 29. Nitrobenzene

415

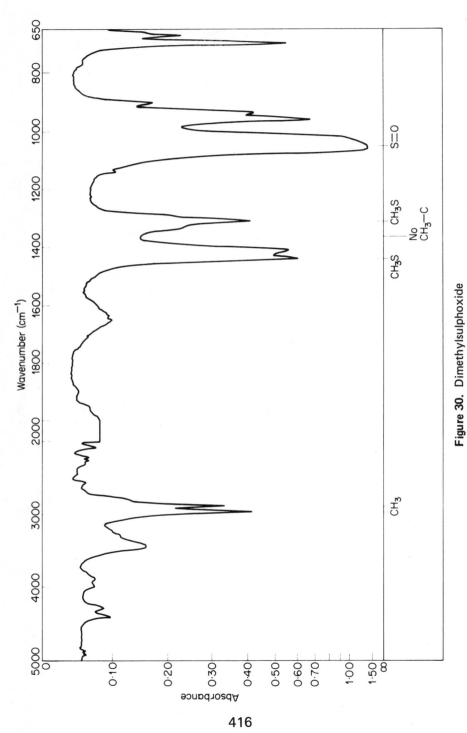

Figure 30. Dimethylsulphoxide

416

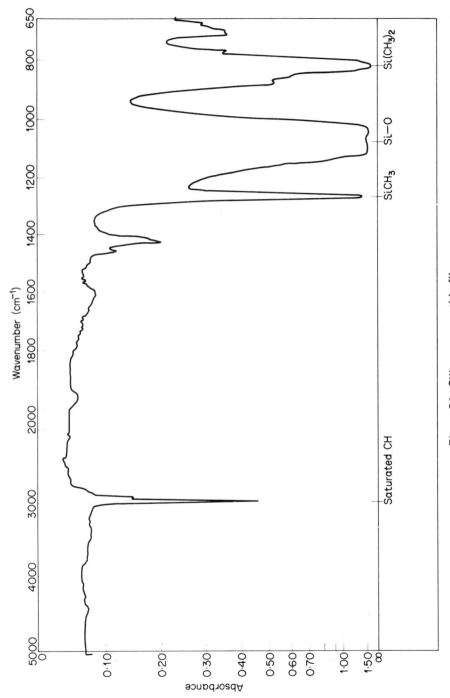

Figure 31. Silicone grease thin film

417

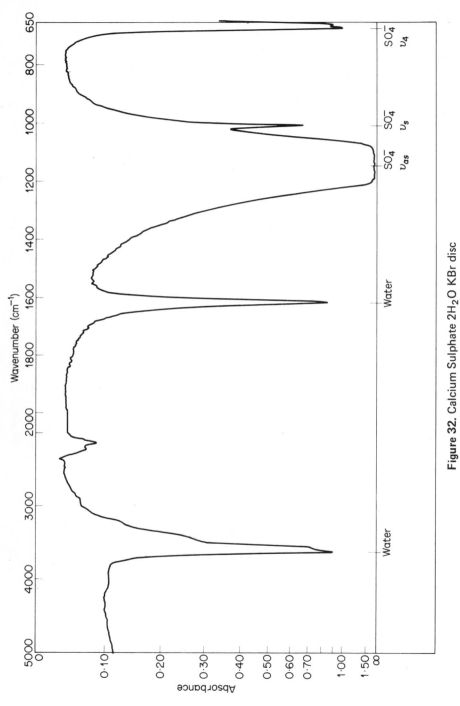

Figure 32. Calcium Sulphate 2H$_2$O KBr disc

Subject Index

Compound Index